Evolutionary Theories of Economic and Technological Change

Evolutionary Theories of Economic and Technological Change

Present Status and Future Prospects

Edited by

P. Paolo Saviotti
and
J. Stanley Metcalfe

University of Manchester, UK

harwood academic publishers
chur • reading • paris • philadelphia • tokyo • melbourne

Harwood Academic Publishers

Post Office Box 90
Reading, Berkshire RG1 8JL
United Kingdom

58, rue Lhomond
75005 Paris
France

5301, Tacony Street, Drawer 330
Philadelphia, Pennsylvania 19137
United States of America

3-14-9, Okubo
Shinjuku-ku, Tokyo 169
Japan

Private Bag 8
Camberwell, Victoria 3124
Australia

Library of Congress Cataloging-in-Publication Data

Evolutionary theories of economic and technological change : present status and future prospects / edited by Paolo Saviotti and Stan Metcalfe.
p. cm.
"Contains a number of papers which were presented at a workshop on Evolutionary Theories of Economic and Technological Change held in Manchester on 21 and 22 March 1989"--Pref.
Includes bibliographical references and index.
ISBN 3-7186-5078-9
1. Economic development--Congresses. 2. Technological innovations--Economic aspects--Congresses 3. Economic history--Congresses. I. Saviotti, Paolo. II. Metcalfe J. S. (J. Stanley) III. Workshop on Evolutionary Theories of Economic and Technological Change (1989 : Manchester, England)
HD73.E88 1991
338.9--dc20 90-21884
CIP

Contents

Preface

This book contains several papers that were presented at a workshop on "Evolutionary Theories of Economic and Technological Change" held in Manchester on 21 and 22 March 1989. The decision to organise such a workshop came to us after some discussions on the present state of development of evolutionary economics. We felt that evolutionary economics is making great progress and is in a very creative stage of its development. Several ideas and concepts that originated in separate research traditions and schools of thought are converging into evolutionary economics. The number of adherents and of sympathizers seem to be growing. It is indeed a very exciting period, opening up many interesting probabilities and avenues for experiment. Following the pioneering work of Nelson and Winter (1982) many articles have been published that have extended and enriched evolutionary thinking. Recently, a book has been published containing a wide range of contributions that can be considered evolutionary (Dosi, Freeman, Nelson, Silverberg and Soete, 1988)

However promising, such a stage is not without its risks. The opening of new perspectives, the introduction of new concepts and their transfer from other disciplines and research traditions can create problems of coherence and of economy of thought, which at some stage become important in the development of a new research tradition and have to be faced by its practitioners. We thought that at this stage it would be particularly useful to focus on those that can be considered the fundamental aspects of evolutionary economics, and we designed the workshop accordingly. We were very fortunate in being able to find several participants who were thinking on similar lines. Some important aspects that should have been of general concern for all the participants were agreed in advance of the workshop itself. We hope that as a result of this procedure, the individual papers have a reasonable degree of coherence and that they make a useful contribution to the progress of evolutionary economics.

In this book, the chapters start with those that deal with the most general aspects, then move to those to try to develop analytical frameworks and finally to chapters dealing with simulations or with policy implications.

The introductory paper by the editors makes clear the variety of research traditions contributing to the evolutionary perspective and the need for a set of organising concepts that can encapsulate variety and change, the two central concerns of evolutionary theory. The locus of evolutionary theory is, of course, the biological sciences and the paper by Mani provides a clear account of the controversies that surround the meaning of evolution in biology and the forces that drive and contain evolution.

Faber and Proops provide different insights into evolutionary analysis by contrasting different aspects of biological, physical and economic systems around the central theme of predicatability and unpredictability of the respective phenomena. They argue that phenotypic evolution is predictable, while the underpinning genotypic evolution is not. This leads them to a discussion of why economics is different from biology: essentially because in the former, genotypic evolution proceeds rapidly relative to phenotypic evolution and because of plausible elements of Lamarckian development in the economic sphere. The relation between Lamarckian processes and the accumulation and dissemination of knowledge is the central theme of Clark's paper. An important emphasis in this paper is given to the open-mindedness and hence, unpredicatbility of knowledge-driven processes.

The paper by Hodgson deals with some of the fundamental difficulties that have to be faced when economic phenomena are treated as evolutionary. Problems exist with respect to the level at which evolution operates and of the kinds of behaviour (rational. habitual etc.) that an evolutionary approach might encompass. He draws attention to the importance of including purposeful behaviour into the analysis of social and economic evolutionary situations and the importance of the institutional contents that temper purpose with routine and habit.

The paper by Dosi and Metcalfe is a survey and discussion of the sources of irreversibility and path dependence in evolutionary processes. They draw a distinction between irreversibility at the micro and macro levels, which give rise to the view that history matters in an essential way for the evolutionary process. Clearly, this reinforces the claim that evolutionary processes are open-ended with respect to their paths of development.

The central theme in the paper by Allen and Lesser is the meaning that should be attached to the idea of variety and different behaviour. They emphasize the crucial point that evolution depends on the prior existence of variety and also generates variety. Their discussion is based on simple simulation models, which are powerful tools for the analysis of evolutionary processes.

The difficult question of how variety in the economic system is to be measured and related to the pace of evolutionary change is addressed in the chapter by Saviotti. He reviews different measures of variety, and relates them to changes in organisational structures, in elementary processes and interactions, such as substitution and competition.

The paper by Stephen Smith also follows the simulation route. Building on the behavioural theory of the firm, he develops a model of search, innovation, finance and selection that captures in detail the process of structural change, which in turn typifies industrial competition.

The empirical study of evolutionary processes is the theme of John Foster's paper. He treats this theme in the context of contrasting econometric methodologies. He argues that, when the empirical purpose is to identify structural breaks and parameter changes in an underlying system, as distinct from identifying the continuity of a system, then new techniques are needed that have yet to be developed.

The final paper in the volume, by Keith Smith, is the only one to address some of the science and technology policy issues implicit in an evolutionary approach. His central point is that policy is not about optimizing, but about doing better, an inherent implication of a world of bounded rationality. Major policy concerns relate to the ability of an economic system to experiment and generate variety, and to the selection process that favours some experiments over others. This latter issue is especially relevant, since the selection processes are myopic and may not identify and support the best long-run sources of innovation in the economy.

Therefore, as the reader can already begin to realise and as it will be evident in the book itself, the themes addressed here comprise a selection of all the possible ones about evolutionary theories. The aim of the book is not to be comprehensive, but to address fundamental issues in evolutionary theories. We would like to separate the motivation to focus on fundamental issues as particularly relevant at this stage from the outcome of the book. We hold that the direction of search oriented towards fundamental issues is right, but that this book clearly constitutes work in progress and that it does not contain a definitive statement about the foundation of evolutionary theories. In organising this book, we hope to have indicated a fruitful search path that other people can follow, thus helping to establish in a clear and rigorous way the foundations of evolutionary theories of economic and technological change.

P. P. Saviotti and
J. S. Metcalfe
Manchester, December 1989

Present Development and Trends in Evolutionary Economics

P.P. SAVIOTTI
Department of Economics, University of Manchester, Manchester M13 9PL, UK
J.S. METCALFE
Department of Economics, and PREST, University of Manchester, Manchester M13 9PL, UK

INTRODUCTION

The recent revival of interest in evolutionary theories of economic and technological change is at least partly due to the dissatisfaction with the way orthodox economics deals with processes of technological change in particular and with any type of change which transforms in a fundamental way the economic system. Attempts to conceive economics in an evolutionary way are not new. The often quoted precedents of Thorstein Veblen (1898) and of other American institutionalists, of Marshall (1890) and above all of Schumpeter (1912, 1943) show that at least the need or the motivation to develop an evolutionary approach to economics had been felt before (Clark, Juma, 1987, 1988; Boulding, 1981). Naturally Schumpeter has gone further than any other economist in developing a theory of economic transformation and structural change, and it is no accident that he is one of the main sources of inspiration for those who are now trying to develop further an evolutionary approach to economics. The fact that Schumpeter himself rejected in his early writings the qualification of evolutionary for his theories is due to a different definition of what constitutes an evolutionary explanation in the social sciences.

At this state of development of an evolutionary approach it is much easier to identify the reasons for the resurgence of interest in it than to define accurately what constitutes an evolutionary approach. This is tantamount to saying that an evolutionary approach is only now beginning to acquire common analytical structures and conceptual tools. It is still far from being a complete theoretical system. One could say that the evolutionary metaphor or paradigm is just emerging and that it will require an extended period of 'normal science' (Kuhn, 1962) to articulate it adequately. At this stage it is therefore particularly important to focus on the foundations and on the basic conceptual tools of the emerging evolutionary approach in order to better facilitate its subsequent development. The adequacy of these conceptual tools depends on their ability to support a wide range of theoretical and empirical investigations while also maintaining sense of coherence in the overall approach.

In order to establish the conceptual foundations of an evolutionary approach it must be realised that a number of research traditions are related to and contributing to it. First, there has been the already mentioned dissenting tradition of some economists, from Veblen to Marshall and to Schumpeter, who were explicitly evolutionary. It must be realised that this dissenting tradition had a limited impact on orthodox economics (Clark, Juma, 1987, 1988). Central here is the orthodox concern with equilibrium states rather than processes of change, together with the overwhelming reliance on the representative agent (firm or household) as the vehicle through which orthodox theory is articulated. By these two devices the essential variety and openness which drives evolutionary change is ruled out of consideration. Second, there is the biological research tradition in which the evolutionary metaphor has been accepted and exploited far more than in economics and in the social sciences and in general, has therefore acquired a more complete conceptual structure. This is not to say that consensus has been achieved on the foundations of an evolutionary explanation of biological change (see for example the chapters by Hodgson and especially by Mani in this book) but that the articulation of the evolutionary paradigm in biology is far more advanced than in economics. In this sense the biological tradition acts as a powerful force of suggestion and of questions to be posed in the economic sphere. Naturally, one has to resist the temptation to think that the structure of the problems is the same in economics as in biology or that answers to given problems can be transferred intact between the two disciplines. It will therefore be quite important to see what similarities and differences there are between the conceptual foundations of an evolutionary approach in economics and in biology. This is particularly important with respect to the role of intention and expectation in the behaviour of economic agents as we explain below. The chapters by Hodgson [Ch. 5] and by Faber and Proops [Ch. 3] address this issue.

The third contribution to a modern evolutionary approach comes from recent developments in chemistry and physics and in particular from non equilibrium thermodynamics. The concepts of chaotic and irreversible behaviour are particularly relevant here, since they illustrate the possibility of strongly path dependent phenomena. From this it follows that history matters and that the cumulation of random 'small' events can fundamentally shape behaviour and options for future choice (Prigogine, Allen, Herman, 1977; Allen, Sanglier, 1978; Allen, 1982a; Allen, 1982b; Arthur, 1983, 1988, 1989; David 1975). Closely related to this contribution is the fourth related research tradition, the theory of complex systems.

Finally, the fifth research tradition is that of the theories of firms and organisations, which has given rise to the behavioural theories of the firm (Simon and Cyert, March 1963), to transaction costs analysis (Williamson) and to some recent applications of these basic ideas to the analysis of technological change (Teece, 1986). Based on these research traditions and following the seminal work of Nelson and Winter (1974, 1977, 1982) a number of economists and other scholars have started to articulate the evolutionary metaphor in economics.

In this chapter we have four objectives. We begin by outlining some of the important research traditions which can shape the future development of evolutionary economics. This is followed by a brief discussion of key elements in an evolutionary approach to economic change. In the third section we select some key issues which must be addressed in future developments. It will be clear to the reader, whether sympathetic or hostile to the case of evolutionary argument, that this volume represents work in progress rather than a definitive statement of results. Much remains to be achieved, the research agenda is open and we anticipate rapid progress. If our deliberations help others to proceed down the evolutionary path of reasoning the contributors to this volume will be well satisfied. To understand what makes the world develop, to be able to better interpret the historical tapestry of variety and change remains our dominant motive.

To explore further these contributions it will be useful to begin with a brief survey of Schumpeter's evolutionary perspective.

1. THE DIFFERENT TRADITIONS CONTRIBUTING TO EVOLUTIONARY ECONOMICS

1.1 Schumpeter

If we concentrate for the moment on Schumpeter it is quite clear that his theories contain several elements embedded in modern evolutionary approaches. To begin with, economic development is defined by the carrying out of new combinations of productive means by entrepreneurs (Schumpeter, 1912, 1934, p66). A broad perspective on new combinations is taken, they can be new products, new processes, new markets, new sources of raw materials and new organisational forms. In more modern terms one would say that Schumpeter attached a great importance to radical innovations as ingredients of economic development. Alternatively one could say that for him *qualitative change* and the generation of economic variety are central to long term economic development. Furthermore, Schumpeter stresses the *non equilibrium*, aspects of capitalist development. The *creative destruction* which 'incessantly revolutionizes the economic structure from within, incessantly destroying the old one, incessantly creating a new one' is one of the fundamental mechanisms of capitalist economic development (Schumpeter, 1943).

Qualitative change can either be gradual or discontinuous. In the latter case, it can lead to gestalt switches changing completely the configuration of the system. As Schumpeter emphasised, no amount of improvement in the horse drawn carriage would have resulted in a transport system competitive with the railroad. The qualitative difference between the two technologies were schismatic. Different knowledge bases, different organisational structures and different working practices were required to exploit the railroad structure. In this sense, Schumpeter's model of evolutionary change was essentially kaleidoscopic. It is therefore understandable that theories of economic and social change concerned with long term developments are more likely to emphasize qualitative change, discontinuities and transitions between different forms than are theories concerned with short term movements and the establishment of equilibrium. The latter would instead emphasize temporary displacements with respect to stable positions. By contrast, Schumpeter and evolutionary economists are concerned with long term developments which entail qualitative and quantitative change in economic structure. The concept of balanced growth in which all economic activities expand at the same rate is quite foreign to this approach. Of course, the classical economists including Marx were also mainly concerned with long term developments in the economy and society and, even if they did not classify their theories as evolutionary, their relevance to modern evolutionary thinking is considerable. As Clark and Juma (1988) point out, the difference between neoclassical and evolutionary economics is therefore one of fundamental goals or, more explicitly, the analysis of processes of change as distinct from the characterisation of states of equilibrium. While neoclassical economics is mostly concerned with what makes the world coherent and ordered, the emphasis in evolutionary theory is upon what makes the world change. This of course does not imply that evolutionary analysis can proceed without some theory of system coherence. As Dosi (1988) has rightly pointed out there is conflict between understanding order and understanding development, a conflict which evolutionary theory must address.

1.2 The biological research tradition

The biological research tradition has now acquired a richness and complexity that is far beyond the boundaries of this chapter to elaborate. However, the role of the discussion in this section is not to give a comprehensive review but to point out some of the main ideas generated within the biological research tradition that can be used, although sometimes in a modified form, in economics and in the social sciences. Furthermore, a more detailed analysis of the main ideas and controversies in biology is contained in Mani's chapter [Ch. 2] in this book.

To begin with the theory of biological evolution in its modern form is generally associated with the name of Charles Darwin. Two of the main aims of Darwin in studying evolution were to establish that all existing organisms are descendants of one or few simple ancestral forms, and to show that evolutionary change was due to natural selection operating on the variations within the population. In other words, the number of surviving species had changed in the course of time with new species appearing and others becoming extinct. Qualitative change was therefore produced by the slow and gradual emergence of new species, or speciation, related to the pre-existing species. This thread of continuity is created by the fact that members of each generation pass on their genetic make-up. Central to the modern synthesis of Darwinian theory are the concepts of phenotype and genotype. The phenotype is identifed by the external appearance of an organism (its characteristics and morphological structure) while the genotype is constituted by its genetic make-up, as embodied in the DNA structure.

According to their genetic make-up, individuals of a species differ from each other in the morphological and behavioural characters and therefore in the degree to which they are suited to their environment. The poorly adapted ones 'perish' whereas the well adapted ones 'survive' and pass on their beneficial genetic underprinting to their offspring. However, these variations characters are generated at random by genetic processes and are then selected on the basis of their relative suitability to the environment in which the phenotypes live. This mechanism of natural selection is at the roots of the controversy between Darwin and Lamarck. As Mani points out in Chapter 2, Lamarck maintained that organisms develop variations in order to adapt to environmental conditions and that such adaptations are passed on to their offspring. Darwin recognised that the environment could affect organisms and species but that effects were limited to the phenotype and that consequently the results of the natural selection process could not be transmitted to their offspring. Very little evidence seems to have been found for a Lamarckian type of biological evolution. However, such an evolutionary mechanism is highly relevant in the case of the social sciences [Hodgson Ch. 5, Faber and Proops, Ch. 3].

Other concepts which can be very relevant for economics are used in ecology. For example, the concepts of species, environment, habitat and niche can be transferred and adapted to economics. Similarly, relevant are the different types of species interactions, such as competition, commensalism and predation (J Maynard, Smith, 1974, p5).

1.3 Non equilibrium thermodynamics and systems theory

Another important influence on the development of modern evolutionary thinking comes from non-equilibrium thermodynamics. A fundamental distinction can be made here between open and closed systems. Open systems are those systems which can exchange matter, energy and information with their environment. Closed systems cannot exchange anything with their environment. The two types of systems behave in a completely different way. Closed systems when left to themselves tend to move towards an equilibrium state corresponding to one of maximum disorder and randomness. The degree of disorder is measured by entropy. Open systems on the other hand do not have equilibria but steady states, in which the time invariance of at least a number of variables characterising the system can be maintained in presence of continuous exchanges of matter and energy with their environment (Von Bertalannfy, 1969). Non-equilibrium thermodynamics applies to open systems. It was developed from the work of Lars Onsager in the 1930s (Prigogine, 1976, 1987; Prigogine Stengers, 1984). To move away from a closed system one can imagine that it is gradually submitted to ever stronger interactions with its environment, which create new constraints on the behaviour of the system. Systems submitted to weak external constraints give linear responses to changes emanating from the environment. Already in the presence of weak interactions the behaviour of the system becomes irreversible and this irreversibility corresponds both to dissipation and to the formation of order. In other words, without some further change in the environment the system cannot recover its previous state.

When a system is submitted to strong external constraints it moves further and further away from equilibrium. In this process the system can undergo a transition to positions where it can choose between a multiplicity of different stationary states. This process can be presented by plotting some variable X, representing the behaviour of the system, against some control parameter, describing the flows into and out of the system, and therefore the distance from equilibrium (Fig 1). For some critical value of the control parameter the system becomes unstable and new solutions emerge. This is a bifurcation point and here the system has a choice between the different branches of the bifurcation diagram (Allen, 1988a, 1988b).

This is an essential indeterminancy in its behaviour in this neighbourhood. Fluctuations in some of the variables can lead to the appearance of new

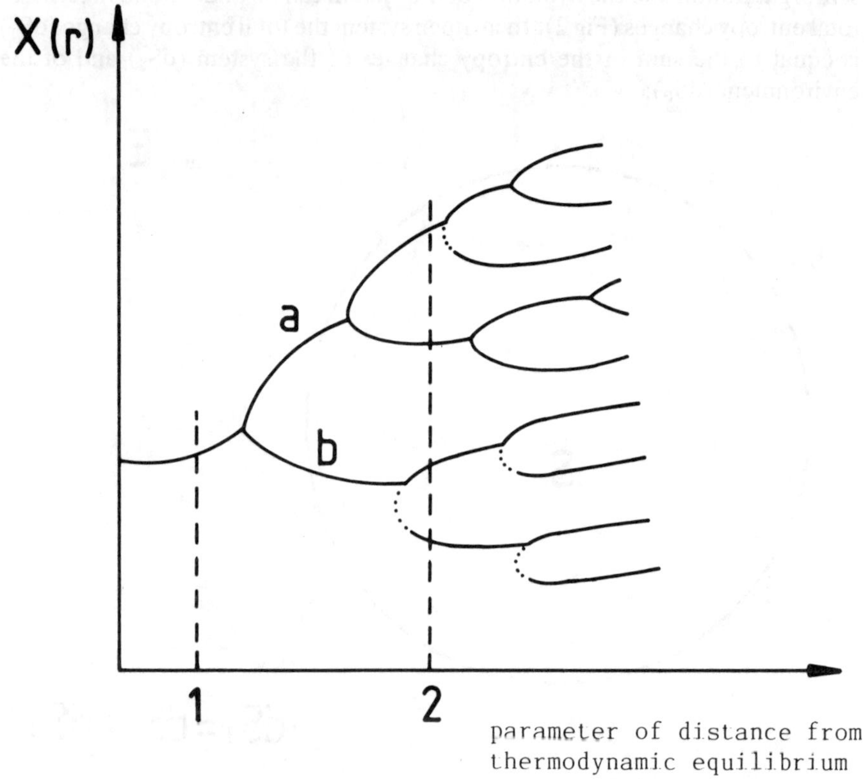

Figure 1. Bifurcation diagram representing the stationary states which are possible for a dissipative structure.

'species' (Allen, 1982a, 1982b, 1988a, 1988b). Far from bifurcation points a reduced description of the system, based on selected parts of it and on their laws of interaction, is valid. However, near a bifurcation point a reduced description breaks down due to the effects of fluctuations, which means of microscopic diversity (Allen, 1987a, 1988a, 1988b). Alternatively, one can describe this change as a transition from one attractor to two different attractors. Non equilibrium can become a source of order: new, more complicated, types of attractors may appear, and give the system remarkable new spacetime properties. The increasing order that the new stable states can display is generally described by the concept of self-organisation (Prigogine, 1976; Prigogine, Stengers, 1984; Allen, 1976, 1988a). The evolution of the system towards states of higher organisation requires non linear mechanisms and is based on a mixture of deterministic mechanisms and fluctuations (Prigogine, 1976; Prigogine, Stengers, 1984; Allen, 1982a, 1982b, 1987a). The existence of self-organisation and the evolution of the system can be understood in terms of total entropy changes (Fig 2). In an open system the total entropy change (dS_T) is equal to the sum of the entropy changes of the system (dS_S) and of the environment (dS_E):

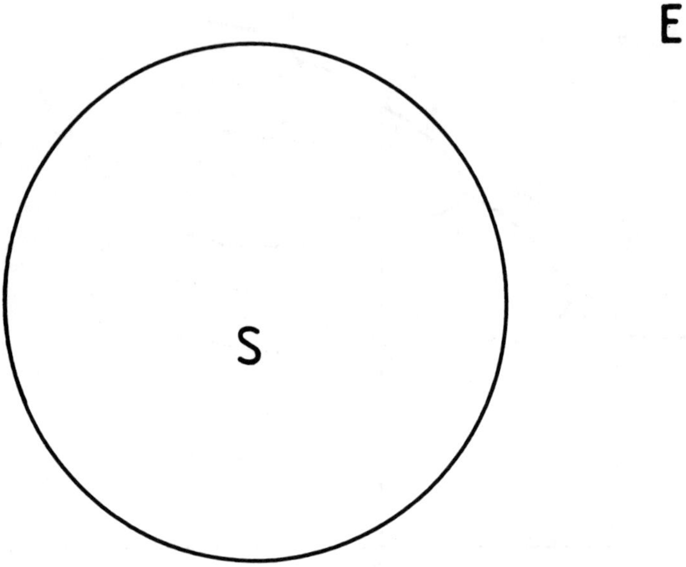

$$dS_T = dS_S + dS_E$$

Figure 2. Entropy changes of a system and of its environment.

It is quite possible for dS_S to be negative, therefore leading to a greater order, if dS_E is positive and sufficiently large to ensure that dS_T greater than 0. In other words, in moving to states which are further and further away from equilibrium the system would be 'exporting' entropy to the environment and becoming internally more ordered and structured.

These concepts, developed for the analysis of physical and chemical systems, can be adapted to biological and social systems. Both biological and social systems are open systems and in both cases evolution tends to lead to a greater degree of order and structure in the course of time (Allen, 1985, 1989). Implicitly related to an open system approach are recent ecological theories of organisational change (McKelvey, 1982; Hannan and Freeman, 1977, 1989) and perspectives on the organisation which focus on its information generating and processing capabilities, what has aptly been termed the organisation as brain (Morgan, 1986).

1.4 Firm and organisation theories

Another research tradition contributing to modern evolutionary thinking focuses on the internal operation of firms and organisations. These theories have been developed from the work of Simon (1947, 1957), Cyert and March (1963), Penrose (1959) and Coase (1937). Some of the fundamental aspects that these theories stress, are imperfect knowledge, satisficing (rather than optimising) behaviour and conflict and bargaining in organisations. Furthermore, it is particularly in theories of organisations that a relatively more detailed characterisation of the external environment has been developed (McKelvey 1982). The fundamental contributions contained in the work of the above mentioned authors have been incorporated by Nelson and Winter into their evolutionary theory (Nelson and Winter 1974, 1982). The knowledge that their firms possess is *local*, which means that it rapidly becomes less effective as the firm moves away from its established area of operation (Nelson and Winter, 1974). Typically, firms have no *direct* and costless access to any other techniques except the one that they are currently using (ibid 1982). Closely related to imperfect knowledge are the concepts of *search activities*, of *routines* and of *decision rules*. Firms will follow established routines and decision rules as long as their targets (eg. percentage return on capital) are exceeded and switch to new routines/decision rules when they are not. The kind of new routines/decision rules which emerge will depend upon the nature of the firm's search activities, that is, on its capacity to learn. Here routines and decision rules function essentially as information ordering and knowledge saving devices. In evolutionary terms the survival of the firm in the face of major environmental changes depends on its ability to learn how to change these internal decision rules.

We may also mention here that imperfect knowledge is incorporated by Paul David into a different type of evolutionary approach. David tries to explain the relationship between mechanisation and labour scarcity in XIXth century America as compared to Britain. He rejects the neoclassical explanation of this factor bias (Fellner, 1961) and develops an alternative 'evolutionary' approach (David 1975; Elster 1983). Of particular relevance in this context is the type of knowledge available to or utilised by firms. According to David, firms at any time have immediate and free access to a small number of practices and any linear combination thereof. Firms behaviour can be represented by a combination of responses to changes in factor price ratios and of local neutral technical progress based on learning by doing. This combination leads to irreversibility, a narrowing of substitution possibilities and path dependence. A change in factor price leads to a switch of production technique, and subsequent improvements due to learning by doing make the new method absolutely superior to any of these existing before the change, and entail the impossibility of going back to the original position even if the hypothesised change in factor price ratio were to be reversed. In short, localised progress based upon focused learning by doing creates irreversible changes in the technique choice set facing firms. Choice becomes a function of history.

The growing awareness of the imperfection of the knowledge used by firms can be interpreted in a number of ways. First, it could be interpreted as a troublesome deviation from an ideal state of perfect knowledge that firms should try to attain in their real behaviour by overcoming frictions in the knowledge generation and dissemination process. Such an ideal world of equal access to knowledge is characteristic of equilibrium theories of competition. Second, knowledge imperfection could constitute the underlying basis of a firm's competitive performance. In this second sense each firm has come to know different things based upon the phenomena of differential creativity, and localised learning. This different aspect of knowledge is encapsulated by the concept of the *knowledge base*, the collective knowledge used by a firm/organisation to underpin its production activity. A knowledge base can never be complete because it is first of all inherently limited in 'span' (coverage) and, secondly, because search activities carried out by firms and other organisations continually redefine the reference framework with respect to which each individual knowledge base can be evaluated. Knowledge is structured, it builds cumulatively and is stored in the organisation's memory to shape its future search activities. Interfirm differences in knowledge base determine the 'asymmetry' of competing firms, which is one of the stylised facts of innovation (Dosi, 1988).

Finally, another distinction between different types of knowledge is useful. Knowledge, even when it is created within individual minds, has to be communicated in order to be used. However, not all human knowledge can be as easily articulated and communicated. Michael Polanyi (1962) analysed extensively the role of knowledge that cannot be readily articulated – tacit knowledge. Tacit knowledge forms a large part of a number of skills such as swimming and landing an airplane. A lot of the skills used in the industrial arts have a high tacit content. On the other hand, there is a type of knowledge, which is called *codified*, which can be expressed in symbolic form and can be easily communicated (Teece, 1981). Real life knowledge can often be placed on a continuum between the totally tacit and the totally codified (Nelson, Winter, 1982). It is possible for the degree of tacitness or codifiability of a given piece of knowledge to change with time. This distinction is very important for the definition of the knowledge base of a firm or organisation and for the possibility that firms can imitate knowledge created outside their boundaries.

Summarising this section one could say that important trends in modern theories of firms and organisations have been on the one hand to stress the nature and not just the degree of imperfect knowledge and on the other hand to point out that it is the very nature of the knowledge possessed by an organisation which determines its performance. The knowledge possessed by organisations is continuously changing but it differs even amongst organisations of the same type, giving rise to persistent asymmetries.

2. KEY ELEMENTS OF AN EVOLUTIONARY APPROACH

We can now summarise the previous discussion in terms of a number of general propositions which are applicable, although in a different way, to the above research traditions and to an evolutionary explanation of economic change. By an evolutionary explanation of economic change we mean one that is driven by two distinct but related mechanisms. The first mechanism generates economic variety and the second selects between those varieties to change their relative economic importance over time. Clearly, selection in the economic sphere also influences the generation of variety. Unlike in the natural world there is no need in the economic world to adhere to theories of blind variation.

2.1 The generation of variety

The fundamental contrast between the biological and the economic world is that in the latter, the generation of variety is purposeful. Firms deliberately seek to differentiate themselves from rivals through a multitude of types of product and process innovations, and while this process undoubtedly contains random elements it is also shaped by the environment in which firms operate. Indeed, this is the central contribution to thinking which is made by the theory of induced innovation. Moreover, firms can anticipate changes in their future selection environment and this anticipatory component is quite foreign to evolution in nature. This of course does not imply perfect knowledge or rationality. As Allen and McGlade (1987) showed, evolution generates diversity and diversity drives evolution. Even though the diversity is costly and suboptimal, over the longer term it is selected for. The mental processes which in social organisations precede actions and outcomes may save a firm from really odd experiments, but too strong a mental selection may result in a failure to innovate successfully. It has already been pointed out that both classical economists and evolutionary economists in general place a greater emphasis on long term development than neoclassical economists. Of course, this does not mean that neoclassical economists have failed to develop theories of technical progress, quite the contrary (Stoneman, 1984, Coombs *et al*, 1987). However, neoclassical explanations proceed within an optimisation framework and generally focus upon improvements in process technology for a given set of commodities. The institutional content of these theories, particularly with respect to the generation and dissemination of knowledge is generally minimal. With its focus on search and selection, evolutionary theory seeks to include an explanation of the processes which generate economic variety through 'product and process speciation'. In this way they are better suited to address problems of qualitative change. The processes by which the institutions, technologies, artefacts, etc, (the economic and institutional furniture) which surround us today, but which were not in existence only one or two hundred years ago, are a central concern of evolutionary economists.

Therefore, economic development consists of the addition to the system of elements (institutions, technologies, etc) which are qualitatively different from those that composed it before. There is a clear analogy here with the emergence of new species and the extinction of some older species in biology. This analogy is one of the most powerful amongst those that have attracted economists and social scientists towards evolutionary explanations. Qualitative change takes place in time and it can be either slow or fast. Evolution was considered by Darwin to be slow but more recently some biologists have challenged this concept and proposed the metaphor of punctuated equilibria (Eldredge, Gould, 1972), [Mani, chapter 2 in this book]. Economic and technological evolution seem to be remarkably fast by comparison with their biological analogue [Faber and Proops, Ch.2]. Here we may also note that Schumpeter was a saltationist, believing that economic change occurred by discrete, punctuated changes. By contrast, Marshall was a gradualist and built his entire theoretical edifice around the concept of incremental change, one thing leading imperceptibly to another.

2.2 Mechanisms of selection

In Darwinian evolution this mechanism is considered to be natural selection. Randomly produced variations of biological organisms have a different suitability to their external environment. The variations which fit better survive while the less well fitting disappear. The randomness with which the variations are produced is the fundamental discriminant between Darwinian and Lamarckian evolution. Lamarck maintained that organisms developed features in order to adapt to their environment and can pass on those features to their offspring. According to Elster (1983), pp. 137–138) in addition to the two previous mechanisms, which can be called natural selection and intentional adaptation respectively, there is a third mechanism called artificial selection. Artificial selection differs from natural selection in that it can take account of the long term consequences of current choice and it differs from intentional adaptation in that it is constrained to accept the small variations generated randomly. While no evidence has been found in biology for Lamarckian mechanism, it is likely to be quite frequent in economic evolution (Hodgson, 1988, p. 143, 1989: Metcalfe, Gibbons, 1987, 1989; Boulding, 1978, 1981; McKelvey, 1982; Nelson Winter, 1982, pp. 134–136).

In the economic sphere the analogue to selection is the process of competition, and, as in the natural sphere competition is myopic. Firms are selected for what they are, not for what they might become. Firms compete by offering products between which customers can choose and selective advantage is associated with offering more 'desirable' products and by producing them more efficiently. The more successful firms not only survive but grow at faster rates relative to their rivals and so increase their relative economic importance or economic weight within the economy. The questions this raises concern the way in which markets and related environments evaluate different sets of technological characteristics, and the abilities of the firms to create and operate with superior technological characteristics. The possibility of Lamarckian evolution now arises because the development of technological characteristics is not entirely random. Firms focus their search activities on promising lines of development where the degree of promise depends upon their judgement of the market environment. Behaviours learned in this way become incorporated within the memory of the firm, through routines and working practices and are thus handed on to determine the firm's future behaviour. Lamarckian inheritance in economics has considerable plausibility. Notice though that learning mechanisms can also become self-limiting and thus undermine the scope of the firm to search when faced with changes in its market environment. Indeed, central to the evolutionary approach is the concept of inertia, that is of imperfect adaptation in response to environmental change. Imperfect adaptation is, of course, a further factor contributing to the generation of economic variety. What is important about evolution is the presence of differential adaptation by firms to prevailing circumstances. In a world of uniform adaptation evolution would be impossible.

2.3 Inheritance, and the transmission of genetic make up

In biology organisms can reproduce (generally sexually) and pass on their genetic make up to their offspring. The meaning of reproduction is much looser in economics but relates to the maintenance of productive competence over time, and integral to this is the generation and storing of information. Organisations and technologies tend to show some continuity in the course of time, although the pace of change can be much faster and more discontinuous than in biological evolution (McKelvey, 1982). What is inherited within the firm is the ability to perform production transformations as embodied in organisational rules and routines (Nelson and Winter, 1982). Here the crucial problem for the survival of the firm is to make its memory independent of the current personnel. Naturally, in economics and in the social sciences the mechanisms of inheritance cannot be sexual although Boulding (1981) considers human artefacts to have a polyparental origin, due to the multiplicity of inputs used to make them.

2.4 Two different layers of reality

The units of observation that we choose (eg. organisms, species, organisations, industries, technologies, etc) have a physical outer appearance, which can be defined by some generalised characters, and a more fundamental part which is less easily visible and more difficult to modify. In biology these two different layers are the phenotype and the genotype respectively. The characters defining the phenotype are subject to environmentally induced changes which are not transmitted to subsequent generations. It is possible to define analogues of phenotype and genotype in the social sciences [Faber and Proops]. For example, Boulding (1978, 1981) considers know-how to be the equivalent of genetic make up in human societies. McKelvey (1982) sees the pool of dominant competence of organisations constituting their equivalent of a genetic make up. Nelson and Winter (1982) consider routines and decision rules as the equivalent of the genetic make up of firms. However, to be able to define an economic or social analogue of the genetic make up of biological organisms does not prevent acquired characteristics from being transmitted. While it is plausible to treat a product and its associated process technology as the basic unit of selection in a market economy, it must be recognised that selection of technologies is not logically equivalent to selection of firms. The multiproduct firm is in principle able to offset adverse selective effects with respect to one product by developing additional products for more favourable selection environments. Moreover, the operation of the capital market provides a level of selection which operates directly on the ownership status of firms and bears only a tenuous relationship to the forces of selection in individual product markets. As in the natural sciences one can recognise a hierarchy of levels at which selection mechanisms operate (Eldredge).

2.5 Population perspective

The concept of population (an aggregation of members of a species) is one of the fundamental concepts in biology. In the social sciences, however, a population perspective has been overshadowed (Hannan and Freedman, 1977, 1989) by a *typological* perspective. The latter is based on the assumption that relevant features of a species/group of study are substantially homogeneous and are captured by a few essential characteristics and therefore one is entitled to use these representative values to identify the species under investigation. In a population perspective on the other hand, the variations internal to a species are extremely important and therefore it becomes relevant to consider not only average values but also variances and their relation to the evolutionary process.

2.6 Systems openness, irreversibility and path dependence

As already emphasised, theories of complex systems and irreversible thermodynamics have important implications for economics and for the social sciences. As several scholars have emphasised, economic systems are open systems and exchange matter, information and energy with their environment

(Silverberg, 1988; Boulding, 1978, 1981). Therefore, they must follow the laws of behaviour of open systems. On the other hand one of the most important features of economic systems, which they share with biological systems, is the increasingly greater order, complexity and variety of institutions, products and technologies to which economic development has given rise. Naturally, this increasing order would not be explicable if economic systems were closed to flows of energy and matter. In this case we would expect these systems to evolve towards equilibria characterised by the lowest possible order or the greatest possible randomness. This would be the implication of the second law of thermodynamics when applied to closed systems.

The possibility of 'evolution', that is, qualitative change in the structure of the system, is linked in a fundamental way to the non-equilibrium nature of biological and economic systems (Prigogine, Stengers, 1984; Silverberg, 1988; Allen, 1982b, 1987a, 1988). If one imagines moving a given system gradually away from equilibrium, for example, by increasing the rate of exchange with its environment, the behaviour of the system changes fundamentally. First, transformations become irreversible. However, in the neighbourhood of equilibrium such transformations are predictable. This occurs in the so called linear regions, in which the flows (eg movements of matter, or heat flows) are a linear function of the 'forces' (eg concentration or temperature gradients) which cause them (Prigogine and Stengers, 1984). In this region the behaviour of the system differs from that at equilibrium but not enough to give rise to evolution.

When the 'distance' from equilibrium, as measured by some control parameter, is sufficiently large the system begins to behave in a completely different way. First, it can undergo a series of transitions to an ever greater number of states, each characterised by a high degree of order and a different structure. This can be represented in a 'bifurcation' diagram (Fig. 1). By contrast with equilibrium systems, in the neighbourhood of a transition point (a branching point in the bifurcation diagram) the behaviour of the system becomes indeterminate and it is not possible to predict which branch of the bifurcation diagram is going to be followed. Random fluctuations are extremely important in determining the outcome of the process. In this situation, the behaviour of each system is extremely specific and not many generalisations are possible (Prigogine and Stengers, 1984). Amongst the few generalisations which can be made is the autocatalytic nature of far from equilibrium processes. In other words one of the products of the process feeds back into the system and accelerates the rate of change of the system as a whole. The mathematical consequence of this is the fact that the process is now described by equations that are nonlinear with respect to the concentration of the 'product' considered. Furthermore, and especially in the neighbourhood of transitions, random fluctuations are extremely important in determining the outcome of a process, for example, in determining to which branch of the bifurcation diagram the transition will take place.

Such far from equilibrium systems can have an ordered structure. One of the first examples of these systems was the so called 'Brusselator', a chemical reaction in which a reagent A is transformed into a product E by means of a series of intermediate steps (Prigogine, 1976; Prigogine, Stengers, 1984; Allen, 1988). A number of substances are involved in the intermediate steps. By changing the concentration of these substances one can move the system away from equilibrium. The Brusselator far from equilibrium can produce what is called a chemical clock. This can be understood assuming for example that the reagent A is blue and the product E is red. Beyond a bifurcation point a limit cycle is established: the reactions proceed completely to the side of the products and then back completely to the side of the reagents. The system is seen displaying a red colour or a blue colour at alternate times. This limit cycle is stable as long as the control parameters are kept at the required values. In other words, such far from equilibrium systems display order structures which are stable as long as the systems themselves are subjected to continuous flows of energy and matter. For this reason they are called dissipative systems. Naturally the behaviour that one would normally expect from a Brusselator at equilibrium would lead to a 'violet' looking system.

These conclusions are drawn from calculations of chemical and physical systems. However, some of them are generalisable to biological and social systems. For example, intrinsic non linearities have been found to be important since the work of Goodwin (1951) who showed that self-sustaining business cycles were only possible in the context of non linear models. Stochastic indeterminancy plays an important role in the models of Nelson and Winter (1974, 1982) and of Arthur (1983, 1988, 1989). Some or all of these aspects are contained in models based on catastrophe theory, on game theory and on replicator dynamics, the last one being based on Fisher's mathematical formulation of natural selection (Silverberg, 1988).

2.7 Elementary interactions

All the different interactions between pairs of species in biology can be classified into three categories (Maynard Smith, 1974, p. 5).

a) Competition. Each species has an inhibiting effect on the other.
b) Commensalism. Each species has an accelerating effect on the growth of the other.
c) Predation. One species, the 'predator', has an inhibiting effect on the growth of the other, the 'prey'; the prey has an accelerating effect on the predator.

Clearly, one cannot expect the same classification of interactions to apply to economics as well. In economics, competition is virtually the only type of interaction which is studied. It is clear that there are different types of competition. Perfect competition is the most commonly mentioned in textbooks but relates to a state of equilibrium not to a process of rivalry and change. It is also considered the 'best' type of competition through its link with efficient resource allocation and other forms are therefore considered to some degree imperfect. Perfect competition is, however, more a state of affairs to be attained after all competitive forces have been eliminated rather than a force leading to change (McNulty, 1968). Of great relevance here is the fact that in perfect competition all firms are equal and therefore they cannot generate any qualitative change in their output or methods or production. Of a different type is the process of competition generated by Schumpeterian entrepreneurs introducing innovations into the economic system (Schumpeter, 1912, 1943, Metcalfe Gibbons, 1989). Undoubtedly, real life competitive situations can be somewhere in between the two extremes defined by perfect competition and innovation competition. As perfect competition is approached the products of different competitors become progressively more similar; alternatively as one moves towards innovation competition the products become progressively more different. Examples of these trends have been found in the evolution of motor car and aircraft technologies (Saviotti 1985, 1988).

On the other types of elementary interactions found in biology, commensalism has a clear analogy with the relationship between complementary products. No comparative economic interaction seems to exist for predation, although Goodwin's model of the trade cycle (1951) is constructed on analogous grounds.

2.8 *External environments*

A systematic analysis of the external environment of organisations has not been developed in economics, except in an extremely oversimplified way. In perfect competition the external environment is constituted by a structure of product and factor prices over which the firm has no influence. A far more complex concept of environment, and one which is quite compatible with an evolutionary approach, has been developed by organisation theorists (Jurkovitch, 1968; Terreberry, 1974). To begin with a difference can be made between the *climate* and the *texture* of an environment (McKelvey, 1982, p120). The climate is constituted by those components of the external environment which have a general effect on the organisation(s) considered. The climate can thus be considered a *non purposeful* milieu surrounding organisations. Organisations are also embedded in a texture of causal relations composed of other (myopically) purposeful organisations and of their interrelations (McKelvey 1982, p.121). Actions by one organisation may have differential effects on the other organisations to which they are related. Each of these components of the environment can be further broken down into dimensions of climates, such as resource capacity (rich–lean), homogeneity–heterogeneity (diversity), stability–instability, resource concentration–dispersion, domain consensus–dissensus (territorial legitimacy) and turbulence (disturbance via increasing rate of interconnection) (Aldrick, 979, pp.63–70). This is only an example of the internal dimensions of the climate. Other authors have created similar classifications (McKelvey, 1982, pp. 120–122).

Environments also influence the structure of successful organisations. Examples of this influence have been found by Burns and Stalker (1961), Woodward (1965), Lawrence and Lorsch (1967) and by a large number of other authors. For example, Burns and Stalker discovered that management styles were specialised into what they termed *mechanistic* and *organic* structure. The former is suited to a stable environment, the latter to a changing environment. Emery and Trist (1965) identified four types or states of causal texture. These are called placid–randomised, placid–clustered, disturbed–reactive and turbulent fields. Starting from the first the degree of interaction between the constituting entities, the complexity and the instability increase. Perfect competition, imperfect competition and oligopoly correspond to the first, second and third of these types of causal texture respectively.

Nelson and Winter (1977, 1982) proposed a 'selection environment', which would be a generalised analogue of the market for the most different types of organisations, such as firms, schools, hospitals, prisons, etc. Their selection environment can be specified by means of four elements: 1) the definition of worth or 'profit' which is operative for firms in the sector; 2) the manner in which consumers and regulatory proficiency influence what is profitable; 3) investment processes; 4) imitation processes. Metcalfe and Gibbons (1989) have developed a characterisation of the selection environment for the analysis of economic growth and competition, which emphasises its capacity to evaluate different product and process technologies, its rate of growth and its homogeneity.

All those characterisations of environments come from different research traditions and the process of integrating them into an evolutionary perspective is far from complete.

3. UNRESOLVED ISSUES

3.1 Fundamental assumptions

As presently developed, evolutionary theories of economic and technological change do not have the formal elegance, the analytical depth or articulation of more orthodox economic theories. Accepting that there is a range of economic and technological phenomena that evolutionary theories can explain better than existing alternatives the task ahead consists of giving these theories better conceptual tools, of improving their analytical depth and their ability to be used as a basis for empirical research.

As it has been shown at the beginning of this chapter, modern evolutionary theorists can draw inspiration from a variety of research traditions, ranging from heterodox parts of economics to non-equilibrium thermodynamics, systems theories and organisation theories. This situation leads to two classes of problems. On the one hand, there is the problem of deriving from the separate intellectual foundations a relatively coherent and unitary set of concepts and tools for evolutionary theories of economic and technological change. On the other hand there is the problem of avoiding coarse forms of reductionism. This implies that while one can take inspiration from biological, thermodynamic, etc, research traditions those same traditions can only provide analogies and suggest problems but not supply answers to economic or technological problems. A continuing attention to the foundations of the inspiring research traditions will therefore be required in future developments. In this context it is relevant to make the distinction between causal, functional and intentional explanations (Elster, 1983). According to Elster, different disciplines tend to be based predominantly on one of these types of explanation. Thus physics used a predominantly causal mode of explanation, biology a functional one and the social sciences an intentional one. In a causal mode of explanation events are explained by means of a regular conjunction with certain causes. On the other hand a feature of an organism is functionally explained if it can be proved to be a part of a local individual maximum with respect to individual reproductive capacity (ibid., p.53). Finally we explain intentionally an action when we are able to specify the future state it was intended to bring about (ibid., p.70). Clearly these modes of explanation are related to our basic perception of the subject matter we are studying. Thus we usually attach a certain degree of intentionality to human behaviour but not to animal behaviour. In particular we consider that technical evolution differs from biological evolution in that the changes are far from random but to some extent directed. These changes are also screened by a mechanism in which human intentionality plays a role (ibid,. p.12).

The separation of these modes of explanation as presented above is somewhat excessive. Thus, in intentional explanation one needs to relate particular actions to a goal or a set of goals that the actions are aimed at bringing about. However, the existence of a goal does not necessarily imply that it will be achieved. Causal mechanisms will have to be invoked in order to explain how the goal itself has been achieved. Therefore, the different modes of explanation are not entirely incompatible and can to a certain extent be used jointly. There are however incompatibilities, such as the impossibility to use intentional explanation in physics.

These considerations are relevant in the present context because the different research traditions which converge in modern evolutionary economics have, according to Elster, different modes of explanation. The transfer and adaptation of ideas and concepts between these research traditions has to be done taking differences in underlying modes of explanation into account. This has two implications: on the one hand, as already mentioned, we can only expect different research traditions to be able to suggest analogies and questions but not necessarily to provide answers for evolutionary economics; on the other hand this imposes a continuing attention to the fundamental aspects of any concept, idea or theory which is integrated into evolutionary economics.

Another important problem is constituted by the presence and the location of intentionality in economics. To have conscious aims or goals that can be generated or change without fluctuations in the external environment is a peculiar feature of human beings and therefore of the social sciences. Different schools of economic thought in the past have stressed the purposeful character of human actions (Hodgson, 1988; and Chapter 5). The Austrian school is perhaps the one that has laid the greatest emphasis on purposeful behaviour. However, they only recognise purposeful behaviour in individuals and not in organisations or institutions. Furthermore, they tend to disregard any influence of the external environment on individuals and on their purposeful behaviour. A number of choices are therefore open for evolutionary theorists to follow: determinism vs purposeful behaviour, determinism vs indeterminacy, individual vs collective (organisations, institutions), environment dependent vs environment independent. Choices of this type are unlikely to be testable but they are important components of any theory and it is important for them to be explicitly articulated in the construction of evolutionary theories. In a similar vein, John Foster (1987) in attempting to establish an evolutionary basis for macroeconomics maintains that scientific materialism and homo economicus form part of the core of orthodox economic theories and that they cannot be refuted empirically. An evolutionary approach will need similarly general foundations.

A further problem of evolutionary theories is constituted by their predictive power. Both fluctuations and qualitative change, which are amongst their important features, can limit their predictive power. This is an important issue because predictive power is considered to be a fundamental property of a well developed scientific theory, and it is important to establish whether there are any limits to the predictive power of a science. Such limitations have been found for physics and they have not compromised the scientific character of the discipline. According to Boulding (1981) prediction is possible only in systems that have stable parameters like celestial mechanics. By his definition, evolution leads to a change of parameters and therefore is essentially unpredictable. Similarly, Faber and Proops [Chapter 3] maintain that qualitative change, which they define as change in the genotype, is essentially unpredictable. At the roots of this problem there is the tension between being and becoming, which has been a fundamental one in western philosophy. In recent times this tension, at least in the context of the natural science, has been solved in favour of being. One of the most important features of the development of modern science in the last two to three hundred years has been the attempt to reduce the different and changing to the identical and the permanent (Prigogine and Stengers, 1984, p 293). Thus all changes were reduced in principle to changes in the position and interaction of the various elementary constituents of the universe. In this attempt a central role was played by the concept of time which became basically 'a geometric parameter that makes it possible to follow the unfolding of the succession of dynamical states (ibid p 293). Naturally, this concept of time implied also the reversibility of processes. The transition to an evolutionary paradigm involves, among other things, a redefinition of the balance between being and becoming. By emphasising 'becoming' instead of 'being', one would largely abandon any hope to explain and above all to predict events and processes. Some degree of continuity and commonality of constituents between past, present and future is required if the world has to make sense and show some predictability. Thus the concept of time and the nature of invariance of the constituents of the system stated are fundamentally related to the reversibility (or lack of) of processes and to the predictive power of science. The acceptance of irreversibility points to a different role of oriented time. Irreversibility is no longer a purely subjective perception but a fundamental characteristic of scientific explanations (Prigogine and Stengers, 1984, p 298). From the outset it has to be recognised that evolutionary processes are inherently open with respect to outcome, and that there is no need within them for any concept of long run equilibrium. Indeed what evolutionary theory demonstrates is that good explanations need not have good predictive power (Scrivens, 1959). That is to say, in a good theory explanatory power need not accompany predictive power. Modern evolutionary ideas allow for a mixture of indeterminacy as a consequence of fluctuations in the vicinity of bifurcation points, and of predictive power far from bifurcations (see Fig. 1). The inherent openness of economic evolution also relates to the inductive nature of the development process. The accumulation of experience is the primary mechanism by which economic knowledge grows and it is this fact which underpins strong elements of irreversibility and path dependence in economic evolution.

The development of neoclassical economics has been dependent on an analogy with classical mechanics (Clark, Juma, 1987, 1988). The same concept of time as a geometric parameter has therefore been used in economics. Furthermore, given the concern of economists with equilibrium states, questions of mutual consistency of given relationships have been highlighted rather than the nature of mechanisms of the corresponding processes which destroy those states. Thus in explaining the sources of economic growth, it is the shifts of the production function and not their timing and origin which are important (Amendola and Gaffard, 1988). It is obviously one of the tasks of evolutionary economics to recover a concept of time which is compatible with becoming, and which gives substantial contrast to the nature of processes. This would supply a much needed connection with history.

This very brief description of some fundamental problems in evolutionary theories does not pretend to be comprehensive. Its aim is simply to point out that problems of this kind are particularly acute at this state and that continuing study and analysis of them if fundamental for the further development of this subject.

3.2 Units of analysis and taxonomy

Like any science, evolutionary economics must have a conceptual apparatus which is compatible with empirical analysis. This enables economic theories to be applied to particular situations, and that empirical analysis can serve as a basis from which to infer further theories. In this sense an important problem for an evolutionary approach is to establish its units of analysis at different levels of aggregation. Given the central role of technology in these considerations one needs to have units of observation for both organisations and technology.

Firms are the most common but not the only producer and user of technology. An example of a generalised definition of organisational species, which encompasses both firms and other types of organisations, has been proposed by McKelvey (1982) on the basis of *dominant competencies*. An organisation has a *primary task* and a set of managerial activities which are directly related to the implementation of the primary task and which constitute the *workplace managerial task*. Of all the competencies of the organisation, the most important ones are those which are *directly* related to the primary task and to the workplace management task. The set of all these competencies constitutes the *dominant competencies* of the organisation (ibid pp 189, 191). The elements of dominant competence are called *comps*. An organisational species is then constituted by population of organisations sharing some dominant competences.

For what concerns the technologies themselves the units of analysis are the products and the processes used to produce them. Examples of possible approaches to these units are those of Sahal (1981a, 1981b) and of Saviotti and Metcalfe (1984). Sahal adopts a systems view arguing that technology is best conceived in terms of its performance characteristics. Consideration of these characteristics is important because variables other than factor substitution can have a profound influence on economic development. Examples of these variables are materials employed, availability of requisite skills, scale of output and type of product. Saviotti and Metcalfe (1984) represent products by means of two sets of characteristics, one describing the internal structure of the technology (technical characteristics) and one representing the services performed for its users (service characteristics). Processes are represented by different characteristics (capital and labour intensity, structure, types of machines, process layout, batch, mass, flow, process, etc). By means of these units of analysis one can represent various types of elementary processes (see next subsection), measure degrees of technological changes and develop a numerical taxonomy.

In addition to a taxonomy of organisations and of technologies one can establish a taxonomy of innovations and innovation strategy. Thus innovations can be classified as product or process, radical or incremental (Freeman, Perez, 1988). Conversely, innovation strategies can be classified as offensive, defensive, imitative, dependent, traditional and opportunist (Freeman, 1982). An innovation taxonomy at a higher level of aggregation and concerned predominantly with institutions rather than with innovations, has been proposed by Pavitt (1984). In his classification industrial sectors are classified as supplier dominated sectors, scale intensive sectors, specialised suppliers and science based sectors. These sectors differ with respect to the source of the innovations that they use (they can be imported into the sector, generated internally or sold to other sectors), with respect to size and capital intensity, to the type of process used (mass, flow, etc) and to the influence of science on the performance of the sector.

A well developed evolutionary theory will require a better developed taxonomy at all levels of aggregation in such a way that the relationships of the various units of analysis within and between each level of aggregation can be analysed. Naturally, this taxonomy is now far from complete and it will need to be extended and refined.

3.3 Elementary processes of technological evolution and other basic concepts

Qualitative change linked to the emergence of new products and services and new organisational forms has previously been indicated in this chapter as one of the fundamental components of an evolutionary explanation. How these qualitative changes are to be measured is a critical problem in evolutionary explanations of economic change.

New products, services, etc, can either substitute for previous ones or coexist alongside them. Depending on the outcome of the substitution process, the economic system will have a changing variety of surviving 'species'. This is obviously a very interesting question because the concept of variety can be the basis for an analytical representation of the qualitative change that is so central both to economic development and to an evolutionary perspective. Therefore, it is important to understand the nature of variety of the economic system and to give a rigorous analytical definition of it. This way it will become possible to map trends in variety in the course of economic development and to analytically relate variety to other important variables characterising the economic system.

As already pointed out new products and services are continuously emerging in the economic system. If the majority of these do not substitute for previous products and services but are simply 'added' to the economy, the variety of the system is likely to increase. Casual observation of the habits of an average household gives the impression that the number of types of goods and services used now is much greater than that which was available only two or three hundred years ago. Product differentiation and diversification have added to this variety. However, it must be remembered that they are essentially a phenomenon typical of mature industrialised societies which began on a large scale during this century. The implications of these considerations are twofold: on the one hand they point to the number of surviving species (products, services, etc) as an important aspect of variety; on the other hand they seem to imply that, at least in recent historical times, the variety of the economic system might have been increasing. As definition of this type of variety at the level of aggregation of the economic system as a whole, is the number of *distinguishable* products, services, etc, in the system. This aspect of variety and its implications for technological and economic evolution is discussed in Saviotti's chapter in this book.

Different types of phenomena, closely related to a population perspective, occur at another level of aggregation in the economic system. During the evolution of a technology or an industry the number and size of firms is likely to change. Such changes are incorporated in the models of Nelson and Winter (1982) and of Metcalfe and Gibbons (1987, 1989). This amounts to a change in the institutional variety of the supply structure of a given industry during its evolution. Consequently, the *variance* of a number of properties of the firms in the industry changes in response to the forces of competition selection, innovation and institution. Other features of the firms in the industry, such as the number of product designs at each time, the nature of the process technologies used, etc, which also contribute to the variety of the system, are likely to change systematically during the evolution of the industry (Abernathy Utterback, (1975). Similar changes in the variance of these properties can be expected.

Thus different aspects of the variety of the economic system exist at different levels of aggregation, and they are not necessarily the only important aspects of variety. Furthermore, they are not independent. The main purpose of these notes, however, is not to give a complete description of the concept of variety, of its aspects and methods of measurement but to suggest that variety is a very important concept which promises to have a fundamental role in the development of an analytical treatment of the processes of qualitative change previously described.

Changes in variety can take place by means of a number of mechanisms. For analytical purposes it would be very useful to be able to represent such complex processes, of which there is a great diversity in the economic system, by means of a small number of *elementary processes*. These would be the common dynamic constituents of complex processes of economic and technological change. To discuss these elementary processes it is useful to adopt the distinction which is often made, although only for didactic purposes, between the act of innovation and its subsequent diffusion. Naturally, this does not imply that these two stages of generation sand utilisation of innovations are independent. Some elementary processes are more closely related to the generations of innovations, other to their subsequent diffusion; amongst these there are technological substitution, specialisation and emergence of completely new products.

Elementary processes can be very useful building blocks in the representation and analysis of complex processes of technological change. Recent research, in the economics of technological change has also uncovered a number of general patterns of evolution of technologies. For example, Abernathy and Utterback (1975, 1978) proposed a life cycle in which technologies evolve from a multiplicity of product designs to a dominant design and in which simultaneously the scale and organisation of the production process change from small and loosely coordinated to large and rigidly coordinated (systemic). In a somewhat similar way, although starting from quite different premises, Nelson and Winter (1977) introduced the concepts of *technological regimes* and *natural trajectories*, Sahal (1981) that of *technological guideposts* and Dosi (1982) that of *technological paradigm*. Apart from their specific features these concepts all point toward some degree of invariance of technological systems with respect to fluctuations in their external environments (Saviotti, 1986) or alternatively a form of self organisation of the system. These developments have therefore redefined the previous need or demand pull/technology push debate (Mowery, Rosenberg, 1979; Coombs et al, 1987) by limiting the influence of demand on technology to changes within the existing technological paradigm.

Similar patterns at a higher level of aggregation have been proposed by Perez (1983) and Freeman and Perez (1988). They talk about *technoeconomic paradigms* as the combination of regularities in the technologies used and in the surrounding institutions. A particularly important role is played for them by the delay with which institutions adapt to the new potential generated by new technologies.

Quite apart from their specific features these concepts imply a considerable discontinuity in economic and technological development and some form of self organisation which gives the technological or economic system relatively stable and ordered structure which only occasionally undergoes gestalt switches to different configurations. The nature of such regularities has so far been proposed but both very little empirical and theoretical research has been done on them. Amongst the first analytical treatments there are those of Metcalfe (1984) and Heiner (1983, 1988). Furthermore, these concepts imply that some features of technology can change continually while others change only rarely. For example, some elements of the knowledge base are common to the firms operating in a given technology and change only when a transition to a different paradigm occurs. Consequently, one can interpret the knowledge base as the equivalent of the genetic make up and the paradigmatic transition as the equivalent of a mutation.

In summary, this section has discussed a number of emerging conceptual tools which were not present in orthodox economics. These concepts have originated recently from the generalisation of findings of studies of technological change. Such studies were generally not carried out according to an explicitly evolutionary framework but were rather eclectic in their methodology and empirical in their approach (Coombs, Saviotti, Walsh, 1987). This eclecticism reflected the unsuitability of orthodox economics for the purposes of the studies and their policy orientation. Although such studies of technological change were not explicitly evolutionary they helped to formulate a number of important concepts which have become an integral part of a modern evolutionary approach because they analysed one of the main ingredients of qualitative change in the system. As already pointed out this level of analysis allows the reconstruction of processes of technological change and not only of their outcome. Those studies of technological change were therefore of a fundamental importance in providing material which one can consider implicit data and evidence for an evolutionary explanation.

From these empirical studies a natural history of innovation is emerging, from which higher level conceptual structures can be developed. In most cases these are still in a qualitative form and are still non analytically founded and non analytically applicable. One can foresee that the next stage of conceptual development will consist of the construction of analytical foundations and structures for basic concepts like those described above. Naturally, this does not exclude that other new conceptual structures can be inferred but it implies that they too will need a sound analytical foundation.

3.4 The role of technology

Technological change plays a very important part in this book. It is one of the fields of study which has given some of the most important contributions to the development of a modern evolutionary approach. Yet technological change is not accorded a very prominent role by the economics profession. It is hardly mentioned in general economics textbooks and it plays a secondary role even in industrial economics textbooks. Courses on the economics of technological change are very rare and specialised options. Policies for science and technology are conceived outside economics and have very limited relationships with the mainstream of economic policy. All this can make sense only to the extent that technological change is exogenous to the economic system and therefore that the economist has only to register its outcomes and proceed to relate them to other aspects of economic behaviour. To the extent that technological change is itself one of the main aspects of economic change and development then it must necessarily play a more central role in economic theory and policy.

This defence of the role of technological change in economic life and theory does not in any way imply that it is the only cause of qualitative change. Schumpeter is often quoted as one of the main proponents of the importance of technological change in economic development but he included new markets and new forms or organisation amongst the main types of change. What is required is not an excessive and one sided emphasis on technological change but a generalised concept of innovation and experimentation and of their relationship to more routine activities.

REFERENCES

W.J. Abernathy, J.M. Utterback, A Dynamic Model of Process and Product Innovation, *Omega,* **3** (6) (1975) 639–656

W.J. Abernathy, J.M. Utterback, Patterns of industrial innovation, *Technology Review*, (1978) 41–7

H.E. Aldrich, *Organisations and Environments*, Englewood Cliffs, N.J. Prentice Hall (1979).

P.M. Allen, Evolution, population dynamics and stability, *Proceedings of the National Academy of Sciences, USA,* **73** (1976) 665–668.

P.M. Allen, Sanglier, Dynamic models of urban growth, *Journal of Biological and Social Structures,* **1** (1978) 265–280.

P.M. Allen, The genesis of structure in social systems: the paradigm of self-organisation, *Theory and Explanation in Archeology*, New York, Academic Press (1982a).

P.M. Allen, Evolution, modelling and design in a complex world, *Environment and Planning,* **9** (1982b), 95–110.

P.M. Allen, Ecology, Thermodynamics and self-organisation: towards a new understanding of complexity, *Canadian Bulletin of Fisheries and Acquatic Sciences,* **213** (1985) 3–26.

P.M. Allen, J.M. McGlade, Evolutionary drive: the effect of microscopic diversity, error making, and noise, *Foundations of Physics,* **17** (1987a) 723–738.

P.M. Allen, J.M. McGlade, Modelling complex human systems: a fisheries example, *European Journal of Operational Research,* **30** (1987b) 147–167.

P.M. Allen, Evolution: why the whole is greater than the sum of the parts, in Wolff, Soeder (Eds), *Contributions to theoretical ecology*, Proceedings of an International Workshop held at the Nuclear Research Centre, Zulich, GDR, 19–30 October (1988).

P.M. Allen, Evolution, innovation and economics, in G. Dosi, C. Freeman, R. Nelson, G. Silverberg, L. Soete (Eds) *Technical Change and Economic Theory*, London, Pinter (1988).

M. Amendola, J. L. Gaffard, *The Innovative Choice*, Oxford, Basil Blackwell (1988).

W.B. Arthur, Competing technologies and lock-in by historical events: the dynamics of allocation under increasing returns, International Institute for Applied Systems Analysis, Paper WP-83–90, Laxenburg, Austria, (1983).

W.B. Arthur, Competing technologies: an overview in; G. Dosi, C. Freeman, R. Nelson, G. Silverberg, L. Soete (Eds) *Technical Change and Economic Theory*, London, Pinter (1988).

W.B. Arthur, Competing technologies, increasing returns and lock-in by historical events, *Economic Journal,* **99**, (1989) 116–131.

A.A. Berle, G.C. Means, *The Modern Corporation and Private Property*, New York, Macmillan (1933).

K. Boulding, *Ecodynamics, A New Theory of Societal Evolution*, Beverly Hills, London, Sage (1978).

K. Boulding, *Evolutionary Economics*, Beverly Hills, London, Sage (1981).

T. Burns, G.M. Stalker, *The Management of Innovation*, London, Tavistock (1961).

N. Clark, C. Juma, *Long Run Economics: an Evolutionary Approach to Economic Growth*, London, Pinter (1987).

N. Clark, C. Juma, 'Evolutionary theories in economic thought', in G. Dosi, C. Freeman, R. Nelson, G. Silverberg, L. Soete (Eds) *Technical Change and Economic Theory*, London, Pinter (1988).

R. Coase, 'The nature of the firm' *Economica*, **4**, (1937) 386–405.

R. Coombs, P. Saviotti, V. Walsh, *Economics and Technological Change*, London, Macmillan (1987).

R.M. Cyert, J.G. March, *A Behavioral Theory of the Firm*, Englewood Cliffs, New Jersey, Prentice Hall 91963).

P. David, *Technical Choice Innovation and Economic Growth,* Cambridge, Cambridge University Press (1975).

G. Dosi, Technological Paradigms and Technological Trajectories: a Suggested Interpretation of the Determinants and Directions of Technical Change, *Research Policy,* **11** (1982) 147–162.

G. Dosi, Institutions and markets in a dynamic world, *The Manchester School,* **56** (1988) 119–146.

G. Dosi, C. Freeman, R. Nelson, G. Silverberg, L. Soete (Eds) *Technical Change and Economic Theory*, London, Pinter (1988).

N. Eldredge, S.J. Gould, Punctuated equilibria: an alternative to phyletic gradualism, in T.J.M. Schopf (Ed), *Models in Paleobiology*, San Francisco, Freeman, Cooper (1972).

J. Elster, *Explaining Technical Change*, Cambridge, Cambridge University Press (1983).

F.E. Emery, E.L. Trist, The causal texture of organisational environments, *Human Relations,* **18** (1965) 21–32.

F.E. Emery (Ed) *Systems Thinking*, Harmondsworth, Penguin Books, (1969).

W. Fellner, Two propositions in the theory of induced innovation, *Economic Journal,* **71** (1961) 305–308.

R.A. Fisher, *The Genetical Theory of Natural Selection*, (Drew, New York, 1958, originally published 1929).

J. Foster, *Evolutionary Macroeconomics*, London, Allen & Unwin, (1985).

C. Freeman, *The Economics of Industrial Innovation*, London, Pinter (1982).

C. Freeman, J. Clark, L. Soete, *Unemployment and Technical Innovation: a Study of Long Waves in Economic Development*, London, Pinter (1982).

C. Freeman, C. Perez, Structural crises of adjustment, in G. Dosi, C. Freeman, R. Nelson, G. Silverberg, L. Soete (Eds) *Technical Change and Economic Theory*, London, Pinter (1988).

R.M. Goodwin, The nonlinear accelerator and the persistence of business cycles, *Econometrica,* **19** (1951) 1–17.

M.T. Hannan, J. Freeman, 'The population ecology of organizations' *American Journal of Sociology,* **82** (1977) 929–964.

M.T. Hannan, J. Freeman, *Organisational Ecology*, Cambridge, Mass., Harvard University Press (1989)

R.A. Heiner, The origin of predictable behaviour, *American Economic Review,* **73** (1983) 560–595.

R.A. Heiner, Imperfect decisions and routinsed production : implications for evolutionary modeling and inertial technical change in G. Dosi, C. Freeman, R. Nelson, G. Silverberg, L. Soete (Eds) *Technical Change and Economic Theory*, London, Pinter (1988).

G. Hodgson, *Economics and Institutions*, London, Polity Press (1988).

R. Jurkovitch, A core typology of organizational environments, *Administrative Science Quarterly,* **19** (1969) 380–394.

T. Kuhn, *The Structure of Scientific Revolutions*, Chicago, Chicago University Press (1962).

P. Lawrence, J.L. Lorsch, *Organisations and Environments, Managing Differentiation and Integration*, Cambridge, Mass., Harvard University Press, (1967).

A.J. Lotka, *Elements of Mathematical Biology*, New York, Dover, (1924, 1956).

B. McKelvey, *Organizational Systematics: Taxonomy, Evolution, Classification*, Berkeley, Unviersity of California Press (1982).

P.J. McNulty, Economic theory and the meaning of competition, *Quarterly Journal of Economics,* **82** (1968) 639–656.

Marshall, (1890), *Principles of Economics*, London, Macmillan, (1890, 1927), 8th Edn.

J. Maynard Smith, *Models in Ecology*, Cambridge, Cambridge University Press (1974).

J.S. Metcalfe, 'Impulse and diffusion in the study of technological change', *Futures,* **13** (1981)

J.S. Metcalfe, Technological innovation and the competitive process, *Greek Economic Review,* **6** (1984) 287–316.

J.S. Metcalfe, M. Gibbons, Technology, variety and organisation, mineo, Manchester University (1987).

J.S. Metcalfe, M. Gibbons, Technology, variety and organization : a systematic perspective on the competitive process, in *Research on Technological Innovation, Management and Policy,* **4** (1989) 153–193.

G. Morgan, *Images of Organisation*, London Sage (1986).

R. Nelson and S. Winter, 'Neoclassical vs. evolutionary theories of economic growth: critique and prospectus' *Economic Journal,* **84** (1974) 886–905.

R. Nelson and S. Winter, In Search of Useful Theory of Innovation, *Research Policy,* **6** (1977) 36–76.

R. Nelson and S. Winter, *An Evolutionary Theory of Economic Change*, Cambridge MA, Harvard University Press 1982).

K. Pavtit, Patterns of technical change: towards a taxonomy and a theory, *Research Policy,* **13** (1984) 343–374.

E. Penrose, *The Theory of the Growth of the Firm*, Oxford, Blackwell, (1959), (1980)

C. Perez, Structural change and assimilation of new technologies in the economic system, *Futures,* **15** (1983) 357–375.

M. Polanyi, *Personal Knowledge: Towards a Post Critical Phylosophy*, New York, Harper Torchbooks (1962).

I. Prigogine, Order through fluctuations in Self-Organisation and Social System, in: E. Jantsch and C.H. Waddington (Eds), *Evolution and Consciousness, Human Systems in Transition*, Addison Wesley, New York, (1976)

I. Prigogine, P.M. Allen, R. Herman, Long term trends and the evolution of complexity, in E. Laaszlo (Ed) *Goals in a Global Community,* London, Pergamon Press (1987).

I. Prigogine, A new rationality? in I. Prigogine, M Sanglier (eds), *Laws of Nature and Human Conduct,* Bruxelles, GORDES, (1987)

I. Prigogine, I. Stengers, *Order out of Chaos,* London, Fontana, (1984).

N. Rosenberg, *Perspectives on Technology*, Cambridge, Cambridge University Press (1976).

N. Rosenberg, *The Black Box: Technology and Economics*, Cambridge, Cambridge University Press (1982).

D. Sahal, Alternative Conceptions of Technology, *Research Policy,* **10** (1981a) 2–24

D. Sahal, *Patterns of Technological Innovation*, Reading, mass., Addison Wesley (1981b).

P.P. Saviotti and J.S. Metcalfe, A Theoretical Approach to the Construction of Technological Output Indictors, *Reseach Policy,* **13** (1984) 141–151

P.P. Saviotti, A. Bowman, Indicators of output of technology, in M. Gibbons, P. Gummett, B. NM. Udgaonkar, *Science and Technology Policy in the 1980s and Beyond*, London, Longman (1984).

P.P. Saviotti, An approach to the measurement of technolgoy based on the hedoic price method and related methods, *Technological Forecasting and Social Change,* **27** (1985) 309–334.

P.P. Saviotti, Systems theory and technological change, *Futures,* **18** (1986) 773–786.

P.P. Saviotti, Information, variety and entropy in technoeconomic development *Research Policy,* **17** (198a) 89–103.

P.P. Saviotti, A characteristics approach to technological evolution and competition, presented at the Conference on Recent Developments in the Economics of Technological Change, Manchester, 22–23 March (1988b).

J. Schumpeter, *The Theory of Economic Develoment*, Cambridge, MA, Harvard University Press, (1934, original edition 1912).

J. Schumpeter, *Capitalism, Socialism and Democracy*, London George Allen and Unwin (1943, 5th Edition 1976).

M. Scriven, Explanation and prediction in evolutionary theories: satisfactory explanation of the past is possible even when prediction of the future is impossible, *Science,* **130** (1959) 477–482.

G. Silverberg, Modelling economic dynamics and technological change, in G. Dosi, C. Freeman, R. Nelson, G. Silverberg, L. Soete (Eds) *Technical Change and Economic Theory*, London, Pinter (1988).

H.A. Simon, *Administrative Behaviour*, New York, Free Press (1947).

H.A. Simon, *Models of Man: Social and Rational*, New York, Wiley, (1957).

P. Stoneman, *The Economic Analysis of Technological Change*, Oxford, Oxford University Press, (1984).

D.J. Teece, 'The market for know how and efficient international transfer of technology', *The Annals of the Academy of Political and Social Science*, (November 1981).

D.J. Teece, Profiting from technological innovation', *Research Policy,* **15** (1986) 285–305.

S. Terreberry, The evolution of organizational environments, *Administrative Science Quarterly,* **12** (1968) 590–613.

T. Veblen, 'Why is economics not an evolutionary science?', *Quarterly Journal of Economics,* **12** (1898) 374–397.

L. Von Bertalanffy, The theory of open systems in physics and bioogy, in F.E. Emery (Ed) *Systems Thinking*, Harmondsworth, Penguin Books, (1969).

J. Woodward, *Industrial Organisation: Theory and Practice*, Oxford, Oxford University Press, (1965).

Is There a General Theory of Biological Evolution?

G.S. MANI
Department of Theoretical Physics, Schuster Laboratory, University of Manchester, Oxford Road, Manchester M13 9PL, UK

1. INTRODUCTION

Evolutionary theories in biology evoke more controversy both among biologists and among onlookers than theories in any other branch of science with the possible exception of social and political theories. Part of this controversy is due to the very diverse range of approaches among evolutionary biologists and the lack of any unifying theory in evolutionary biology. In recent years the experience has been more and more like Alice's experience with the Cheshire Cat. The more we go along the reductionist path the harder it becomes to recognise any synthesis in biological evolution.

It could be argued that the neo-Darwinism and the extension of it to the 'modern synthesis' provide a unifying model for the evolution of the biological systems. Unfortunately there exists no unique description of what the modern synthesis implies. The modern synthesis is a set of ideas which is sufficiently broad and variable to accommodate a multitude of truths and sins. The early advocates of modern synthesis, especially Dobzhansky (1937) and Simpson (1944), were pluralistic and expansive in their approach. With the passage of time these same authors hardened their attitudes and had become more inflexible (Dobzhansky, 1951; Simpson, 1953). An analysis of this transformation to a more rigid attitude by Simpson between 1944 and 1953 is discussed by

Gould (1980). A large number of critics of the neo-Darwinian model use the works of Dobzhansky (1951), Simpson (1953) and Mayr (1963) to define the neo-Darwinian theory. This has been unfortunate, since the deficiencies in the models proposed by the above mentioned authors have then been highlighted by the critics as positive proof that Darwinian theory is inadequate in explaining biological evolution. Various garbled versions of such criticisms by eminent biologists have been used by creationists and by popular media to question whether evolution in biological systems did at all occur. In this chapter I shall discuss some of the problems encountered in the study of biological evolution and indicate why it is not feasible at present to formulate a theory of biological evolution that takes into account evolution from the molecular level through the organismic level to the global ecosystem level. It may be that no that no single theory can be constructed to describe biological evoultion at all the various levels of hierarchy. But, as we shall see in the following sections, within each level, evoultion can be understood on the basis of Darwinism, provided out application of Darwinian theory takes into account the constraints in the system arising through laws of physics and chemistry. It is our inability to incorporate such constraints into the model and to bridge the gaps between the various levels of hierarchy that has led to the present tendency among some biologists to reject Darwinism outright and yet provide no tangible model to replace it.

2. LAMARCK, DARWIN AND THE RISE OF NEO-DARWINISM

What we define as *Lamarckism* today is a much modified version of the views regarding evolution held by the French biologist Lamarck (1744–1829). According to this theory, an organism, during its life, adapts to the enviromental conditions to which it is exposed and such adaptations can be passed on to its offsprings. If this could occur, then it would contribute to the evolution of new and improved adaptations. Lamarck further held the view that organisms had an inherent drive to evolve into higher and more complex forms. Charles Darwin in the *Origin of Species* (1859) aimed to establish two basic axioms. First, that all existing organisms are descendants of one or a few simple ancestral forms, thus arguing that evolution has in fact occurred. Secondly, Darwin claimed that the mechanism for evolutionary change was natural selection operating on the variations in the population and that the origin of the variations was non-adaptive. Darwin had no mechanism for the origin and maintenance of variation and he thought that if the organisms acquired characteristics through *use and disuse* during their lifetime, these would influence the nature of the offspring and thus produce variations. In this sense, Darwin was a Lamarckian. But Darwin believed that natural selection was the primary cause of evoultion, thus rejecting the Lamarckian thesis that organisms had an intrinsic drive towards evolving into more complex forms.

The next major advance in the theory of evolution came from August Weismann (1834–1914). Weismann totally rejected the Lamarckian theory of inheritance of acquired characters. Instead he proposed that the development of the fertilized egg involved two independent processess of cell division, leading to the 'soma' cells and to the 'germ' cells. Acquired characters affect only the 'soma' cells while the germ line leads to gametes that start a new generation. The 'soma' cells would eventually die while the 'germ' cells are potentially immortal. Interestingly, as Maynard Smith notes, Weismann was led to this model more through his insight into the fact that the process of inheritance is a process of information flow, rather than through empirical evidence. Weismann could not conceive of a mechanism of 'reverse translation' whereby the hypertrophied muscles of a blacksmith could be translated into large muscles in the next generation (Maynard Smith, 1982). As Weismann remarked 'if one came across a case of the inheritance of acquired character, it would be as if a man sent a telegram to China, and it arrived translated into Chinese' (Maynard Smith, 1989). Weismann's theory provides a distinction between the phenotype and the genotype. The somatic cells, adapting to the environment are phenotypes while the germ cells form the genotypes. Information flows from the germ cells to the somatic cells defining the form and the character of the organism while the reverse is not possible. This is the 'central dogma' of evolutionary biology and as we shall see later, has its basis in molecular biology.

Weismann's postulate greatly strengthened Darwin's theory. Natural selection becomes the only process that can yield evolutionary changes. Since natural selection can only operate on the variations in the population, one still had the problem of the origin and maintenance of variation. This question was resolved with the rediscovery of Mendel's laws during the first decade of the 20th century and with the formulation of the chromosome theory of heredity. The rediscovery of Mendel's laws started a controversy regarding the importance of mutation that raged with greater ferocity than equivalent controversies today such as that between the punctuationalists and the gradualists. The Mendelians saw mutations as the origin of variation and hence as the potential starting point for the formation of new species. The biometricians, on the other hand, considered mutations to be irrelevant and rapidly lost through selection. To the biometricians continuous variation was the essence of evolution when acted on by the forces of natural selection. To the Mendelians, mutations being discrete, evolution could not be continuous and thus they considered themselves to be non-Darwinians! We have come full circle with the present controversy on whether evolution is continuous or jerky with the proponents of punctuationism, Eldredge and Gould, considering themselves as non-Darwinians!

The early controversy between the Mendelians and biometricians was resolved by the work of population geneticists Fisher, Haldane and Wright.

The continuous variation studied by the biometricians was shown to arise from a number of alternative genes at may loci. The discipline of population genetics became well established with robust mathematical arguments and a series of experiments, mainly done in the laboratory. There was a feeling of euphoria that biology, like physics, can be turned into a *hard science*. Since the mathematics becomes intractable when one considers systems with more than two loci, almost all mathematical attempts were concentrated on single locus models with very few alternative genes at that locus. Such simplistic models led to the formulation of Fisher's so called 'Fundamental Theorem' which states that the rate of increase of the average fitness of a population is equal to the population variance in fitness. As we saw above, in the Darwinian theory of natural selection, selection is assumed to operate on the variation within a population to improve the average fitness of the members of the population. Fisher's result encapsulates this traditional view of natural selection and was thus taken to be a mathematical vindication of Darwin's theory. Apart from the various difficulties in the interpretation of the Fundamental Theorem which I shall not discuss here, Fisher's theorem is a static result since it implies that natural selection would eventually produce a static population with no variation and hence no evolution. Thus for evolution to proceed there must be a continuous injection of variation through mechanisms such as mutation, migration and stochastic effects arising from finite populations.

As mentioned earlier, the characters such as height, size etc. which vary continuously within a population were explained in terms of population genetics through the action of a large number of genes at a large number of loci, with each locus contributing a small effect to the eventual phenotypic character. Since the mathematics of a multilocus systems is not easily amenable to analytic considerations, the theory for such continuous characters, called quantitative genetics, was mainly developed at the phenotypic level, based on assumed variation in the character space acted on by natural selection. Thus, though there was a reconciliation between the biometric and the Mendelian approach, the theory of quantitative characters was not easily reducible to the underlying genetic field as the theory of discrete characters.

Population genetics describes a 'bottom to top' dynamical system connecting the genetic space to the phenotypic space. In general it does not consider the ecological interactions such as competition, prey–predator interations and mutualism. Its dynamics is reversible and thûs, by itself, it provides no mechanism for birth and extinction of species. An application of population genetics to the problem of phenotypic evolution can be seen in the rapid increase of the melanic or darker forms of the peppered moth in the industrial regions of Britain. This is an oft-quoted example of the occurence of Darwinian evolution. The phenomenal increase in the melanic forms of this moth from the middle of the 18th century to the earlier part of this century has been attributed to differential bird predation. With change in background colour in the environment due to industrial pollution, the lighter or the typical variety is more visible than the melanic form, and hence suffers a relatively high rate of predation. Population genetics models have had reasonable success in explaining the change in the melanic frequency as well as its distribution over England and Wales from 1850 to 1960 (Cook and Mani, 1980; Mani, 1980, 1982, 1989). Since 1960, with the advent of the Clean Air Acts and with the replacement of domestic coal fires by central heating, the pollution levels have steadily decreased and the frequency of the melanic forms has also shown a steady decrease; and this change in the peppered moth has been predicted with considerable degree of success using population genetics models (Cook *et al.*, 1986; Clarke *et al.*, 1985; Mani, 1989). Thus the population genetics model, acting upon the variation in the population of genotypes, can produce reverse evolution in phenotypic forms, provided the variant genotypes are not lost. In other words, at this level of sophistication, population genetics models have no 'arrrow' in time. Long term biological evolution has a direction in time and thus any evolutionary model incorporating the mathematical structure of classical population genetics must also have features that allow for non-reversibility in long term evolution.

The work of population genetics led to the amorphous 'theory' called the 'modern synthesis' of evolutionary biology developed during the period 1930–1950 by, among others, Dobzhansky, Simpson, Stebbins, Mayr, Ford and Julian Huxley. The essentials of the theory can be summarized as follows: (1) the ultimate source of variability is point mutations; (2) natural selection acts on the variability to produce evolutionary changes; (3) rates and direction of changes are controlled by natural selection alone; (4) the selective process leads to adaptation. The modern synthesis emphasized that the neo-Darwinian mechanisms of natural selection in Mendelian populations were sufficient to explain the evolutionary process observed in nature and that it could also explain the observed geographic variation between and within species as well as the origin and formation of species.

In this section I have traced the origin and the evolution of the neo-Darwinian theory. This, as we saw, was based on the concept of natural selection in Mendelian populations and on the rejection of the Lamarckian process through the 'central dogma' of Weismann. Are there any cases in biological evolution where the Darwinian model is invalid or inapplicable?

There are a few cases in which the neo-Darwinian assumption appear to be dubious. For example, the members of a clone of the water flea *Daphnia* can have different morphologies that are adaptive in response to different environmental stimuli and once occurred, are transmitted through the egg. These are assumed to be caused by changes in gene activation rather than through changes in the sequence of DNA and therefore may be the result of a system of flexible response incorporated through natural selection. Similarly, flax plants are know to acquire morphological changes in the presence of high levels of fertilizer and these are seen to persist for a number of generations, though not indefinitely. The changes again are caused not by changes in DNA base sequence but through a process of producing multiple copies of the same gene, called gene amplification. In ciliated protozoa, newly acquired patterns of cilia are transmitted through binary division with no apparent change in the underlying DNA. This process is not still understood. Finally, cultural evolution, and animal learning do get propagated with no hereditary mechanism. The famous example is the transmission of the information concerning the change in milk bottle tops from one blue tit generation to another. The process of cultural evolution, (and hence economic and technological evolution), is Lamarckian. Recently, E. J. Steele, in his book *Somatic Selection and Adaptive Evolution* (1979) has proposed a mechanism whereby acquired characters might be transmitted to the offspring, which led a leading British newspaper to splash across its centre page that Darwinism is dead! Steele's mechanism involves the DNA-coded information from selective cells to be incorporated in RNA viruses, which are then carried to the germ cells and there incorporated into their chromosomes, resulting in the genetic trasnmission of an acquired character to the offspring. This is quite a plausible model since one knows that there are viruses, called retroviruses, which can transcribe their RNA information into DNA, thus slightly denting the 'central dogma'. The AIDS virus is one such virus. An experimental confirmation of Steele's hypothesis was reported (Gorczynski and Steele, 1980, 1981) but many attempts to reproduce the results of the experiment by others (Howard, 1982) have failed and thus Steele's model for Lamarckian transmission of characters is yet to be confirmed.

Why is the Lamarkian process, except for the case of cultural inheritance, so uncommon in biological evolution? The answer partly lies in the fact that there exists the gentic system which prevents reverse translation and thereby reduces error propagation. Phenotypic changes, (except learnt ones) are non-adaptive and thus a hereditary mechanism by which these are transmitted to the offspring would not be favoured by natural selection.

Are there any other mechanisms by which the variability in a population can be altered without the force of natural selection? The answer is yes. Kimura, a Japanese theoretical biologist and others have developed a robust mathematical model, called the neutral model (Kimura, 1984) by which they show that the variability in a population can be maintained without natural selection. In the neutral model, variability is generated through mutation and is lost through effects of finite population size, called genetic drift. The rate of loss depends upon population size and there is a dynamic balance between loss through genetic drift and gain through mutation. Thus the variability is maintained and altered without the intervention of a selective mechanism. There is ample evidence to show that at the molecular level the neutral model does operate, though not exclusively, but at the organismic level natural selection is the dominant force. One of the observations supporting the neutral model is the existence of an approximately even ticking molecular clock, which favours the conclusion that most changes in structural genes arise from neutral substitutions rather than from a grand averaging of various types of selection over time. In the past two decades, there has been very strong and sometimes acrimonious debates between the neutralists and the selectionists and occasionally these arguments have gone beyond the bounds of scientific thinking into the political and social arena. Selectionists are considered to be the product of capitalist free market philosophy evolved from the English gentry class. Neutralists, on the other hand, are products of Marxist thinking and socialist philosophy. Could the recent tendency for these two groups to converge towards each other's point of view be a reflection of the corresponding convergence between Thatcherists and Reaganists on one side and the socialists and Marxists on the other side? There indeed exists noise at the molecular level and adaptation at the individual level does prevail. What is needed is a unifying approach that can exhibit neutral effects and at the same time be amenable to selective forces. The extreme points of view developed by either the neutralists camp under the leadership of Kimura and Nei or by the selectionists such as Richard Dawkins in his well publicised book *The Selfish Gene* contribute little to the real understanding of evolution.

3. ECOLOGY, POPULATION BIOLOGY AND EVOLUTION OF COMMUNITIES

The population genetics models discussed in the previous section do not include selective interactions at the species level. The very complex nature of the inter- and intra- species interactions as well as the interactions with the environment, precludes the inclusion of such effects within these models. So the study of the populations at the species level has traditionally been divorced from the reductionist genetics models and the discipline of evolutionary ecology had emerged. The models in evolutionary ecology are based on the Darwinian theory that the evolution of adaptation takes place as a result of natural selection but the non-inclusion of genetics implies that Mendelian mechanisms as well as effects of mutations are neglected. It would be a great step forward in theoretical biology if a coherent and consistent model of evolutionary biology could be constructed through a marriage of genetics and ecology. The lack of such a 'Grand Unifying Model', to borrow a phrase from physics, has led Lewontin (1979) to bemoan the fact that ecology and genetics 'remain essentially separate disciplines, travelling separate paths while politely nodding to each other as they pass'.

Theoretical ecology has its origin in the work of Lotka and Volterra on prey–predator interactions. The simplest dynamical equations of the prey–predator system is based on (1) an exponential growth of prey in the absence of predator; (2) a rate of increase of the predator population that is proportional to the product of prey and predator densities; (3) a constant death rate for the predators. The resulting equations closely resemble the equations for the oscillations of a pendulum in the absence of friction. The prey and predator densities are analogous to the kinetic and potential energies in the case of the pendulum and exhibit cyclic changes with time. This basic set of equations is biologically unrealistic since it assumes that the prey density is controlled by the predator density alone and that of the predator by prey density alone in a linear fashion. Also the model is structually unstable since small perturbations could significantly alter the dynamics. A large industry has grown in constructing various modifications of the Volterra equations, introducing more realistic biological features. The Lotka-Volterra equations can also be modified to include other types of intra- and inter- species interactions such as competition and mutualism. It has been shown by May (1974) and others that the stability of the ecosystem is dependent upon these interactions. Such stability arguments have been based on complex models involving the network of food-webs in the ecosystem. These types of ecological models which analyse the conditions of stability of eco-systems have also been applied with varying degree of success in the evolution of economic systems.

The main direction of study in recent ecological models has been to understand the mechanism of coevolution of species. In the simplest version of competition between two species, when restricted to a single resource, one is led to the concept of competitve exclusion, which states that two species cannot coexist in the same niche provided there is only a single resource available. If, on the other hand, they compete for a range of resources, then it is possible that they partition the resources between them and both will survive and coexist. The ecological models suggest that two similar species will usually evolve so as to utilize different resources, resulting in the two species becoming morphologically more different. This process is usually referred to as 'character displacement'. Character displacement can also occur in the absence of resource competition, for example, with selection for mating isolation. In real ecosystems, any species would interact with a large number of competitors and parasites. In such cases, the models of coevolution assert that evolution will cease in a physically stable environment once the coexisting species attain their maximun fitness. This obviously is not borne out by observations in real ecosystems. Real ecosystems are rarely, if ever at all, in stable environments. As each species evolves, it alters the environment in which the other species coexist. This led Van Valen (1973) to propose the so called 'Red Queen' hypothesis. The name arises from the Red Queen explaining to Alice in Lewis Carroll's *Through the Looking Glass* 'here, you see, it takes all the running you can do, to keep in the same place. If you want to go somewhere else, you must run at least twice as fast as that!'. In this Red Queen model, evolution would continue indefinitely, even in a constant environment, as each species evolves to meet the changes in others. A model involving the Red Queen hypothesis and based in Darwinian theory has been proposed by Stenseth and Maynard Smith (1984). An analysis of the model shows that evolution in a physically constant environment may continue for ever at a constant rate in what is presumably in a species-rich systems but would cease in a species-poor system. The results are important to the extent that in a species-rich system, Darwinian evolution does not necessarily cease in a physically constant environment. Ecosystems are not physically stable over a period of time and in this case both species-rich and species-poor ecosystems exhibit continued evolution. The long-term behaviour of the ecosystem depends on the nature of the ecological interactions and on the stability of the physical environment. Thus on the basis of the Darwinian theory, an ecosystem could be in a fixed state or slowly changing for long periods of time and then suddenly have an accelerated evolution. Thus the observation of stasis followed by punctuation in fossil records need not be non-Darwinian as claimed by Eldridge and Gould. I have discussed at some length these types of ecological models since I feel that they would be the ones most useful in any theory of evolutionary economics. It would be interesting to apply the concepts of 'character displacement' and Red Queen behaviour to economic models.

There has been in biology a large amount of interest in understanding the evolution of behaviour and the evolution of sex, mostly based on the game theory concept 'Evolutionary Stable Strategy' (ESS) initiated primarily through the work of Maynard Smith (1976, 1982). Since these models discuss only short-term evolution, they will not be discussed here. The ESS models are basically devoid of genetics and hence could have useful application in economic systems.

4. THEORIES OF SPECIATION AND PALEONTOLOGY

In the previous two sections I have described genetic and ecological models that explain the dynamics of the changes that occur in biological evolution on the short or ecological time scale. A true understanding of biological evolution can only occur when we consider the mechanisms for formation of species and the rate at which they are formed and go extinct.

Even the most casual observer of nature could not but be fascinated by the enormous diversity of organisms which live in the world around him. Further he would not fail to recognise that this set of diverse organisms at any one place could be classified into unique and discrete subsets. The members within each subset are more or less morphologically connected while the individuals from different sets are most often morphologically distinct with, in general, no continuity among the subsets. In other words, the morphological differences from one subset to another are discrete with no subset having intermediate characteristics, and he would recognise these subsets to be species. He would, for example, be able to recognise two individual dogs from their differences in morphology but would classify them both as dogs since such differences are continuous while the morphological differences between cats and dogs are so discrete and noncontinuous that he would not confuse one with the other.

Thus even at the level of the most casual observer, the various organisms can be classified in terms of species and his limited observation is found to be true for a wide range of animals and plants. Formation of such discrete groups is seen to be almost universal whether one looks at animals or plants or at structurally simple or complex organisms. Any adequate theory of biological evolution should thus be able to explain the formation of such discrete species as well as the maintenance of continuous variation within each species.

This leads us to the question of the definition of species. In the case of sexually reproducing organisms, one can define species as being groups of organisms that are isolated from one another through mating incompatibility. Since mating and Mendelian segregation more or less randomize the genetic structure of each individual in the population, it is not surprising that there exists a continuous distribution of characters within the species population. The concept of species becomes more ambiguous when we consider organisms that reproduce asexually. In this case one can define species in terms of their morphological as well as functional differences. Since these differences are also more or less reflected at the molecular level, one could also use the study of DNA and protein sequences in resolving some of the ambiguity. Even at the level of the viruses, which can only reproduce with the aid of DNA in the host cell, one can easily distinguish the species that cause influenza from that, for example, that causes AIDS. In both cases it is known that at the molecular level there exists large diversity and yet the morphological and functional differences are sufficient to separate them into two different species. I shall not enter here into the debate concerning the concept of species, and I shall assume that in the large majority of cases we can classify organisms unambiguously into species.

The true extent of organic diversity can only be surmised at present. According to Mayr, there are at present around one and a quarter million known species in the animal and plant kingdom. Of these, one million belong to the animal kingdom. In the animal kingdom the largest abundance of species is found among insects, being around three quarter of a million. These estimates are those that have been recorded and the question arises as to how close these values are to the actual number of species existing today. Estimates between 10 to 80 million species has been put forward by various authors, based on the rate at which new species are being observed, especially in rain forests. When we consider that over 99 percent of all species since the origin of life are extinct now, one realises the enormous diversity in the living system. This leads to the interesting but unanswered (and maybe unanswerable) question of whether the diversity has increased or remained constant over a large part of the evolutionary time, namely, 2.7 billion years. I would venture the opinion that at least in the past 0.5 to 1.0 billion years the species diversity has been more or less constant with some periods when there has been episodic increase in the diversity. This arises from the fact that if the number of species in the past one billion years is anywhere between 1 to 2 billion, then the species turnover must be very rapid yielding on the average a reasonably constant value for the index of diversity. Thus there exists no evidence that biological evolution has proceeded in the direction of increasing species diversity, at least in the past half a billion to a billion years. In the very early stages of evolution the diversity must have been small as I shall discuss later. What then was the mechanism that triggered the enormous diversity observed in the past billion years? A partial answer to this question is given in the next section.

The species can be classified into higher hierarchical order by clustering those species that have more common morphological, taxonomic, behavioural and molecular characters. This process of classification can be continued to higher orders and one obtains a phylogenetic tree showing the direction of speciation. Such classifications are not unique and many possible trees can be obtained. The ambiguity increases with the number of species involved and with the complexity of the pattern. In general one chooses the most parsimonious tree. This criterion of parsimony is invoked because we cannot empirically have knowledge of the ture historical pattern but this does not imply that the hypothesis is true and that it mirrors the true historical pattern. Parsimony is accepted 'not because nature is parsimonious but because only parsimonious hypotheses can be defended by the investigator without resorting to authoritarianism or apriorism' (Wiley, 1975). In general one attempts to obtain consistent patterns using such varied data as morphological characters and changes at the molecular level and this, when coupled with paleontological study, yields an acceptable picture of the process of species evolution. Such a model need not necessarily reflect the historical pattern since most weight is given to the existing species, which as we saw earlier, are only a small sample of the total number of species that has evolved since the origin of life.

I shall now discuss why there exists such a large diversity in organic life, and why such diversity can be clustered into converging nests of higher hierarchical orders? We have already seen that species diversity has remained more or less constant over long periods of time, implying that species are formed and go extinct at a reasonably constant rate. This probably is what occurs at the lowest level with a regular turnover of species within the lowest group due to small accumulation of mutation and adaptation in agreement with the model of the neo-Darwinists. This is probably true, since at the lowest level the members within a group usually have very similar morphology at the embryonic stage and the differences are magnified in the adult stages. The large diversity is a reflection of the enormous range of environmental conditions over the earth.

Species which are well adapted to large environmental changes would then show greater diversity, with the caveat that this diversity would decrease with increasing body size, since due to energy considerations, the number of large body-sized animals must be restricted. Such evolutionary changes cannot be seen in fossil records which are only a extremely small sampling of the totality of species that has existed. Thus paleontology yields a very crude average of the evolutionary process. In recent years, based on paleontological studies, many biologists, primarily Gould and Eldredge have asserted that evolution proceeds in jerks with long periods of stasis followed by a sudden accelerated speciation, claiming that such a process is non-Darwinian. The sudden appearance of a new species in a geographical region need not necessarily be due to evolution but may be due to migration. Fossil records cannot often distinguish between the two possibilities. It is true that there are many cases when such punctuationism has occurred but as we saw earlier, this could be explained on the Darwinian basis. In general, most mutations have small effects but occasionally there could be a macro-mutation that produce a large change. How one can reconcile such mutations with large effects with adaptation to large-scale changes in the structure of the organism is still a matter of debate. Some biologists believe that the answer lies in the mechanism of development and my own sympathy is with them. This could still be adaptive and there is no reason why Darwinian theory should be discarded. I am not claiming that all morphological changes have an adaptive reason but the general evolutionary trend is consistent with Darwinism.

In any geographical region, the fact that energy comes from solar radiation through photosynthetic plants to the rest of the living system in that region implies a constraint on the amount of biomass that can be supported. Within this constraint, different species utilize the available resources through a food chain network. The stability of the ecological community in the region depends on the stability of the food chain and evolutionary changes take place through changes in the food chain caused by environmental fluctuations. When we consider the global situation, it is selfevident that the environmental conditions are infinitely variable and hence could produce large species diversity. It may be that the constancy of species diversity over evolutionary time arises from the restriction on the photosynthetic efficiency in converting the constant solar energy into biomass. But why in spite of speciation and extinction, the diversity is apparently maintained constant is not fully understood.

Up to now I have discussed the diversity of life and the various mechanisms that are invoked to explain the maintenance of this diversity. I now turn to the question of how species originate. Following on the views of Darwin as expressed in the *Origin of Species*, the modern synthesis has used the accepted model of adaptive geographic variation as a paradigm for the origin of species. Populations get geographically isolated with no gene flow between them. Under this condition mutational and adaptive changes accumulate at a slow rate, and after sufficient divergence between the populations they become reproductively isolated. Thus successful speciation is a cumulative and sequential process driven by selection through a large number of generations. This mode of speciation is called *allopatric speciation*, the allopatric mode is an extension of the standard population genetic model for microevolution to the species level. It is certainly true that many species have originated this way. But recent studies show that allopatric speciation may not be the only, and perhaps not the most dominant, mode for the origin of species. Species occupying the same geographical area could be isolated by many other mechanisms, such as low fitness of hybrids and temporal isolation through, for example, delay in emergence time and thus provide a barrier for gene flow. These could then accumulate changes that eventually produce new species. Such modes of speciation are called *sympatric speciation*. The inherent assumption in the allopatric mode as described by modern synthesis is that the population within a geographical region is large and randomly mating. In reality the physical space is divided into demes within which random mating can occur. The gene flow between the demes are often too weak to overcome selection and other intrinsic processes within local demes (Ehrlich and Raven, 1969). And it is thus possible that local demes are sufficiently independent for potential speciation. If this be the case, then the distinction between allopatric and sympatric speciation becomes cloudy. If demes are largely independent, new species can originate anywhere within the geographical range of an ancestral form. White (1978) believes that speciation could also occur between populations in continual contact, provided the gene flow can be overcome by strong selection or through sheer rapidity of potential fixation of major chromosomal variants. Even Mayr (1963) acknowledged that sympatric formation of new insect species could arise through switch in plant preference for host-specific forms. In recent years, many authors have pointed out that chromosomal alterations could provide an isolating mechanism. Also there have been suggestion that speciation may be more a matter of gene regulation and rearrangement than changes in structural genes that adapt local populations to varying environment. We thus see a multiplicity of mechanisms that induce new speciation. Many biologists have remarked that since the allopatric mode of speciation through accumulation of small changes is not always observed and since speciation can occur on the ecological time scale, the premise of the exponents of the modern synthesis cannot be true and thus the processes must be non-Darwinian. I cannot see the logic of this argument and as I have argued earlier, Darwinian theory can explain such sympatric speciation and one need not invoke any new mechanism. Natural selection would still operate at the species level, though some of the assumptions in the formulation of population genetics and evolutionary ecology may have to be altered. But then, science proceeds through such a process of evolution of theory.

Another question one can ask is whether biological evolution proceeds towards greater complexity? In many phyletic lines of animals, the general tendency is towards larger size. This general rule works more often than it fails and is call 'Cope's rule'. The opposite phenomenon of gradual size decrease is very rare. In general when such size decrease occurs in a lineage, it leads to extinction. Gould (1983, pp.314) cites two examples in which a decrease in body size was followed by extinction. (This article of Gould's is a humorous essay on the phyletic size decrease in the products of human manufacture!) The reason why Cope's rule operates in biological system is not well understood. But associated with Cope's rule is the general belief that evolution does proceed towards systems of greater complexity. Since evolution is not teleological, why it should move towards large size and perhaps greater complexity is an open question.

5. OXYGEN REVOLUTION, THERMODYNAMICS AND COMPLEXITY

The two most important steps in biological evolution have been the innovation of photosynthesis and respiration. Evolution would have come to a dead end but for these two evolutionary landmarks. In the primordial world, the atmosphere had very little oxygen and the surface of the Earth was exposed to the harmful ultraviolet radiation from the Sun. Around three billion years ago, biochemical evolution had proceeded far enough for discrete heterotrophic organisms to appear. As their name implies, these organisms derived their energy and nourishment from externally formed organic molecules. They were probably found in deep waters like seas and lakes to avoid the harmful effects of ultraviolet radiation. The oceans were a very dilute, virtually oxygen-free, broth of organic molecules. In the absence of oxygen, the main energy source for these primal organisms was through the process of fermentation, in which energy was derived by breaking organic molecules and rearranging their parts. The most familiar example of such a process is fermentation of sugar by yeast to produce alcohol. In this chemical reaction, 1 gramme of sugar is converted into 0.49 grammes of carbon dioxide and 0.51 grammes of alcohol, releasing around 110 calories of energy. If on the other hand, sugar can be completely oxidized to produce carbon dioxide and water, as happens in respiration, then 1 g of sugar would produce around 3900 calories of energy! Fermentation is a very inefficient process for energy production compared with oxidation. Also, in the fermentation process various poisonous waste products such a alcohol, lactic acid and formic acid are produced. Thus, these early primitive, and basically inefficient forms of life consumed the organic compounds in the oceanic broth to live, grow and reproduce and in so doing, created vast amount of poisonous pollutants in the environment. Such a process of utilization of limited resources with the consequent production of self-destructive waste products would have brought evolution and perhaps life to a halt.

The reasons why these primordial organisms faced an evolutionary dead end can be summarized as follows: (1) They were consuming the dilute organic materials from the surrounding water envrionment; (2) they were accumulating poisonous waste products in their environment; (3) their energy production was very inefficient; (4) they were not in equilibrium with the environment; (5) they were unable to colonize the land mass because of the harmful effects of ultraviolet radiation. All these points have extreme relevance to human cultural and technological evolution. Fortunately the waste product carbon dioxide saved the situation, at least partially. This gas was entering the ocean and the atmosphere in ever increasing quantities. Some time before these early heterotrophic organisms had exhausted the organic compounds in the environment, they had managed successfully to evolve new organisms which were capable of utilizing the carbon dioxide produced by their predecessors. In this process, the energy of sunlight was made use of in processing organic molecules from carbon dioxide and water and yielding oxygen. For example, using 3900 calories of solar energy, they were able to produce one gramme of sugar molecule. Though photosynthesis made the organisms independent of the limited external supply of organic molecules, energy for the metabolism and reproduction was still being produced through the inefficient process of fermentation creating poisonous waste products. The next large step in the evolution solved this problem. The oxygen from the photosynthetic process was entering the atmosphere in ever increasing quantities, changing the composition of Earth's atmosphere. Also the oxygen, rising to higher altitudes was helping in the formation of the ozone layer through photochemical reactions. The ozone layer grew to sufficient thickness to shield the Earth from ultraviolet radiation making it possible for life to spread from oceans to land and air. It was during this period that a radically new evolutionary stage, namely respiration, as a method of energy production was developed. In this process sugar was reduced to carbon dioxide and water, releasing around 35 times as much energy from unit mass of sugar as was possible with the fermentation process. The waste products were no longer harmful, and the carbon dioxide released is recycled through photosynthesis. The oxygen–carbon dioxide cycle can be considered as a giant wheel pumping solar energy into the living system. The new atmosphere, composed mainly of nitrogen and oxygen, maintained in equilibrium through the dual process of photosynthesis and respiration, possessed a stability that has lasted for the past billion years. There exist in the biosphere many other cycles such as the hydrological (water) cycle, the nitrogen cycle, the sulphur cycle, the phosphorus cycle and others. These cycles are interconnected through the living system and the external environment.

Apart from the energy requirement satisfied through the carbon dioxide–oxygen cycle, living systems require various other materials such as sulphur, phosphorus, nitrogen etc. as building material. Since these materials are finite resources, for evolution to be maintained they need to be conserved through recycling. As remarked above, the material cycles involve both the physical environment and the layer of life. All material cycles have three basic pathways in common. The first of these is growth, taking food matter from the environment. Involved in this growth phase are all energy-requiring processes of the living organisms, including motion, perception, reproduction and body maintenance. Plants and other autotrophic organisms absorb material from the environment and convert it into usable organic compounds. Animals derive their raw materials from plants or other animals and after processing these raw materials, the waste products are released into the physical environment. The second path in the cycle is the process of death of the organism and the accumulation in the environment of waste products and dead organic matter. These are made up of complex organic molecules that need to be degraded before they can re-enter the food chain. In the third step the metabolic waste products and dead organic matter are decomposed into simpler chemical compounds that can be reused to start the cycle again. This decomposition is accomplished by a specialized and diverse group of micro-organisms called decomposers.

There are two points to note in this basic cyclic structure that exists in the biosphere. Firstly, matter assimiliated by organisms is eventually returned to the environment in its original form for further usage. Thus there is no such thing as a waste product in the broad ecological context. Waste exists only in terms of specific species. The waste product of one species forms the food input of another. The self-regulatory equilibrium that exists in the biosphere depends on this recycling of waste products and thus has deep significance for the stability of the ecosystem. Secondly, the cycle can only be maintained through birth and death of organisms. Thus birth and death form a necessary condition for the long term stability of evolutionary systems. Living systems are thermodynamically far from equilibrium, consuming negentrophy from the sun to produce highly complex ordered systems. The various interconnecting cycles discussed above are an essential feature for the maintenance of the system over large periods of time. There is a slow, small input of materials into the biosphere through the weathering of rocks and volcanic eruptions. The amount that is thus gained is balanced through sedimentation in the oceans. This part of the material cycle is very slow compared to the recycling time involved in the biological cycle. The material cycle can be considered as made up of two interacting flywheels, a slow wheel based on geochemical mechanisms and a fast wheel turning through the biochemical process. The energy to power the slow wheel comes from the Sun as well as from the energy stored in the interior of the Earth. The energy from the interior of the Earth is used in the uplifting of crustal rocks. Winds and rain that produce weathering of surface rocks derive their energy from the Sun. The biochemical cycle utilizes solar energy to maintain its motion. The solar energy is stored as potential energy in molecular bonds of organic molecules and is released during various metablic processes that take place in the organism.

I shall now describe briefly the mechanism by which cells transform energy during respiratory metabolism, since this process gives some clue to the question of large-scale conservatism at the molecular level. Surprisingly, oxygen is destructive to all forms of carbon-based life. Molecular oxygen reacts spontaneously with reduced organic compounds. There are, for example, very sensitive anaerobes that cannot tolerate even 1–2% increase in the oxygen concentration above the present levels found in the atmosphere. It is the toxic nature of oxygen that has necessitated the evolution of a very complex mechanism for respiratory metabolsim. The evolutionary process had to contend with the following criteria. The energy requirment for higher forms of life can only be met with oxidative metabolsim. Yet the burning of oxygen in the furnaces of organisms had to be at a low temperature to prevent cellular damage. Nature's solution for satisfying these two criteria was to endow molecular oxygen with the role of an electron sink or equivalently, a hydrogen acceptor in biological oxidation. Biological oxidation proceeds not so much by the addition of oxygen as by the removal of hydrogen. This process of dehydrogenation is brought about by the action of enzymes which remove the hyrogen atom from the substrate molecules that function as hydrogen carriers. The energy released in the oxidation process is harnessed to regenerate the molecule ATP from the molecule ADP and phosphate. The ATP-ADP cycle can be thought of as a fule cell for energy storage. This method of storage is utilized by both photosynthetic and respiratory processes, exhibiting the economy of usage in nature. The respiratory cycle, known as Krebs cycle consists of a large number of stages, such that the free energy of the glucose molecule is transferred in smaller units to thirty six ATP molecules. The hydrogen in the glucose molecule ends up as water and the carbon atoms as molecules of carbon dioxide. The efficiency for the conversion of the free energy stored in glucose to the 'organic fuel cells' ATP is around 66%. When this energy finally gets converted into muscular work, the overall efficiency drops to around 30%. But is worth noting that a large part of the energy stored in ATP molcules is used in the direct biosynthesis of macromolecules like proteins without conversion first to mechanical or heat energy.

A modern steam generating plant is capable of converting around 30% of energy input into useful work and thus compares favourably with the efficiency for the production of mechanical work by biological cells. But to achieve this, the steam temperature has to be as high as 600°–800° C. Thus evolution has been able to realise overall energy efficiency comparable to, if not better than, that obtained in human technology but at room temperature. The price that nature had to pay for such a low temperature, high efficiency energy conversion system was to evolve an extremely complex chemical techonolgy.

As pointed out earlier, the evolution of photosynthesis and respiration made it possible for the living system to utilize the constant energy source of the sun, yielding a mechanism for an increase in the specific energy available. This thus paved the way for evolving multicellular systems. Also the creation of the ozone layer increased the physical environment in which life can exist. The main constraint on the maximum possible biomass is the amount of sunlight that is incident upon a region and the very low photosynthetic efficiency of 1–2%. Thus the conditions were optimum around 500 million years ago for an explosive increase in species diversity. This period, called the Paleozic era, was also marked by large scale continental drifts, producing vast swampy regions. The decaying biomass of this era was turned into bituminous coal over geological time scale. A large fraction of the present day coal deposits, almost 60%, is bituminous coal that originated around half a billion years ago. When we indiscriminately use our fossil fuel resource, it might be worth while reflecting on the sequence of favourable events that helped to accumulate this heritage for us.

6. MOLECULAR AND BIOCHEMICAL EVOLUTION

Since the discovery of the helical structure of the DNA by Watson and Crick in 1953, there has been an enormous information explosion in molecular biology. The success of molecular biology in explaining many aspects of the genetics of the organism has led to a strong and widespread belief among molecular biologists that biology can be reduced to the molecular level and nothing new can be gained by studying the properties at the level of the organism. There exists at least among some molecular biologists the feeling that biological systems are no more than a collection of atoms and molecules oriented in specific fashion through chemical and physical laws and that all aspects of biology can be ultimatley explained in terms of the physics and chemistry of macromolecules. It is true that organisms are made up of atoms and molecules, but they are highly complex patterns of form and structure of these atoms and molecules. In short, living processes are highly *improbable* patterns of physical and chemical processes. In view of the claims by some molecular biologists, it would be interesting to ask what the study of the organism at the molecular level tell us about biological evolution.

The DNA, as it exists in the chromosomes of cells, consists of two polymeric fibres wound around each other in a helical way – the *double helix.* Each fibre consists of a backbone of sugar-phosphate groups to which are attached four types of molecular units, called nucleotides, in some specific order along its length. The chromosome is composed of the long DNA polymer coiled into a complex three dimensional structure and various proteins needed for the basic functioning of the DNA molecule. The number of nucleotides in the DNA can vary from a few thousand to hundreds of millions depending on the organism. Along the length of the DNA molecule are regions of finite length, called coding regions, whose beginning and end are recognised through specific sequence of nucleotides. The coding region contains the information that is translated into proteins through a complex mechanism involving intermediaries called RNA. The protein molecule is made up from molecular units called amino acids. There are twenty different types of amino acids and these are almost universally used by all living systems in producing proteins. In the coding region of the DNA, the amino acids are recognised through specific triplet sequences of the nucleotides. Thus the coding regions are made up of such triplets defining the protein they code for. The genes we have alluded to earlier are the coding regions and we could assume that each gene corresponds to a single protein, though this is not universally true. The mechanism of translation is one-way, namely from DNA to RNA to proteins. This then is the molecular explanation of the 'Central Dogma' discussed in an earlier section. Though there are a few instances in which an RNA can be translated back into DNA, the retroviruses being one of them, in no case has one found proteins to be reverse translated to DNA. It is this central dogma that prevents a Lamarckian mechanism for biological evolution. The proteins together with the environmental interaction produce phentoypes while the coding regions are the genotypes that are passed on more or less faithfully to the offspring. The DNA also has the ability to self replicate with the help of enzymes that are coded by it. Thus when a cell divides, each of the resultant cells contains an almost exact copy of the original DNA. In man, for example, there are ten million cells and each of these carry a complete copy of the DNA in their chromosomes. The coding regions, producing proteins and enzymes that determine the structure, form and functions of cells and the organisms, are called structural genes. There are also coded into the DNA regulatory genes that determine the expression of structural genes both spatially (in different cells) and temporally.

At the molecular level one sees an enormous conservatism in biological evolution. Many of the chemical elements such as oxygen, carbon and hydrogen are found in very similar proportion in almost all living organisms. The DNA structure is universal and the DNA code is also almost universal. There exist some differences in what are called the mitochondrial DNA that regulates energy production. All organisms are based on cellular structure with the cells falling into two general patterns, namely, the prokaryotes and the eukaryotes. All organisms utilize the same twenty amino acids for constructing proteins. It is even more interesting that most of the proteins are found in an extremely large range of organisms. The number of novel types of proteins in the organisms of any species is very small indeed. The initiator and terminator sequences in the coding regions of the DNA are almost universal. Quite often, instead of producing new and novel types of proteins, the organisms use homologous proteins to perform different functions. Finally almost all macromolecules involved in the living system have a specific chiral structure. Many chemical substances exist in two dissymetric molecular forms, the right (D) and the left (L) isomers, whose structures are the mirror reflection of one another. In most organisms, the principal constituents of protoplasm, particularly the amino acids, are represented exclusively by L isomers. Certain bacteria do have D type amino acids but these are mostly componenets of antimetabolites, toxic to competing organisms. The essential proteins of these bacteria do contain exclusively L isomers. The D isomers are also found among some earthworms and insects. Before stepping through the mirror into the nonsense world behind the looking glass, Alice said to her kitten 'How would you like to live in Looking-Glass House? I wonder if they'd give you milk in there? Perhaps Looking-Glass milk isn't good to drink'. In fact it follows from what has been said above regarding the asymmetry in the molecules of living beings that to drink mirror-image milk is, at best, useless.

What is the cause for such a high degree of conservatism and how can we reconcile this with the extreme diversity observed at the phenotypic and species level? A partial answer to this question can be found in the earlier discussion of thermodynamics. We have already seen the type of constraints required for maintaining respiration. The ATP and ADP molecules may have been present from the early stages of pre-biotic evolution. The need for the mechanism of electron transport in converting the free energy in glucose into the energy-rich ATP molecules puts a severe constraint on the chemical mechanism. Once the Krebs cycle has been established, it would be difficult to alter the process without seriously affecting the whole stability of the system. The extraction of energy from the ATP molecules for both the construction of macromolecules as well as for mechanical energy imposes restrictions on the type of enzymes needed and the various metabolic pathways become more or less fixed. The various other cycles that pass through the living strata produce similar constraints. Thus for a large class of proteins, mutational changes that do not alter the form and hence the function of the proteins are the only ones that can, in general, survive. Thus evolutionary changes at the molecular level are, for large part, neutral and the claims by the neutralists are justified. The conservatism and hence the neutrality must have occurred during the period when the photosynthetic-respiratory mechanism got established. Thus the long term evolutionary stability through recycling of materials imposes conservatism at the molecular level. I shall discuss the significance of this to technological evolution later.

If molecular evolution, at least of the coding regions and hence of enzymes and proteins, is neutral and conservative, how can we understand the large diversity seen at the phenotypic and species level? The answer to this lies in the fact that the eventual structure and form of the organism depend on the spatial and temporal distribution of proteins controlled by regulatory genes especially in the early stages of development. Different cells produce different amount and types of proteins at different stages in the development. In spite of the vast strides made by molecular biologists, the mechanism for the control and production of proteins at the right time and at the right place is still a mystery. The adaptive ability of the phenotype thus formed would determine which would survive. Thus the Darwinian process is more applicable at the organismic level and at the molecular level the various constraints have established an almost conservative system. If this be true, then one should see the effects of selection more on regulatory genes, especially at the early stages of development than in structural genes. Developmental pathways controlled by the protein production in space and time, thus determine the structure and form of the adult experiencing Darwinian selection. Thus it is not unreasonable to predict that mutational changes that affect the spatial and temporal expression of structural genes could produce large scale evolutionary change, provided the pre-adult and the adult stages can adapt to the environment. Most of these mutational changes would be harmful but favourable changes could occur at infrequent intervals.

A surprising feature of the DNA molecule in higher organisms is the occurrence of a large amount of repeated sequences, occupying betwwen 30 and 90% of the polymer length, with apparently no function. Many biologists regard these 'junk' regions of DNA as *selfish DNA* replicating selfishly with no advantage and presumably, no disadvantage to the organism. This, they claim, is a clear example of non-Darwinian evolution with no adaptive significance. I believe that the highly repeated, apparently functionless regions do have an important function. The replication time for a DNA molecule is related to its length. In complex organisms, the intrinsic time scales, for example of metabolism, heart beat etc., must all be tuned such that the system can function efficiently. This tuning process need not be very critical. The situation has close analogy with complex manufacturing systems, where the timing of various inputs and outputs are very important for the efficient functioning of the system. Interestingly, in mammals it is well established that the time required to complete 50% of its growth is about 3% of its lifespan, independent of body size. Similarly, it is found that 1.5% of a mammal's life span is required for gestation no matter how big it is. Both respiratory cycle and heart beat cycle occupy nearly size-independent fractions of its lifespan. As a consequence, every mammal can expect to live for one third of a billion breath cycles and for 1.5 billion heartbeats (McMahon and Bonner, 1983). These examples clearly indicate that there are constraints on the various intrinsic time scales in the system. DNA replication is associated with cell division which occurs in order to increase the number of cells during the developmental stage or to partially replace dead cells in developed adults. Since both these processes are related to the various time scales in the system, it should not be surprising that the length of the DNA is adjusted so that its replication time is in conformity with the rest of the system. I do not believe that this is a very severe constraint. The replication time scale can probably vary by a fair amount. Also, the constraint is on the whole of the genome in a statistical way. But this constraint still provides an adaptive significance for the existence of junk DNA. How could such junk DNA been introduced into the organism's genome? The answer lies in the fact that there are entities called plasmids and transposons that are bits of DNA which can attach themselves at specific points in the organism's DNA. Multiple copies of this transposed DNA material can be produced through various mechanisms present in the genome. Such mechanisms would be adaptively suppressed or would be entirely absent in lower organisms such as viruses and bacteria. In the case of the higher organisms multiple copies would be produced. The number of copies would depend upon what happens in other regions of the genome. This will not be a static process, but a dynamic one with copy loss at some regions and copy gain at other regions. Thus one would expect a large variability in the copy number of any specific repeat unit. If the transposed element lodges itself in the middle of a coding region, then it has to be removed in the process of translation to proteins. This could be a time and energy consuming process and natural selection would tend to keep the copy number down. Such regions are know to occur in the coding sections, especially in higher organisms, and are called *introns*. Since introns,probably caused by transposable elements, do occur from time to time in a random way, the protein manufacturing process has developed mechanisms to recognise such regions and remove them at the stage of protein formation. The translation of the DNA coding region into proteins passes through a large number of intricate intermediate steps and hence it may be more efficient to excise these introns towards the end of the manufacturing process. This also may be due to the fact that the DNA with junk and introns is passed on to the germ line and thus the incisions at the DNA level during translation may not be efficient. If the mechanism described above for the occurrence of junk DNA is valid, then this is compatible with Darwinian mechanism at a statistical level over the whole genome. There is no need to invoke non-Darwinian, non-adaptive 'selfish genes'.

I have already remarked on the conservatism that exists at the molecular level. The form and structure of proteins are very important in determining their function and this implies that some amino acids in the protein molecule can be replaced by other amino acids such that the protein function is not inhibited. Proteins are composed of a few hundred to a few thousand amino acids. Thus within the constraints of form and function, there could be considerable variability in the amino acid structure of proteins caused through mutations at the DNA level. Since a triplet code is used to define an amino acid, one has a mapping of 64 triplets onto 20 amino acids. Thus the code is degenerate in the sense that some amino acids are defined by more than one triplet sequence. Thus one would expect even greater variablity at the DNA level. Both these types of variability have been observed and it is not surprising that the variability can be well explained on the basis of the neutral model. Since mutations at the DNA level occur, in general, randomly, the variation in the DNA coding region or in the amino acid sequence of the proteins yield a time direction for the evolution. One can construct phylogenies using this data and such phylogenies are not incompatible with those arrived at through phenotypic characters and fossil data. The discrepancies that one does observe may partly be explained by the fact that there is no unique one to one mapping between proteins and phenotypes. Given the two sets of phylogenies, one arrived at through macroscopic analysis and the other from analysis at the molecular level, provided the macroscopic data is reasonably complete, biological evolution would be better described by the macroscopic model. Unfortunately, the data at the macroscopic level are often inconsistenet and incomplete. On the other hand, at least for the present living organisms, there is an explosive increase of data regarding DNA and proteins. This gives one the misplaced feeling that the study of the variation at the molecular level is all that is necessary for describing the mode of biological evolution.

As we have seen, the most fascinating aspect of a living system is the strong conservatism at the molecular level and the extreme diversity at the species level. I have argued earlier that this molecular conservatism implies that molecular biology alone cannot provide all the answers to biological evolution. On the other hand, such a strongly conserved system could provide the clues to the origin of the primordial molecular self-organizing system. With our present knowledge regarding molecular organization in living cells, we could conjecture the following physico-chemical steps in the transition from non-living to the living molecular complex. We could recognize three major phases in the early development of life. The first phase can be termed the period of chemical evolution. It is fairly well understood how under the conditions of the primitive earth, the precursors and monomers of biological macromolecules, such as amino acids, nucleotides, phosphates, saccharides etc. could have formed. There must have existed a large diversity in the chemical species, since all conceivable chemical substances that conform to the laws of chemical thermodynamics and kinetics could, and probably did, form spontaneously. There is experimental evidence to indicate that these early monomers could condense spontaneously to biological marcomolecules leading to almost random sequences and many of the resultant proteins thus formed are seen to possess some catalytic activity at a low level. It could be thought that the nucleation of life could occur once a sufficiently large number of these primordial catalytic proteins were present. But as shown by Eigen and Schuster (1979), Kuppers (1983) and others, the transition from such random sequences to self-organizing biological marcomolecules which formed the precursors to living organisms could only have occurred through a process of selective optimisation. The next phase in the march towards biological evolution could be called the phase of molecular self-organization. In the early prebiotic condition, proteins were more easily formed than sequences of nucleic acids that are needed for information storage and transmission. Molecular selection and evolution presupposes a capacity for self-replication, a property possessed by nucleic acids and not by proteins. Since RNA molecules fold themselves into specific three dimensional structures, unlike DNA molecules, the RNA molecules are capable of expressing themselves phenotypically and thus interact selectively with the environment. The evolution of such molecules in this early stages of life on earth was severly limited by the rate at which error propagates. It can be shown that such molecules could not have been much more than 100 bases (nucleotides) long, which is approximately the size of the present day tRNA molecules. For further evoultion to take place, one would require highly optimized proteins for highly accurate copying and information transmission. But the RNA were too short to carry information for such proteins. This appeares as the classic chicken and egg example. The problem was solved by Eigen through postulating the catalytic hypercycle in which the proteins and the nucleic acids were cyclically coupled. A hypercycle has information capacity much higher than that which existed in the early shortlength RNA's and is thus well equipped to further the evolution of the complex machinery of replication and translation. Eigen showed that the hypercycles were optimized through natural selection. They have nonlinear growth properties and hence led to the 'once-for-all-time' decision in their structure and function when a particular species had reached macroscopic population numbers. Thereafter they could not be disloged by the few selectively advantageous competitors that were present in the environment. It has been shown that even if several code schemes are equally likely to be found in the environment, the nonlinear growth will reinforce fluctuations and thus ensure that only one hypercycle becomes established with all the others dying out. Even though this model is very attractive, it is not clear why different hypercycles could not have occurred at different geographically separated regions. The final phase in the early biological evolution can be called compartmentalization. Since compartmentalization provides a selective advantage to hypercycles, the formation of protocells was inevitable. This saw the end of nonlinear selection and the era of divergent evolution had begun. Further evolution of differentiated cells could only occur when the information-carrying capacity is increased and this led to the evolution of the DNA molecule. The process through which this transition took place is still not understood.

In this brief review I have omitted many interesting problems in evolution, the most outstanding of these being the evolution of sex. It is well known that sexually reproducing organisms have a two-fold disadvantage compared to non-sexual systems. Why then is sex so successful and how can we understand it within the Darwinian frame work? There are many theories attempting to explain the evolution and maintenace of sex but at present there is no consenus among biologists. Though most biologists would agree that sex evolved and became universal through a process of natural selection, there is little agreement on what this process may be. The other major aspect in evolutionary biology that I have only alluded to is developmental biology. Though both these subjects have important relevance to biological evolution, I have not discussed them since they are still in the early stages of understanding.

7. CONCLUSIONS

In this review I have argued that the extremely long term dynamic stability in biological evolution is a consequence of the evolution of resource recycling. The only resource that is consumed is solar energy which eventually returns to the environment as low grade heat. Molecular conservatism is a consequence of this evolutionary constraint. The mode of converting solar energy into ordered living systems has been responsible for freezing the mechanisms by which energy and materials are utilized by living organisms. This then implies once the balanced system has been perfected, the amount of biomass in the biosphere could not have altered very much since biomass production ultimately depends on the total solar radiation incident on Earth and on the photosynthetic efficiency. The latter again could not alter very much because of molecular conservatism. I would suspect that this situation has lasted since the Cambrian period. The approximate constancy in the total biomass has been the reason for the constant species diversity over the past billion or so years. In this context I would like to stress that the constancy I refer to does not imply that there have not been large fluctuations. What I assert is that on the average both biomass and species diversity have been reasonably constant. Thus the long term 'economic growth' in the biosystem has to be measured through species turnover rather than through an ever increasing biomass production and species diversity. Some evolutionary biologists claim that the fundamental law of evolution is that evolution proceeds in the direction such that all available ecological niches are occupied. I prefer to recast this in terms of the constancy in the biomass and in species diversity.

The living world can be described in terms of various hierarchical levels such as the molecular level, the genotype level, the phenotype level, the population level, the species level, the ecological community level etc. There exist a large amount of overlaps between the levels but still such categorization is useful since there exists at present no Grant Unified Model for the description of biological evolution at all levels. I have shown in this review that at each of these levels evolution, in the main, proceeds through Darwinian mode in terms of the macroscopic variables pertaining to the level. But as we have seen, natural selection is often modified by stochastic effects and by constraints in the system. Many of the controversies among biologists arise in tryng to understand the nature of evolution at any hierarchical level in terms of models specifically formulated for the level above or below.

What can economists learn from a study of biological evolution? Unlike biological evolution, economic evolution is Lamarckian in character. The rate of economic evolution is also dependent on the variablity that is present and it is possible to formulate an equivalent Fisher's Fundamental Theorem. The variability in economic systems arises through competition and occassionally through new discoveries, the 'hopeful monsters'. Both these can be incorporated into a model for economic evolution which parallels much of the ecological theories in biology. For example, one has the concept of competitive exclusion for two products (species) occupying the same economic niche. The variability in economic systems could be thought of as arising through the Red Queen effect and 'character displacement'. Free market economy, encouraging more competition, would thus generate more variability and hence higher rates of evolution than a controlled economy. The great difference between biological evolution and economic evolution is in resource utilization. Economic systems, particularly capitalist economy, are often tuned to the concept of growth which, in the ultimate analysis, implies increasing resource utilization. Economic systems, like biological systems, are far from equilibrium. Thus if economic systems are to enjoy long term stability, it needs to adopt the biological constraints, namely resource recycling. In this case the concept of economic growth has to be considerably modified. Pure free markets cannot be stable for long periods and the constraints we need to impose are not the ones that present socio-political systems apply. Whether we can achieve such a balanced economic evolution depends on our collective political will. Since the economic system is strongly coupled to the biological world, unless we could bring about such a balance, we shall be also strongly perturbing the stability of the biological system. It would be an irony if the biological system that has enjoyed more than a billion years of stability faces instability through the actions of the species that has been endowed with intelligence and means to utilize it more than any other species in the long journey of biological evolution.

Acknowledgements

I am very grateful to Professors B.C. Clarke FRS, A.J. Cain FRS, and J.C. Wilmott and to Drs. L.M. Cook, A. Wallace, and A. Hillel and to my wife Jean Mani for reading the manusrcipt and for many helpful suggestions. Many of the errors that still exist in the text are due to my own pig-headedness in ignoring some of their comments.

REFERENCES

Cook, L.M. & Mani, G.S. (1980). A migration–selection model for the morph frequency variation in the peppered moth over England and Wales. *Biological Journal of the Linnean Society*, **13**: 179–198.

Cook, L.M., Mani, G.S. & Varley, M.E. (1986). Post-industrial melanism in the peppered moth. *Science*, **231**: 611–613.

Clarke, C.A., Mani, G.S. & Wynne, G. (1985). Evolution in reverse : Clean air and the peppered moth. *Biological Journal of the Linnean Society,* **26**: 189–199.

Darwin, C. (1859). *The Origin Of Species*, Murray, London.

Dobzhansky, T. (1937). *Genetics And The Origin Of Species*, Columbia University Press, New York, (1st ed.).

Dobzhansky, T. (1951). *Genetics And The Origin Of Species*, Columbia University Press, New York, (3rd ed.).

Eigen, M. & Schuster, P. (1979). *The Hypercycle : A Principle Of Natural Self-Organization*, Springer Verlage, Heidelberg.

Ehrlich, P.R. & Raven, P.H. (1969). Differentiation of populations. *Science* **165**: 1228–1232.

Gorczynski, R.M. & Steele, E.J. (1980). Inheritance of acquired immunological tolerance to foreign histocompatability antigens in mice. *Proceedings of National Academy of Science,* **77**: 2871–2875.

Gorczynski, R.M. & Steele, E.J. (1981). Simultaneous yet independent inheritance of somatically acquired tolerance to two distinct H-2 antigenic haplotype determinants in mice. *Nature,* **289**: 678–681.

Gould, S.J. (1980). G.G. Simpson, paleontology and modern synthesis. In: *The Evolutionary Synthesis*, Mayr, E. & Privine, W.B. (eds.), Harvard University Press, Cambridge, Mass.

Gould, S.J. (1983). *Hen's Teeth And Horse's Toes*, Penguin Books.

Howard, J.C. (1981). A tropical volute shell and the Icarus syndrome. *Nature,* **290**: 441–442.

Kimura, M. (1983). *The Neutral Theory Of Molecular Evolution*, Cambridge University Press, Cambridge.

Küppers, B-O. (1979). *Molecular Theory Of Evolution : Outline Of A Physico-Chemical Theory Of The Origin Of Life*, Springer Verlag, Heidelberg.

Lewontin, R.C. (1979). Fitness, survival and optimality. In: *Analysis Of Ecological Systems,* Horn, D.J., Stairs, G.R. & Mitchell, R.D. (Eds.), Ohio State University Press, Columbus.

McMahon, T.A. & Bonner, J.T. (1983). *On Size And Life*, Scientific American Book Inc., New York.

Mani, G.S. (1980). A theoretical study of the morph ratio clines with special reference to melanism in moths. *Proceedings of the Royal Society of London (B),* **210**: 299–316.

Mani, G.S. (1982). A theoretical analysis of the morph frequency variation in the peppered moth over England and Wales. *Biological Journal of the Linnean Society,* **17**: 259–267.

Mani, G.S. (1989). Theoretical models of melanism in *Biston betularia* – A review. *Biological Journal of the Linnean Society,* **39**: 355–371.

May, R.M. (1974). *Stability And Complexity In Model Ecosystems*, Princeton University Press, New Jersey.

Maynard Smith, J. (1982). *Evolution And The Theory Of Games*, Cambridge University Press, Cambridge.

Maynard Smith, J. (1982). *Evolution Now : A Century After Darwin*, Nature and Macmillan Press Ltd., London.

Maynard Smith, J. (1989). *Evolutionary Genetics*, Oxford University Press, Oxford.

Maynard Smith, J. & Price, G.R. (1973). The logic of animal conflict. *Nature,* **246**: 15–18.

Mayr, E. (1963). *Animal Species And Evolution*, Harvard University Press, Cambridge, Mass.

Simpson, G.G. (1944). *Tempo And Mode In Evolution*, Columbia University Press, New York.

Simpson, G.G. (1953). *The Major Features Of Evolution*, Columbia University Press, New York.

Steele, E.J. (1979). *Somatic Selection And Adaptive Evolution*, Williams – Wallace International and Croom Helm, Toronto and London.

Stenseth, N.C. & Maynard Smith, J. (1984). Coevolution in ecosystems : Red Queen evolution or stasis?. *Evolution,* **38**: 870–880.

Van Valen, L. (1973). A new evolutionary law. *Evolutionary Theory,* **1**: 1–30.

White, M.J.D. (1978) *Modes of Speciation*, Freeman, San Francisco.

Wiley, E.O. (1975). Karl R. Popper, systematics and classification : A reply to Walter Bock and other evolutionary taxonomists. *Systematic Zoology,* **24**: 233–243.

Evolution in Biology, Physics and Economics: A Conceptual Analysis

MALTE FABER
Alfred Weber-Institut, Universitat Heidelberg, Grabengasse 14, D 6900 Heidelberg, FRG
JOHN L.R. PROOPS
Department of Economics and Management Science,
University of Keele, Staffordshire ST5 5BG, UK

1. INTRODUCTION

Many branches of conventional science tend to conceive of their objects of study as timeless; they therefore tend to represent their findings, in 'eternal laws', such as the Laws of Classical Mechanics. The application of such a kind of science leads easily to the belief that future events are predictable. This predictability would even have been complete for that ideal scientist, Laplace's demon. In contrast to this approach we start from the assumption that the objects and their relationships which science examines are intrinsically characterised by the complete or partial emergence of novelty in the course of time. This leads us to approach the question of predictability in a new way; we shall need to develop new concepts to answer this question. A key notion for our approach is 'evolution'.

It is known that the objects of study in biology at any period of time are the result of an evolutionary process. We shall show that this is also true for other sciences, in particular for economics, and even for physics. Our broad task is to develop a general framework for conceptualising evolution. We shall proceed in such a manner that biology, physics and economics can be contained within it. In this way we also intend to contribute to the furthering of interdisciplinary research between these three disciplines. For example, physics, biology and economics are all of vital importance for environmental analysis, where the simultaneous and interacting evolution of physical, biological and economic systems is involved.

Evolution is a very encompassing concept, and is used in many different contexts. For example, one speaks of physical, biological, social, political and economic evolution. Thus for example key concepts for a definition of evolution in biology are mutation, heredity and selection, while in systems theory one employs the concepts of self-evolution, self-sustained development, and self-reference. For our conceptual framework it is useful to begin with the broadest possible starting point, for otherwise there is the danger that evolution is seen in too narrow a sense. Such a narrow conceptualisation risks the exclusion of important aspects of evolutionary processes.

In our view the key characteristics of evolution are (1) change, and (2) time; we consider them also to suffice as elements for the following starting point: Evolution is the *changing* of something into something else *over time*.

Our aim is to specify and operationalize this conceptualization. To develop our conceptual framework we use concepts originally developed for biological systems.

In particular we shall use the notions of genotypic evolution and phenotypic evolution. Since we employ them for very different sciences than that for which they were originally developed, they are useable only with more or less fuzziness. We willingly accept this fuzziness as long as it does not endanger our endeavour.

We shall argue that the evolutionary process, as we understand it, takes place in many different areas of science and human experience, albeit with different velocities. The evolution of our present physical laws and fundamental constants took place a long time a go and within a very short period of time (see Section 4). The biological evolution of species lasts millions or thousands of years and proceeds continually (see Section 3). The evolution of social and of economic institutions takes place over centuries or decades. As the present evolution of techniques shows, it may even have a much faster pace at certain periods of history (see Section 5).

We are interested in particular in those changes in the course of time which are related to the emergence of novelty, because it is the latter which restricts the area of validity of predictability.

The degree of predicatability of phenomena vary in different areas of science (see Section 2). Clearly, there are unpredictable processes in physics, nevertheless compared to physics the degree of predictability is very low in economics, since the latter is characterized by the emergence of novelty during very short time periods, e.g. invention, and the behaviour of people through their own volition.

Before proceeding briefly to summarise the contents of our paper, we wish to acknowledge that there is a considerable, and growing, literature in economics which uses, to a greater or lesser extent, evolutionary ideas drawn from biology. Examples of such work include Marshall (1890), Veblen (1902), Penrose (1952), Cyert (1963), Hirshleifer (1977), Schelling (1978), Boulding (1981), Nelson and Winter (1982), Norgaard (1984) and Witt (1991).

In Section 2 we distinguish between two types of process in time: predictable and unpredictable processes. While it turns out that the former may be of an 'equifinal' nature, the latter are not. Section 3 is pivotal to this paper, in that in biology the concepts of evolution, particularly the notions of phenotype and genotype, have been developed and applied. Here is laid the conceptual foundations for our evolutionary framework. In Section 4 and Section 5 we apply these evolutionary concepts to physics and to economics, respectively. Section 6 shows why physics is 'easy' and economics is 'difficult', with biology 'in between', because of the different relationships between genotypic and phenotypic evolution in the three disciplines. In Section 7 the importance of ex post and ex ante considerations for conceptualising evolution are discussed. Section 8 offers some conclusions for economic analysis, and suggests some directions for future research.

2. PREDICTABILITY AND THE EVOLUTIONARY PROCESS

To allow us to proceed to our conceptual analysis of evolution we distinguish between two types of processes in time; we call these 'Predictable' and 'Unpredictable' processes. Before discussing these processes it is appropriate to note that actual processes lie on a continuum with regard to their predictability. That is to say certain processes are predictable with great certitude, while others are entirely unpredictable; many processes fall into neither of these categories, but are predictable with a greater or lesser degree of exactitude. However, for the time being it is convenient to employ the above rough categories.

2.1 Examples of predictable processes in physics, biology and economics

We shall illustrate predictable processes with five examples, from physics, biology and economics. We have ordered them according to their degrees of predictability.

(i) Consider a ball falling on to a pavement. Using the laws of classical mechanics one can predict with great exactitude the movement over time of the ball. After a certain amount of time the ball will come to rest, and a mechanical equilibrium will be established.

(ii) The dynamics of the launching of an Earth satellite offer a different type of predictable mechanical process. The satellite initially follows a path determined by the rocket propelling it, but eventually achieves a stable and predictable orbit. In contradistinction to example (i), in this case the predictive equilibrium outcome achieved is dynamic rather than static.

(iii) An example from thermodynamics is the case of an ice cube exposed in a warm room. Depending on its volume and the temperature it will melt in a predictable way, and be converted into a puddle of water and atmospheric moisture. If the room is sealed, an equilibrium will be established between the liquid and gaseous water. However, in terms of the molecules that constitute the water, predictability in an exact sense is impossible. The description of the behaviour of the water molecules at thermodynamic equilibrium can be at best stochastic.

(iv) The life-cycle of an organism is generally predictable. The life of a chicken begins as a fertilised egg, continues as a chick developing within the eggshell, progresses to the newly hatched chick, and thence the mature chicken, the old chicken, and the dead chicken. In this example, while the general pattern of the process is well known, the detailed development of any particular chicken will be impossible to predict with any accuracy.

(v) For an economic example of a predictable process we can consider a simple market which is initially not at equilibrium; that is , the price of the commodity does not equate supply with demand. Assuming that market mechanisms exist to allow adjustment of supply and demand which use price signals, we would expect that eventually a market equilibrium would be established. The exact time path of the adjustment would, of course, depend on the exact nature of the supply and demand curves and the adjustment process involved.

From these five illustrations we conclude that a predictable process is one whose time path can be described as being deterministic or stochastic.

An important further characteristic of many predictable processes is that they are also 'equifinal'. We call a process equifinal, if, given certain boundary conditions, the predictable outcome is dependent of the initial conditions governing the process. For example, if a small ball is placed anywhere on the inner surface of a large bowl, then under gravity the ball will eventually settle at the unique base of the bowl. We see that the final position of the ball is predictable, and is also independent of the initial position of the ball. It is therefore clear that any equifinal process whose dynamic behaviour is known is predictable, though, of course, not all predictable systems are equifinal.

2.2 Unpredictable processes

We now give three examples of unpredictable processes, also from physics, biology, and economics.

(i) The first example we use may surprise the reader somewhat, but it illustrates an important point to which we wish to return later, There is a growing literature suggesting that what we now regard as Universal Physical Constants (eg. the gravitational constant G, Planck's constant h, etc.) were far from constant in the first few microseconds after the Big Bang which brought the universe into being. At the time of the Big Bang it seems that the constants could have ended up with values different from those finally assumed. Modern theory therefore suggests that the whole nature of the universe was, in principle, unpredictable at the time of the Big Bang (Hawking 1988), (for more details see Section 4).

(ii) The whole history of the evolution of species demonstrates unpredictability. Surely no reasonably intelligent dinosaur would have expected that this very successful and long established group of species would not only be extinguished, but replaced by mammals.

(iii) The modern world we inhabit is largely the product of cumulative human inventiveness. By its very nature inventiveness is unpredictable, for correctly to predict the nature of an invention is similar to making that invention.

2.3 Predictability and ex ante theory

As noted above, predictable and unpredictable processes are rough catergorizations. When speaking of 'predictability' in the following, we will often mean in *principle* predictable. By this we mean that it is possible to get to know the dynamics of a system, which allows one to make predictions concerning future developments. It is furthermore useful to note that, when we say that a process is 'predictable in broad terms', this only means we can say at least something about its behaviour, be it that it is equifinal, unstable, chaotic, stochastic, etc. This implies that whenever we are able to conceive an ex-ante theory about an empirical subject, we are able to obtain 'predictable' results, at least in 'broad terms'. In this sense our paper deals with the question of in what areas one can hope to develop ex-ante theories. Wherever a field is in an essential way subject to the emergence of novelty, an ex-ante theory cannot be developed. To give a simple illustration, it would not have been possible to conceive a theory on the behaviour of lions at a time when lions had yet evolved as a species.

In this Section we have illustrated and discussed predictable and unpredictable processes. In Sections 3, 4 and 5 we shall develop a new framework to analyse these processes from our evolutionary point of view.

3. CONCEPTUAL ELEMENTS OF BIOLOGICAL EVOLUTION

In this Section we shall draw upon some well developed concepts from biology; in particular we shall use the notions of phenotype and genotype employed in the Darwinian framework of the evolution of species. We shall relate these evolutionary concepts to the notions of predictable and unpredictable processes as developed in Section 2. In particular we shall contrast predictable and equifinal processes with unpredictable processes. As the concepts of phenotype and genotype are central to all our discussions below, we first define these terms as they are generally used in biology. An organism has a certain appearance, capabilities, characteristics, etc. The appearance that it presents to the world is known as its phenotype. The phenotype displayed by an organism results from the interplay of two factors. The first factor is the potential inherited from its parents; i.e. its genetic make-up, or genotype. The second factor is the environment of the organism; for example, organisms that are otherwise identical (e.g. indentical twins) will grow to different sizes and exhibit different capabilities if they are subjected to widely different nutritional regimes.

It is generally accepted, in biological systems, that the genotype affects the phenotype directly, but there is no such direct influence of phenotype on genotype. For example, a chance mutation may cause the genotype to alter, such that the phenotype becomes, for example, longer-necked. Further, this characteristic of being longer-necked may be inherited by that creature's offspring. On the other hand, an animal might, through its feeding habits, stretch its neck, thus altering its phenotype. This alteration will not influence its genotype and therefore this will not be a heritable characteristic.

From the above discussion it is clear that the genotype reflects the 'potentialities' of the organism. The phenotype represents the 'realisation' of these potentialities, in as far as is permitted by the environment of this organism. In all the discussion that follows we shall be concerned with how the potentialities of biological, physical and economic systems evolve, and also how the realisations of these potentialities evolve. As a convenient short-hand, and to reflect our desire to discover a fundamental conceptualisation of evolution, when we refer to 'realisations' we shall use the term 'phenotype', and when referring to 'potentialities' we shall use 'genotype'.

3.1 Organism, species and biological system: macro-phenotype, macro-genotype and the unit of evolution

At this stage it will be useful to clarify what is meant by 'organism', 'species' and 'biological system'.

An organism is the basic unit of independent life, whether plant or animal, that contributes to reproduction.

A *species* is the set of all organisms whose genotypes are so similar that they allow interbreeding (e.g. horses, dogs, primroses, etc.).

A *biological system* consists of a set of interacting species, each species being made up of distinct but genetically similar organisms. (e.g. Within a pond there will be various photosynthetic plant organisms, which will be foodstuffs for various insects and fishes. These herbivorous animals may, in turn, be preyed on by carnivorous fishes, mammals, birds, etc.). The phenotype as described above refers to an individual organism. As organisms, of their various species, interact with each other in large and heterogeneous biological systems, it is useful to consider an overall description of such a biological system. Such a description would be a listing of the phenotypes of all the orgnisms within the biological system in terms of the species represented and their relative abundances. We term this description a 'macro-phenotype'.

Similarly, the genetic potential of a biological system of organisms could be listed to give that population's 'macro-genotype'. This list comprises all genotype information of all organisms of the corresponding population and is independent of the relative frequencies of the organisms.

The advantage of descriptions of biological systems in terms of macro-phenotype and macro-genotype is that it gives explicit recognition to the fact that the evolutionary process deals with populations rather than with individual organisms, as is more fully discussed in Sections 3.4 and 3.5 below.

There has been a long-running debate in evolutionary biology as to whether the unit of evolution is best conceived as being the individual organism, the species, or the entire heterogeneous population (Levins and Lewontin, 1985).

We have noted above the usefulness of describing populations in terms of the macro-phenotype and macro-genotype. It is clear from our discussion that an individual organism can only exhibit phenotypic evolution, but not genotypic evolution, because for the latter at least two successive organisms have to exist, to allow the recognition of genotypic evolution. In contrast, a species can exhibit not only phenotypic but also genotypic evolution. The genotypic evolution of a species, however, is influenced by that of other species. It is therefore not possible to study the evolution of one species in isolation. For this reason it is useful to consider the biological system as our unit of evolution.

3.2 Phenotypic evolution: predictable equifinal processes

Macro-phenotypic evolution relates to the relationships between, and the relative abundances of, a number of species in mutual interaction in a given environment, where the genetic make-up, or macro-genotype, of the mixture of species is take to be given. For example, the process of ecosystem succession subsequent to some traumatic event may lead to systematic changes over time of relative species (macro-phenotype) abundance. For instance, a major fire in an area of woodland will destroy the majority of the vegetation, particularly large trees. In the years subsequent to the fire there will first be rapid growth by small plants; later the fast-growing trees will become established, and their shade will reduce the growth of the already established low-growing plants. Ultimately, slower growing but bigger and more robust trees will themselves shade-out the faster growing trees. It is clear that in such circumstances the proportions of the species, i.e. macro-phenotypes, will change radically but, systematically and predictably, over time. In this case the phenotypic evolution will also be equifinal, to the 'climax' structure of that ecosystem. It is in this sense that we consider phenotypic evolution to be a predictable process.

One might illustrate the concept of phenotypic evolution by using a simple two-pheontype model, with the dynamics of their interaction shown on a phase diagram. In Figure 1 is plotted the relative abundances of two biological species, as they vary over time. For example, phenotype 1 may be a herbivore and phenotype 2 its foodplant. The arrows indicate the direction of the

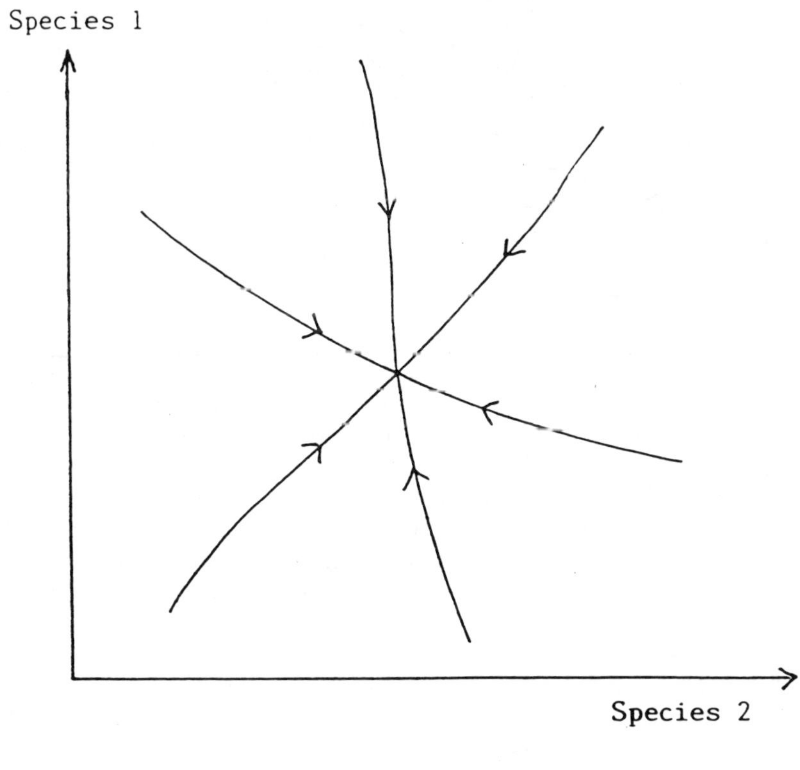

Figure 1 Predictable and equifinal phenotypic evolution.

evolutionary process from any particular starting point along the corresponding phase path. It is clear that a unique, stable equilibrium exists at point A. In the terminology of Section 2, we could say that the process described is predictable and equifinal.

Figure 2 shows a common alternative relationship between two phenotypes.

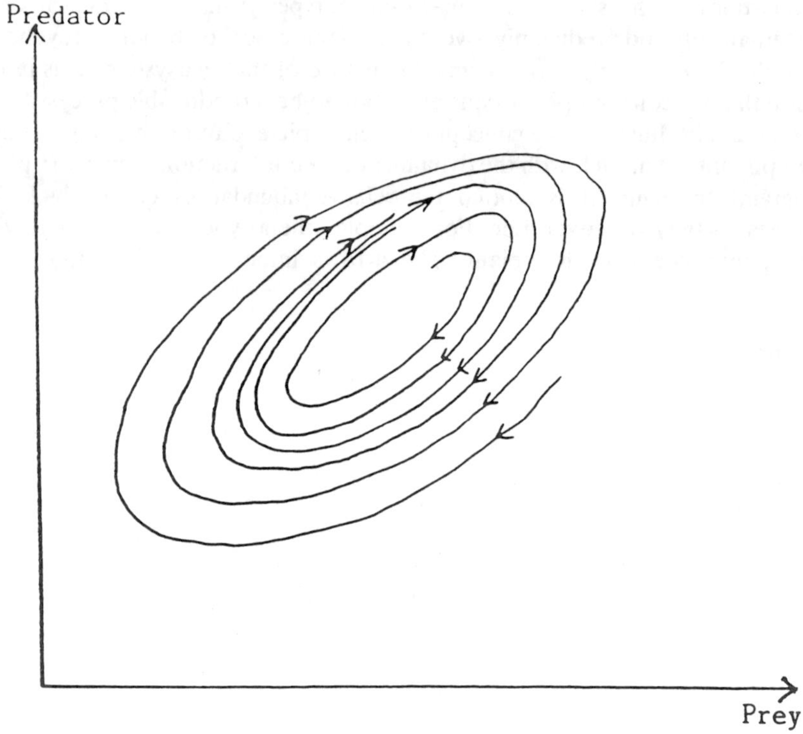

Figure 2 Predictable phenotypic evolution with a limit cycle (1).

Phenotype 1 in this case is a carnivorous predator and phenotype 2 is its prey animal (refrence). Here we see the long-run outcome is a limit-cycle, so that the observed abundances over time in the long-run of these phenotypes will be out-of-phase oscillations over time, as shown in Figure 3. Though this process is not equifinal it is predictable.

The reader will, of course, recognise that in an ecosystem with many species and organisms the nature of the equilibria, and the details of the adjustment dynamics, may be extremely complicated. For example recent work has suggested that the dynamics of some ecosystems may be 'chaotic', as the dynamics are dominated by a 'strange attractor' (Beltrami, 1987). However,

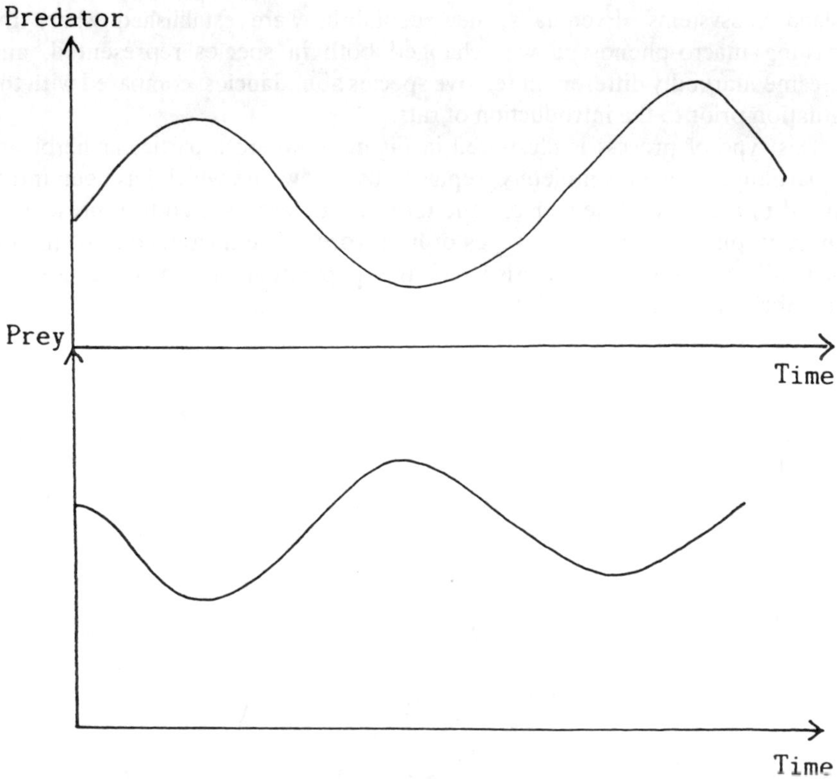

Figure 3 Predictable phenotypic evolution with a limit cycle (2).

even such complicated systems will exhibit behaviour which is systematic and reproducible, at least in its broad outlines, between similar ecosystems. It is in this sense that we may say that it is predictable in 'broad terms'.

Phenotypic evolution may also occur when the system is disturbed, for example by an alteration in climate of some other change in the environment. Another source of disturbance leading to phenotypic evolution, that has been well documented, is the introduction of a new genotype, with its corresponding phenotype, into an ecosystem (Crosby, 1986). The effect of such an introduction will be to shift the equilibrium of the ecosystem, leading to a process of adjustment from the old equilibrium to the new one. For example, the European rat was unknown in many Pacific and Atlantic island ecosystems until the arrival of European explorers and traders. The rat was introduced accidentally from the European ships, so these ecosystems received a new phenotype. These rats rapidly established themselves on these islands and proved to be fierce competitors for resources with the indigenous fauna; over a period of decades they became important, even dominant, components of the island ecosystems. Eventually, new equilibria were established where the existing macro-phenotype was changed both in species represented, and became markedly different in relative species abundancies, compared with the situation prior to the introduction of rats.

This type of process is illustrated in Figure 4, where a particular herbivore is partially, or even completely, replaced by a new one which has been introduced into the ecosystem. Here the term 'herbivore' is used to represent an entire population of various types of herbivores. The introduction of the rat markedly alters the characteristics of this population, and hence causes the equilibrium to shift.

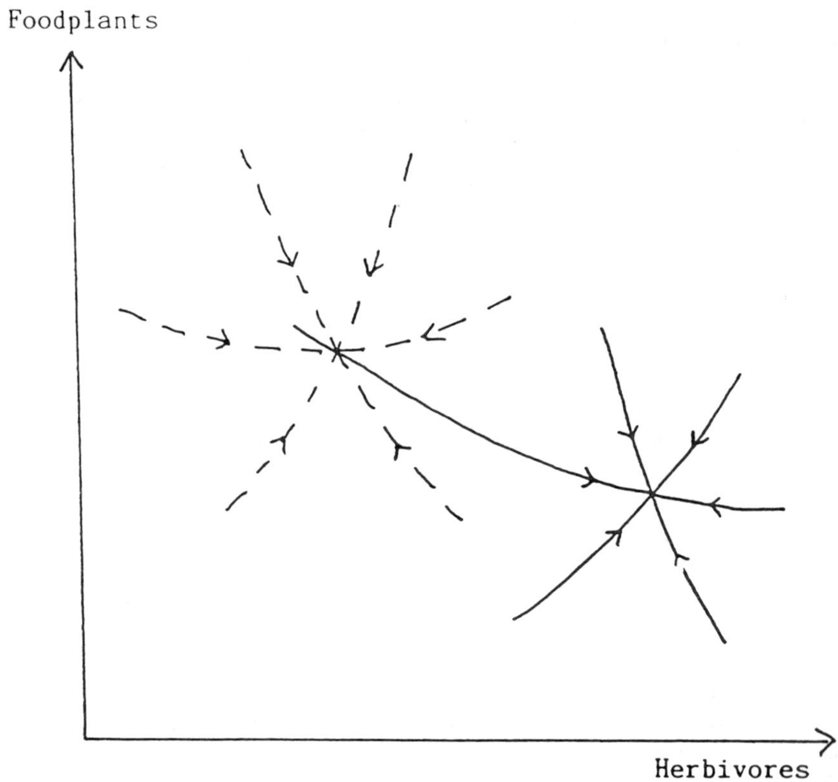

Figure 4 A shifted phenotypic equilibrium.

From the above discussion we see that phenotypic evolution involves alterations to the relative abundances of species, but no alteration to the underlying potentialities of these species, as embodies in their genotypes. Therefore such evolution does not involve the emergence of novelty, as the species are taken as given. Further, in principle phenotypic evoultion is susceptible to description in terms of dynamic models which give rise to well-specified, and predictable, outcomes. In particular, these outcomes are equilibria, which indicates that phenotypic evolution exhibits predictability.

3.3 Genotypic evolution: unpredictable processes

Let us recall the meaning of the term genotype. For a biological organism the genotype is the underlying description of its potential for development, in interaction with its environment. In organisms the genotype is embodied in the genetic material, which mainly resides in the nuclei of the organism's cells. Changes to the genotype of an organism will be inherited by its progeny, and this new genotype may give rise to a new phenotype. In biological organisms alterations to genotypes typically occur through the effects of chemical or ionising radiation (e.g. gamma rays, X-rays, ultraviolet light, etc.) on the reproductive organs. As we shall discuss below, changes to genotype for economic systems correspond to, e.g., changing available techniques of production.

Genotypic evolution may be defined as the alteration of the genotypic description of an organism and its progeny over time. For example, the chance impact of a cosmic ray upon an animal's genetic material may alter that genotype so that the animal's progeny have, say, a slightly longer neck. The first-generation progeny may, in turn, produce second-generation progeny which are also longer-necked. It is well understood that random alteration to the genetic material of organisms only rarely produces mutations which 'improve' the capabilities of that organism's progeny. Therefore the process of changing the genotype is usually extremely slow, although there have been occasional cases of it being recorded in recent times (e.g. the appearance of melanic [dark coloured] moths in nineteenth century industrial Britain) (Kettlewell, 1973). Only long and patient study by palaeontologists has given us a still partial understanding of the genotypes that existed on earth over the past two billion years. Such study has also indicated the relative abundances of the genotypes and how, over very long periods, the appearance of new genotypes led to changes in the relative abundance of the genotypes.

Unlike phenotypic evolution which is in all cases, at least in broad terms, predictable where its dynamics are know, genotypic evolution is unpredictable. This unpredictability arises from three sources. (1) The mutating influence (e.g. radiation) cannot be specified beforehand. (2) The effect of the mutating influence of the genetic material cannot be predicted. (3) The influence of this genetic change on the phenotype is also unpredictable. In economic terminology this means that we are confronted with true uncertainty if genotypic evolution occurs.

3.4 The interaction of genotypic and phenotypic evolution in biological systems

It must be stressed again that changes to biological genotypes take place extremely slowly. In contrast, the relative abundance of phenotypes may change relatively rapidly, as discussed in Section 3.1. The full scheme of biological evolution is thus seen to be the interplay of extremely slow and unpredictable changes to macro-genotypes with relatively rapid but predictable changes to macro-phenotypes.

It is worth noting that the very different rates of phenotypic and genotypic evolution in biological systems allow the predictable and unpredictable parts of the whole evolutionary process to be distinguished. (As we shall discuss in Sections 5 and 6, in economic evolutionary processes it is usually much more difficult to separate out the predictable from the unpredictable processes than in biology, and particularly in physics).

To illustrate the interaction of genotypic and phenotypic evolution in a biological system, let us imagine an ecosystem composed of a given set of macro-genotypes. We suppose that this system has been left undisturbed sufficiently long for it to achieve an 'equilibrium' state (such 'equilibria' may be stationary, limit cycles or chaotic). Now let us suppose that there is a significant alteration of one of the genotypes in the system, and we further suppose that this new genotype is reflected in improved, but unpredictable, 'competitive effectiveness' of its progeny. The result of this appearance of a new and more competitive phenotype will be a predictable macro-phenotypic evolution, leading to altered relative abundances of these phenotypes, and hence of the embodies genotype. The outcomes of the macro-phenotypic evolution will therefore be a changed relative frequency of species genotypes.

The long-run genotypic evolution can be illustrated by reference to the observed evolution of certain predator and prey species (Bakker, 1983). We begin by considering such an interaction with a given pair of genotypes and with an equilibrium phenotypic relative abundance. Let us suppose a random genetic mutation causes the prey phenotype to become longer-legged and swifter. Supposing further that the genotype of the predator is unchanged, then the relative abundance of the phenotype will evolve towards a new equilibrium as showin Figure 5.

Point A1 represents the equilibrium for the phenotypes of the original genotypes. With the introduction of the new genotype of the prey a new phenotypic equilibrium is established at A2. Phenotypic evolution takes place as the system moves from point A1, which is now no longer an equilibrium, towards A2 along the corresponding phase path.

Any mutation in the predator genotype which increases its swiftness will in turn generate yet a further new equilibrium at A3 so that a further phenotypic evolution would take place, as shown in Figure 6.

This series of mutations of genotypes will lead to successive periods of phenotypic evolution.

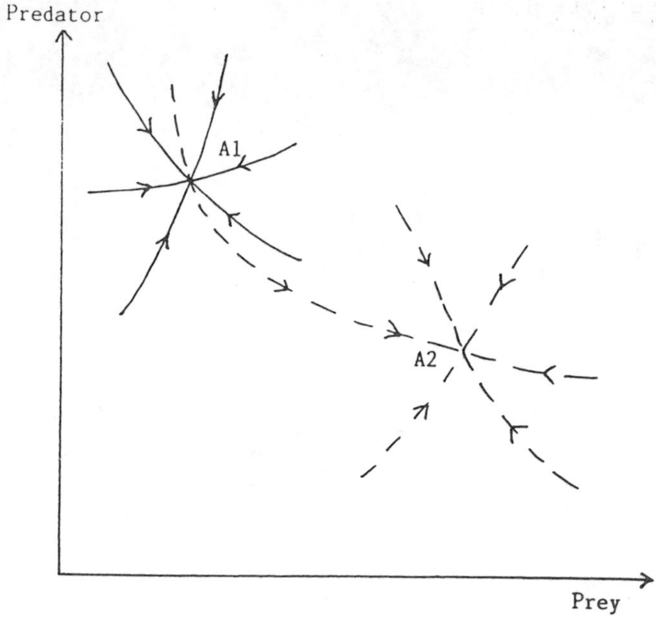

Figure 5 Shifting phenotypic equilibria subsequent to genotypic evolution (1).

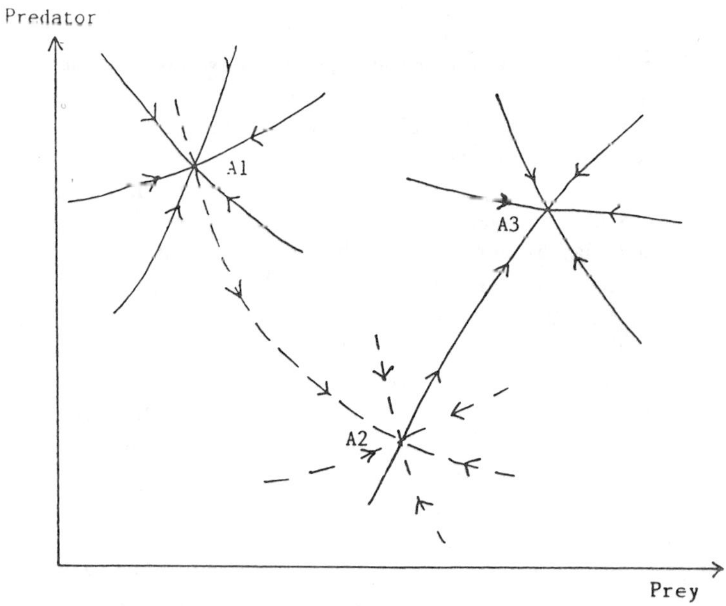

Figure 6 Shifting phenotypic equilibria subsequent to genotypic evolution (2).

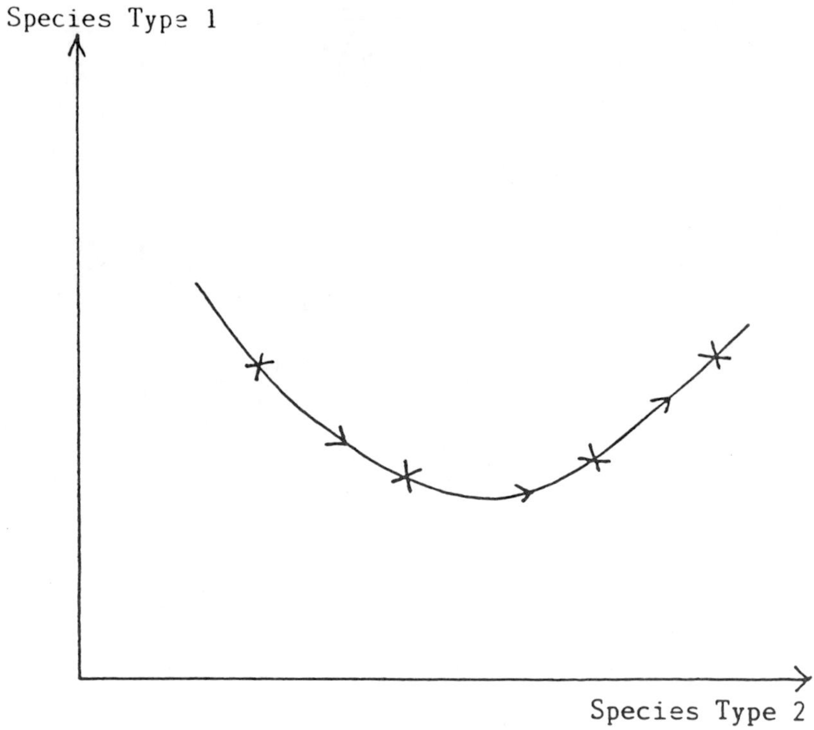

Figure 7 Path of long-run phenotypic equilibria with genotypic evolution.

Now if, as is the case in biological evolution, genotypic mutations which are 'beneficial' take place at intervals which are long compared with the time taken in phenotypic evolution to re-establish 'equilibrium', then, in the long-run time scale, the observed macro-genotypic evolution can be represented by the locus of the successive macro-phenotypic equilibria as shown in Figure 7.

Figure 7 can itself be interpreted as an evolutionary phase diagram. But unlike the phase diagram in phenotypic evolution, as genotypic evolution is unpredictable, the phase path can only be constructed ex post, i.e. once the genotypic/phenotypic evolutionary process has occured. This is in contrast with phenotypic evolution, which is predictable; knowledge of the dynamics of the system allows the full phase diagram to be constructed ex ante. We shall return to a fuller discussion of the role of ex post and ante analysis in Section 6.

It follows from the above discussion that while the notion of an equilibrium, of whatever type, is central to the analysis of phenotypic evolution, there is no way of introducing the concept of equilibrium into genotypic evolution. Thus while the notion of equifinality is important in assessing relations between phenotypes in the relatively short-run, equifinality plays no part at all in the

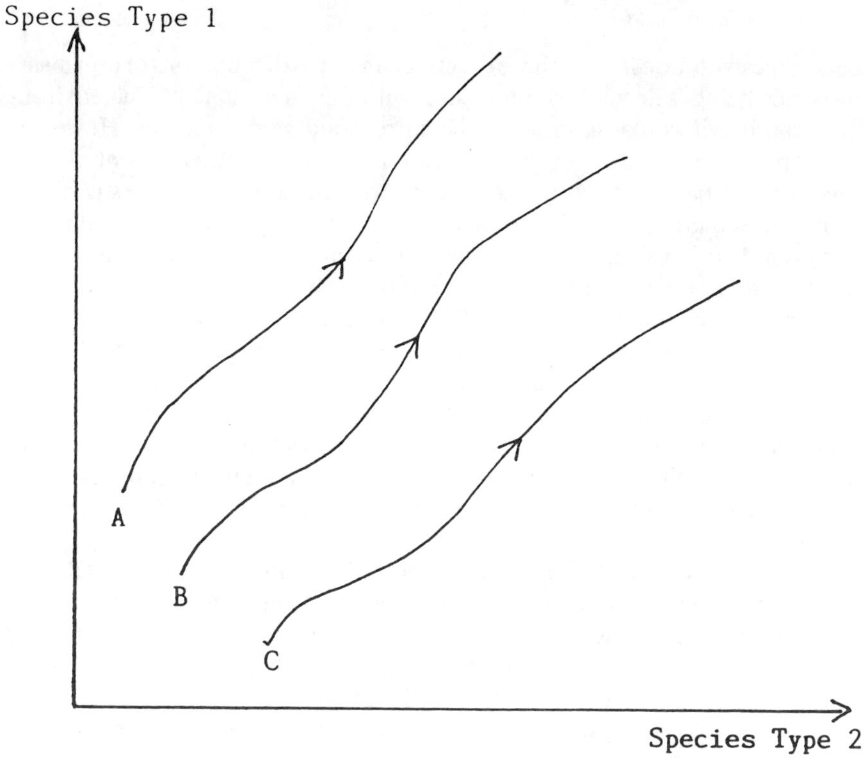

Figure 8 Phase-paths for genotypic evolution.

notion of genotypic evolution in the very long-run. For example a simple two species system might have a genotypic evolutionary phase diagram, as shown in phase Figure 8.

It is clear from Diagram 8 that the initial specification of the macro genotype leads to a completely different unique evolutionary path and there is no meaningful concept of an 'end' in genotypic evolution. For example, if in an isolated ecosystem (e.g. a large island) the system were in state A initially while other isolated ecosystems were in states B and C, there will be no reason to expect that long-run genotypic evolution would lead to similar outcomes.

To illustrate this, consider the cases of Australia and New Zealand which were isolated from what is now the Asian landmass approximately forty million years ago, and from each other shortly thereafter (in geological terms). In Australia there has evolved a rich fauna of marsupials, such as kangaroos and wombats, while in New Zealand the fauna came to be dominated by flightless birds such as the kiwi and the moa (Keast, 1981).

3.5 Coevolution and niches

Genotypic evolution is sometimes described in terms of mutations of organisms in species being 'selected' by interaction with the environment. The environment can be viewed as a 'filter' of the corresponding phenotypes. However, every species, and indeed every organism, is part of the environment of each organism in a biological system. As described in Section 3.3, where successive genotypic evolution of a predator with its prey species was discussed, it is less than useful to consider the genotypic evolution of any species in isolation. Much more useful is to think of the evolution of the entire population; that is, to consider 'coevolution' of many interacting species (Thompson, 1982).

From the point of view of any individual organism, or species, only a limited amount of the full environment actually interacts with the organism or species. This limited range of interacting environment is known as the species' 'niche'. For example, the niche of the honeybee comprises a limited range of vegetation and airspace, while the niche of the fox comprises a relative limited range of rural, or even urban, land and the flora, fauna and some waste materials it contains.

In coevolution, the niche of every species is periodically disturbed by a genotypic evolutionary change to other species. Sometimes a niche may be enlarged or enriched, leading to a greater abundance of a species. At other times niches may be reduced, or even completely eliminated, leading to the extinction of species. However, it is worth noting that an increase in the complexity of a biological system, through the emergence of new species, will tend to lead to a greater number, and thus to greater diversity of niches, and thereby open up possibilities for further genotypic evolution. This is so because any new organism that evolves can act as both a prey (or forage) species for external predation, and as a host to parasites. Therefore the process of macrogenotypic evolution, towards ever greater diversity, has the potential to sustain itself, and may lead to greater complexity and diversity, through the continual opening up of new niches.

We shall indicate the importance of the concepts of coevolution and niches for economic evolution in Section 5.3 below.

We now turn to a discussion of the role of phenotypic and genotypic evolution in physical systems, before we move on to an analysis of evolutionary concepts in economics.

4. EVOLUTION IN PHYSICAL SYSTEMS

In our discussion on biological evolution we have stressed that the potentialities of a system, its genotype, interacts with the environment to give a realisation of these potentialities, in the phenotype. The systems with which physics is mainly concerned have their potentialities defined by the fundamental constants and laws of nature. The realisation of these potentialities is the observed behaviour of that physical system.

We therefore define the genotype for a physical system to be its potentialities; i.e. the fundamental constants and laws of nature. We similarly define the phenotype of a physical system to be the realisation of those potentialities; i.e. its observed physical behaviour. As was discussed under the heading of 'predictable change' in Section 2.1, predictable change is often exhibited by physical systems. As these changes are in the nature and appearance of the physical world, in its realisation, we term such change phenotypic evolution. To illustrate how the potentialities (the genotype) of the physical world may, through interaction with the environment, give rise to a phenotype and to phenotypic evolution, we present the following simple example. A ball held a certain distance above a pavement has potential energy because of its position (i.e. its environment), and because of the laws of nature, its potentialities or genotype. If the ball is released, the interaction of its environment and its genotype (the laws of nature) cause it to behave in a certain predictable manner. The system tends towards a new physical equilibrium, so that is exhibits phenotypic evolution. From the above discussion it should be clear that physical systems, like biological systems, may exhibit phenotypic evolution towards an equilibrium state. However, unlike different biological systems, physical systems all share a unique genotype, comprising the physical constants and the laws of nature, which is unchanging.

We mentioned in Section 2.2 that modern thought suggests that the universe was unpredictable in its first few moments, as the physical constants attained their present values. One could say therefore, that even physical systems have experienced one period of genotypic evolution, albeit for a vanishingly short length of time. The key notions of our evolutionary framework are the genotype, i.e. potentialities, and the phenotype, i.e. realizations. From the above discussion of evolution in physical and biological systems, it is apparent that there exists a hierarchy of these concepts, in the following sense. Realizations in physics, e.g. the state of the earth, are prerequisites for the formation of genotypes in biology. In the same way realizations in physics and biology are prerequisites for the emergence of potentialities in economics.

5. EVOLUTION OF ECONOMIC SYSTEMS

5.1 Economic genotypes and phenotypes

Just as in biology and in physics, as discussed in Sections 3 and 4 respectively, we can conceptualise economies in terms of potentialities and their realisations. The former we will call economic genotypes, and the latter economic phenotypes. Before proceeding we want to point out that in practice it may often be not easy to distinguish these two types from each other. We first consider economic genotypes. The genotype of an economy comprises the following elements: preference orderings of the economic agents (including religious and ethical norms), the technology (i.e. the set of all techniques which are known in that economy), the legal system, the economic and social institutions[1]. The nature of these elements defines the potentialities of a particular economy, i.e. defines its genotype. The economic phenotype of, for example, a market economy is a realisation of these potentialities. This phenotype consists of the following elements: the techniques of production employed, the types of capital goods employed, and their amounts, the types of capital and consumption goods produced, the quantities of these goods, and their prices, the distribution of consumption, income and wealth among the economic agents, and the market structure.

From the above definitions it is clear that, just as in biology we defined the unit of evolution to be the whole interacting biological system, so for economics we define the unit of evolution to be the entire set of interacting economic agents and their institutions and artefacts. We note that over the course of history the unit of economic evolution has tended to increase in size. For a self-sufficient hunter-gatherer society the unit would be the band, while at the present time the high level of integration of national economies suggests that the unit of evolution should be regarded as the global economy. We suggest that the great bulk of dynamic economic modelling can be regarded in terms of the evolution of economic phenotypes. For example, comparative static international trade theory can be conceptualized as phenotypic evolution. In the classical model England shifts its production towards cloth, while Portugal shifts its production towards wine. The equilibrium pattern of trade that is eventually established is mainly dependent upon the structure of production, tastes, resources and techniques available. Pursuing this example further, were there to be a disruption to this equilibrium pattern, and this disruption was then removed, the equilibrium pattern would be reestablished in due course.

Very often the assumption of given constraints to behaviour by economic agents, and a general expectation of optimising by these agents, will give us equilibrium outcomes; i.e. the system will be predictable and equifinal. This tendency in economic analysis is most clearly seen in the continuing struggle by general equilibrium theorists to discover 'reasonable' models of economic activity which give rise to a unique and stable equilibria in price-quantity space. In our view, this working through of the dynamics of economic systems, where tastes, techniques and resource availability are taken as fixed, can justifiably be termed economic phenotypic evolution.

Genotypic evolution in economic terms we would see as being changes in tastes and techniques as well as in economic institutions. For example, the process of industrialisation of the past two hundred years has been one of continually changing economic institutions, techniques of production, and a continually expanding list of natural factors which can be regarded as economic resources. For instance, discoveries in physics stimulated the engineering development of nuclear power, which rendered uranium an economic resource, whereas previously it had been regarded as a material of only limited usefulness. The development of microelectronics and the corresponding increase in economic valuation of germanium (used in transistors) is another example of this process of economic genotypic evolution.

Just as in the biological illustration above, economic genotypic evolution cannot involve the notion of an equilibrium outcome. Unlike the case of biological evolution, however, economic genotypic evolution may take place very rapidly, as is particularly evident in the evolution of the microelectronics industry. Indeed, the rate of change of the genotypic specification (the technical capabilities) in this industry is so rapid that successive changes in specification occur before a new phenotypic equilibrium can be established. The very notion of an equilibrium in such rapidly evolving industries seems to us therefore to be of limited usefulness. The complexity of the phenotypic response to such rapid genotypic evolution is illustrated in Figure 9.

In Figure 9 rapid genotypic change moves the equilibrium of the phenotype from point A to point B, and subsequently to points C and D. The rate of phenotypic evolution towards these successive equilibria is, for

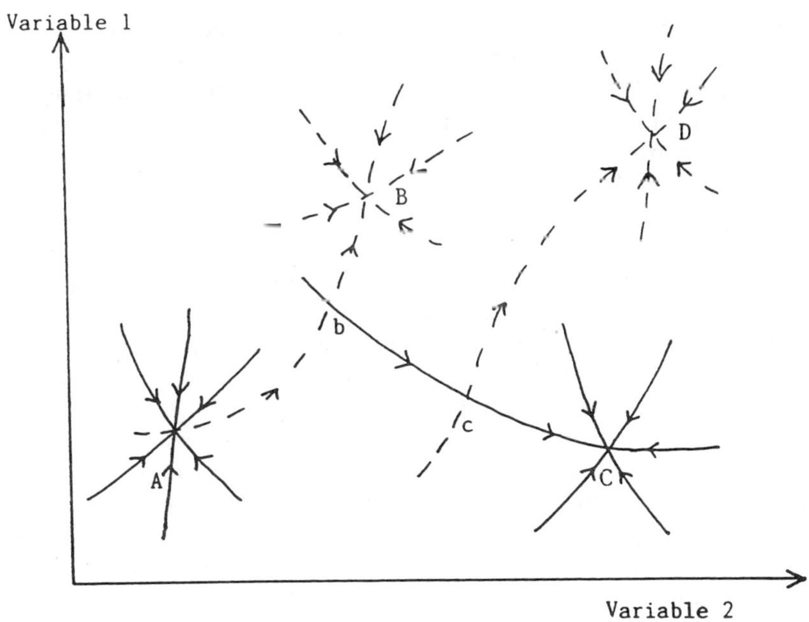

Figure 9 Continuation of phenotypic and genotypic evolution for an economy.

economic systems, of the same orders as the rate of genotypic evolution. The result is that phenotypic evolution proceeds along the ray towards the stable equilibrium B which passes through point A. However, the genotype changes before equilibrium B is achieved, and the phase path shifts to C, the new equilibrium. At point B on the original phase path, the old phase path is replaced by a new phase path leading towards the new equilibrium. This new phenotypic evolution is also incomplete, as by the time the system achieves point C, the phenotypic equilibrium has shifted to D.

It is important to note that in Figure 9 the observed trajectory of the system in phase space is a combination of several rays associated with different equilibria, and except for the starting point, no phenotypic equilibrium state is attained, or even very closely approached.

5.2 The long-run interaction between invention and innovation

In this Section we shall restrict our analysis to one main area of evolutionary economics, the interaction between invention, innovation and the natural world. To introduce these concepts we need first to define the term 'technology'. The technology of an economy, at any moment in time, is the set of techniques which are known, though not necessarily all used. By invention we mean the becoming known of a new technique, which thus expands the technology. Innovation is the process of introducing a technique of the technology which is not currently being used. While invention occurs at some point in time, it takes time to innovate a new technique, because the introduction of a new technique makes it necessary to construct and establish the corresponding necessary capital goods, as well as to train the workers. It is obvious that this innovation gives rise to many adjustment processes over time throughout the economy. (For fuller discussions see Faber and Proops (1986, 1989, 1990)).

The long-run relationships between these elements may be illustrated as in Figure 10.

Recalling the discussion at the beginning of Section 3 the phenotype is our short-hand for what it realised, and the genotype for a system's potentialities. Using the notions of phenotypic and genotypic evolution, we can characterise the process of innovation as economic phenotypic evolution, and the generation of invention as economic genotypic evolution.

Just as in biological evolution, economic genotypic evolution leads to economic phenotypic evolution. And, again as in biological evolution, the phenotypic nature of the economic system offers further potentialities for genotypic evolution. In the case of a biological system, the potentiality lies in niche exploitation and destruction, as discussed above. In an economic system, niche destruction, in the sense of the long-run depletion of natural resources, is a principal motivation for seeking an improvement in economic technical potentialities; i.e. in seeking economic genotypic change. This genotypic

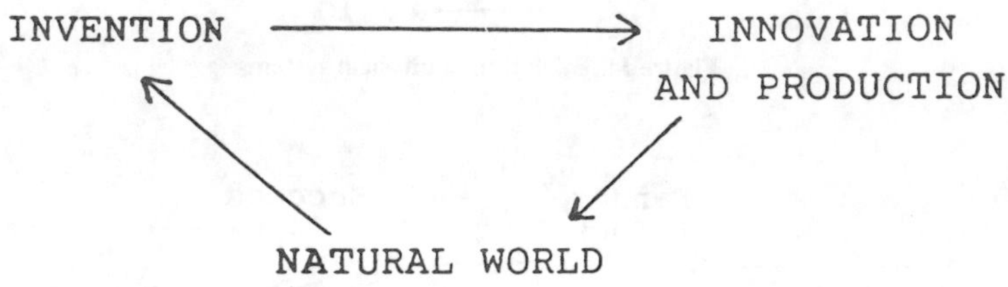

Figure 10 Long-run interaction between the economy and the natural world.

change leads, in turn, to niche production through the bringing into use of new natural resources.

We may illustrate the interplay of phenotypic and genotypic evolution as follows. Consider a resource using economy, which, through the workings of its market system, has established a predictable equilibrium; i.e. relative prices and relative quantities produced are stable and overall output of the economy is growing at the rate of interest. This would be equivalent to the phenotypic equilibrium of an ecosystem, as discussed above in Section 3.1 (see e.g. Figure 1). If the resources used are steadily depleted by economic activity, then this will initially cause marginal adjustments in relative prices and quantities, and a reduction in the overall rate of growth. This phenotypic evolution can be compared with the phenotypic evolution of an ecosystem subject to, say, a slight but continuing alteration in climate over a long period, such as the general warming in northern latitudes after the last ice age. However, long-run depletion of resouces by an economy is likely to initiate a search for new and unpredictable techniques of production, which use less of the diminishing resource (resource-saving invention) or make use of alternative resources (resource-substituting invention). The new techniques give rise to a new and unpredictable genotypic description for the economic system; the new economic genotype in turn produces a new phenotype which must now, through market operations (phenotypic evolution), establish a new, predictable equilibrium economic structure.

6. WHY PHYSICS IS EASY, ECONOMICS IS DIFFICULT AND BIOLOGY IS IN BETWEEN

We start our discussion by representing physical biological and economic evolution in a diagramatic way. In Diagrams 11, 12 and 13, G_i and P_i (i = 1,2) denote the genotype and the phenotype, respectively, of population i. G_i^p and P_i^p (i = 1,2) denote the perturbed genotype, and its corresponding phenotype.

We recall from Section 4 that the complete genotypic physical evolution of the universe occured in the very first few moments of its beginning i.e. in less that a second. Thereafter, there was no more genotypic evolution, but only

$$G_1 \longrightarrow P_1$$

Figure 11 Evolution in physical systems.

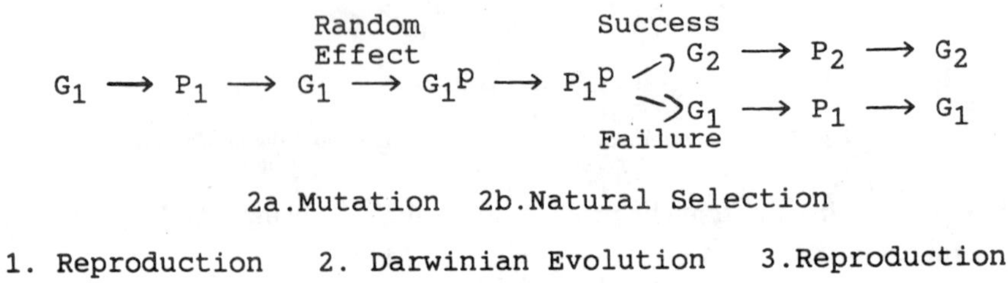

Figure 12 Evolution in biological systems.

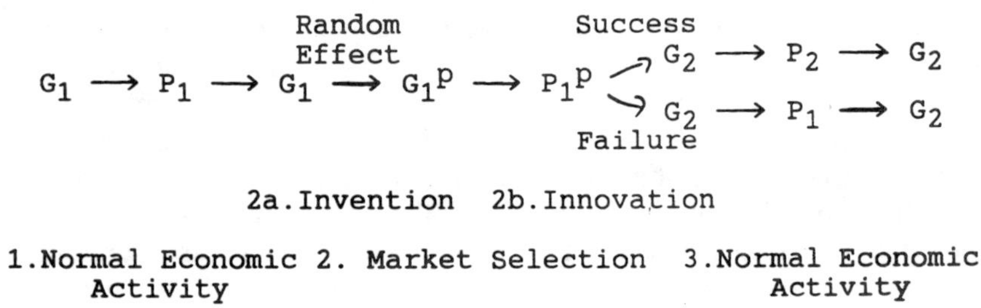

Figure 13 Evolution in economic systems.

phenotypic evolution. Hence physical evolution is simple to characterise in terms of phenotypic and genotypic notions, as shown in Figure 11. It is important to note that there is no recursion between genotypic and phenotypic evolution in physics, unlike biology and economics.

In physics there is no longer genotypic evolution, i.e. no emergence of novelty of potentialities, and physics is only concerned with phenotypic phenomena. This stability of the potentialities of physical systems has allowed physics to deal only with in principle predictable processes, enabling physicists and mathematicians to seek and to find numerous simple and elaborated methods of dynamics. This ability to concentrate on problems which have predictable dynamics is, we believe, the major source of the enormous success physics has enjoyed over the past three centuries. It might have been just this feature of physics that led Planck to comment to Keynes that he gave up the study of economics in favour of physics because he found the former too difficult!

We now turn to evolution in biology. As we see in Diagram 12, the interaction between genotypic and phenotypic processes is much richer than in physics. It may be useful for the reader to explain this diagram in some detail. We see that the representation consists of three phases.

Phase 1 shows the genotype and phenotype of a biological system at an equilibrium; here the genotype and phenotype are recursively maintaining each other through the Reproductive Process.

Phase 2 illustrates Darwinian Evolution; this consists of two parts: 2a. Mutation, and 2b. Natural Selection. In the first part, Mutation is initiated by a random effect which perturbs genotype G_1 to G_1^p, which in turn produces a corresponding new phenotype, P_1^p. In the second part, Natural Selection, the environment acts as filter for the new phenotype P_1^p. If this new phenotype is successful in its environment then the new genotype, G_1^p, becomes established, and it is no longer provisional we rename it G_2. If the new phenotype is unsuccessful in its environment, it is eliminated and the original genotype, G_1, remains.

Phase 3 is a further reproductive phase. If the perturbed phenotype, P_1^p, is unsuccessful then the reproduction is still of the old phenotype and genotype, as for Phase 1. If the mutation is successful, there will be a period of reproduction where the potentialities of the new genotype, G_2, are being realised in the new phenotype of the biological system, P2. This will consist of altering species abundances, and therefore constitutes a period of phenotypic evolution. Supposing the mutation is successful, once a new phenotypic equilibrium is established in Phase 3, one could say that genotypic evolution has occurred.

Finally, we turn to evolution in economics, which we represent in Diagram 13. The reader will notice that this diagram has almost the same structure as the previous one, and is also divided into three phases.

In Phase 1 the economy is in a stationary or steady state, which we have termed 'Normal Economic Activity'. The potentialities (i.e. knowledge base) of the economy are being realised in productive economic activity, and this production in turn allows the maintenance of the knowledge base. Therefore, as in the biological case, the genotype and phenotype are in a recursive, self-maintaining relationship.

Phase 2 is Market Selection. At the beginning of Phase 2 an Invention (2a). changes the genotype of the economy; the market conditions will act as a filter for whether this invention will be realised. However, unlike the biological case, this increase in the potentialities of the system is not so often lost. In a biological system, if natural selection rejects the mutation, the mutation disappears from the macro-genotype of that biological system, as the only host for the mutation is the genetic material of the phenotype of that organism. However, in an economy knowledge is not restricted to any individual or group. It can be stored and made available of a wide range of economic actors. For this reason it is much more difficult to assert that an invention is rejected by the economic evolutionary process than to assert that a biological mutation has been rejected by the biological evolutionary process.

If the market judges the invention to be successful, at any subsequent time, the invention will be innovated in Phase 2b, leading to a restructuring of the economy. That is, economic phenotypic evolution takes place.

Phase 3 is a return to Normal Economic Activity. The new economic genotype, G_2, and the new phenotype P_2, become established as the new economic phenotypic equilibrium of the economy. If the invention is unsuccessful, then the original phenotype, P_1, remains in force. However, as the new invention is a permanent addition to the technology of the economy (the economic genotype), it is the new genotype, G_2, which is maintained. We note once more that, unlike in biological systems, G_1 remains part of the corresponding economic genotype even if it is not used any more.

The above discussion has shown one major reason why economics is conceptually more complex than biology. A second reason has been mentioned already in Section 3.3. and 3.4 above, where we noted that successful genotypic changes in biology occur only infrequently. Biological genotypic evolution can be regarded as movements between successive phenotypic equilibria. For economies, on the other hand, it is obvious that in modern times scientific and technical knowledge has been increasing rapidly, and at an accelerating rate. This has led to a tremendous amount of economic genotypic change, which takes place at a rate comparable to market Adjustments (phenotypic change). Therefore economic genotypic evolution cannot be regarded as movements between successive phenotypic equilibria.

We now offer a third reason for why economics is conceptually more difficult than biology. As we have pointed out at the beginning of Section 3, in biology the phentoype cannot affect the genotype. In contrast to this, in economies the phenotype not only maintains the genotype (its potentialities, i.e. the knowledge base), it also generates it, through research and development activities. That is, the phenotype does affect the genotype in economics.

In summary, while we see great similarities between the evolutionary processes exhibited in biology and economies, we consider economics to be conceptually more difficult than biology for the following three reasons:

1. In economies, additions to the genotype can be preserved even when they are not being realised in the phenotype.
2. In economies, genotypic evolution often takes place at a similar rate to, and sometimes even at a faster rate than, phenotypic evolution.
3. In economies, the phenotype can effect the genotype.

7. THE IMPORTANCE OF EX POST AND EX ANTE CONSIDERATIONS FOR CONCEPTUALISING EVOLUTION

Considering again Diagram 10, where the interaction of invention, innovation and the natural world is illustrated, it will be clear to the reader that historical experience supports this conceptualisation. For example, industrialisation in eighteenth and nineteenth century Europe can be understood, ex post, in terms of the substitution of relatively abundant fossil fuels for the rapidly diminishing woodland in industrial processes. The rapidly rising price of charcoal in the eighteenth century led to intensive research into finding a technique of smelting iron using coal as fuel rather than charcoal. The key invention was the transformation of coal to coke, which is nearly pure carbon and can act as a good substitute for charcoal in iron-smelting. This invention dramatically reduced the price of iron and led to enormous changes in the structure of everyday life; an economic phenotypic evolution of great proportions was initiated by this economic genotypic change.

With the massive expansion in iron making went an equally massive expansion in coal mining. Just as early iron making encountered natural limitations when the genotype demanded the use of charcoal, so coal mining encountered limitations when the genotype demanded the use of solely human and animal labour. The genotypic change that occurred as a result of further search for invention, to counter the flooding of deep coal mines, was the introduction of steam engines driving water pumps. (For a fuller discussion of this see Faber and Proops 1985:608–609).

Ex post this analysis of the invention, innovation and natural world relationships is clear. Phenotypic evolution depleted the natural resource and thereby stimulated a search for a new genotype. With the benefit of hindsight the invention of smelting with coke, and of the steam engine, seem both timely and inevitable. Prior to their invention, however, industrialists were far from sanguine about the prospects for the iron and coal industries. It must ever be thus prior to an invention.

The conceptualisation of the evolutionary process must recognize the very different statuses of the past and the future (Faber and Proops 1986, 1989). The concept of novelty is an ex ante concept, and by its nature novelty can only occur in the future, and by the nature of the future it cannot be known in detail. Further, if the perspective of the future is sufficiently long, even the broadest notions of future novelty that actually comes to pass may be impossible to envisage. Here, uncertainty and unpredictability in its starkest form is encountered. This distinction of the ex ante and ex post in systems which exhibit novelty, and are therefore subject to genotypic evolution, is in marked contrast to a system without the emergence of novelty, where only phenotypic evolution may occur. Here we stress again that phenotypic evolution is closed and in many cases equifinal and therefore accessible to prediction, while genotypic evolution is open, not equifinal, and not accessible to prediction.

From the above discussion it will be clear that phentoypic evolution may be considered both ex post and ex ante, while genotypic evolution may properly be understood only ex post. As we have noted above, for biological systems the distinction between phenotypic evolution is generally clear because of the enormously different time scales on which these processes operate. However, evolution in economic systems will generally exhibit phenotypic and genotypic evolution which not seldom operate on similar time scales, making the discrimination between the phenotypic and genotypic evolutionary processes a source of some difficulty in economics.

Of course we know that different industries have different rates of technological change (i.e. of the occurrence of novelty embodied in new genotypes). This rate is very slow in long-established industries, such as electricity generation by fossil fuel power plants, and steel production. On the other hand, industries such as electronics and aerospace exhibit rapid rates of invention. The evidence even suggests that the rates of invention in these two industries is still increasing. These observations suggest that conventional economic analysis can be used in those industries which exhibit a low rate of genotypic evolution, but that it is only of limited value for industries with rapid rates of genotypic evolution[2].

As a further example of the importance of the distinction of genotypic and phenotypic evolution in economics, we turn to environmental economics. We note that there are a multitude of models in resource economics, particularly relating to the depletion of fossil fuels, which seek a market oriented, and therefore phenotypic characterisation of economic activity in the long-run. The fact that it is apparent that complete depletion of fossil fuels with a fixed technology of production (i.e. a fixed genotype) inevitably leads to the collapse of the economy, is dealt with by invoking a more or less arbitrary back-stop technology, which becomes available in the long-run (see e.g. Dasgupta and Heal, 1979) and leads to a new and unique genotype and thence a new phenotype. Thus a limited form of genotypic change is invoked, through as the change is to a technology which is considered to be known ex ante (i.e. predictable), we cannot consider this to be genotypic evolution in its full sense, as we have discussed above. Indeed, as discussed above, by the nature of genotypic evolution it is unpredictable and therefore, in principle, not susceptible to modelling ex ante.

This is not to say that speculations about the long-run evolution of economic systems is either impossible or worthless. Discussion of the types of inventions (i.e. changes to genotypes) offer, through the constructions of scenarios, prognostications about economic development that may be both interesting and valuable. However, we feel it important to be stressed that predictions based on models of phenotypic evolution (e.g market dynamics) have a completely different status to models based on notions of possible genotypic evolution (e.g. invention). The nature of the possibility for phenotypic and genotypic predictions are quite different. While for the former, specific and concrete predictions can be made, this is not possible for the latter.

8. CONCLUSIONS

In this paper we have offered a general and broad conceptualistion of evolution. Because of the novelty of our approach it is obvious that much further work has to be done on this topic.

Here we wish to summarise our discussion so far. We began by suggesting that evolution should be regarded as a very general notion. We remind the reader that we conceptualised it as follows: evolution is the changing of something into something else over time.

Biology offers useful insights into this kind of conceptualisation of the general nature of evolution, as the notions of genotype (potentialities) and phenotype (realisations) are most clearly delineated in this area.
The former allows for the emergence of novelty, while the latter does not. We saw that phenotypic evolution allows for equifinality, and hence for equilibrium solutions, while genotypic evolution does not. This in turn implies that the former allows for prediction, while the latter does not. This further implies that phenotypic evolution may be understood (i.e. it is accessible to analysis and in particular to modelling which can be used for forecasting) in both ex ante and ex post terms, while genotypic evolution may only be understood ex post. A summary in the form of a juxtaposition of the characteristics of genotypic and phenotypic evolution is given in Table 1.

Table 1

Figure 1: Predictable and Equifinal Phenotypic Evolution
Figure 2: Predictable Phenotypic Evolution with a Limit Cycle (1)
Figure 3: Predictable Phenotypic Evolution with a Limit Cycle (2)
Figure 4: A Shifted Phenotypic Equilibrium
Figure 5: Shifting Phenotypic Equilibria Subsequent to Genotypic Evolution (1)
Figure 6: Shifting Phenotypic Equilibria Subsequent to Genotypic Evolution (2)
Figure 7: Path of Long-Run Phenotypic Equilibria with Genotypic Evolution
Figure 8: Phase Paths for Genotypic Evolution
Figure 9: Combination of Phenotypic and Genotypic Evolution for an Economy
Figure 10: Long-Run Interaction Between the Economy and the Natural World
Figure 11: Evolution in Physical Systems
Figure 12: Evolution in Biological Systems
Figure 13: Evolution in Economic Systems

In particular, we have seen that economies are characterised by an extremely complicated interaction of genotypic and phenotypic evolutionary processes. Since the rate of genotypic change in the modern global economy is so rapid it is often not possible to distinguish between the consequences of phenotypic and genotypic changes. Since the changes of the boundary conditions are not considered central to the method of explanation of conventional economics (Faber and Proops 1989), conventional economics can be expected to be of only limited help for understanding economic evolution.

In summary, predictable, phenotypic economic evolution may be subject to ex ante analysis, allowing sensible predictions of future economic realisations. However, genotypic economic evolution, such as invention of techniques, or the emergence of new institutions, is by its nature unpredictable and therefore amenable only to ex post analysis.

Since physics is entirely concerned with phenotypic evolution, this explains why this discipline has been so extremely successful, and also why so many other disciplines have tried to emulate its method. As our analysis shows, however, this emulation can be successful only so far as other disciplines are concerned with phenotypic processes.

As we have noted in the introduction, the key characteristics of evolution are change and time. While we have concentrated in this paper on change, for further research in this area we consider it to be useful to tie in concepts of time more closely into our analysis than we have done up to now.

In particular, we think we will gain additional insights into the nature of evolution by incorporating various work which has already been done on the nature of time in economic processes. For example, methods of dynamics employed in physics and economics, which would be applicable only to phenotypic evolution, may be usefully studied in this context. Particularly important, we consider, is to marry concepts of irreversibility in time, for example as it involves the emergence of novelty, with the concept of genotypic evolution in economies (cf. Faber and Proops 1986, 1989, 1990).

*We are grateful to Alexander Gerybadze, Peter Michaelis, Armin Schmutzler, and Gerhard Wagenhals for helpful comments and in particular to Reiner Manstetten for extensive discussions.

NOTES

1. The endowments of natural resources is determined both by the naturally occurring materials, and by the technology available. A material can only be regarded as a resource if it can be exploited (e.g. Uranium before 1940 was not a valuable resource, but simply an available type of material).
2. We intend to examine this hypothesis by evaluating the literature on industrial economics.

REFERENCES

Alchian, A.A. [1950], 'Uncertainty, evolution, and economic theory', *Journal of Political Economy*, Vol. **58**, pp.211–222.

Bakker, R.T. [1983] 'The deer flees and the wold pursues: incongruencies in predator–prey coevolution', in: D.J. Futuyama and M. Slatkin (eds.), *Coevolution*, Sinauer, Sunderland, Mass.

Beltrami, E. [1987], *Mathematics for Dynamic Modelling*, Academic Press, New York.

Boulding, K. [1981], *Evolutionary Economics*, Beverly Hills, London, Sage.

Crosby, A.W. [1986], *Ecological Imperialism*, Cambridge University Press, Cambridge.

Cyet, M. [1963], *A Behavioral Theory of the Firm,* Englewood Cliffs, Princedon-Hall.

Dasgupta, P.S. and Heal, G. [1979], *Economic Theory and Natural Resources*, Cambridge University Press, Cambridge.

Faber, M. and Proops, J.L.R. [1985], 'Interdisciplinary research between economists and physical scientists: retrospect and prospect', *Kyklos*, Vol. 38, pp.599–616.

Faber, M. and Proops, J.L.R. [1986], 'Time irreversibilities in economics: some lessons from the natural sciences', in: M. Faber (ed.), *Studies in Austrian Capital Theory, Investment and Time,* Springer-Verlag, Heidelberg.

Faber, M. and Proops, J.L.R. [1989], 'Time irreversibility in economic theory: a conceptual discussion', *Seoul Journal of Economics*, Vol. **2**, pp.109–129.

Faber, M. and Proops, J.L.R. [1990], 'Economy–environment interactions in the long-run', *Ecological Economics,* Vol. 2, pp. 27–55.

Hawking, S.W. [1988], *A Brief History of Time*, Bantam, London.

Hirshleifer, J. [1977], 'Economics from a biological viewpoint', *Journal of Law and Economics*, Vol. **20**, pp.1–52.

Keast, A. (ed.) [1981], *Ecological Biogeography of Australia*, Junk, The Hague.

Kettlewell, H.B.D. [1973], *The Evolution of Melanism*, Oxford University Press, Oxford.

Levins, R. and Lewontin, R. [1985], *The Dialectical Biologist*, Harvard University Press, Cambridge, Mass.

Marshall, A. [1890], *Principles of Economics*, Macmillan, London.

Matthews, R.C.O. [1984], 'Darwinism and economic change', *Oxford Economic Papers*, Vol. **36** (Supplement), pp.91–117.

Nelson, R.R. and Winter, S.G. [1982], *An Evolutionary Theory of Economic Change*, Harvard University Press, Cambridge, Mass.

Norgaard, R. [1984], 'Coevolutionary development potential', Land Economics, Vol. **60,** pp. 160–173.

Penrose, E.T. [1952], 'Biological analogies in the theory of the firm', *American Economic Review*, Vol. **42**, pp.804–819.

Schelling, T.C. [1978], *Micromotives and Macrobehavior*, Norton, New York.

Thompson, J.N. [1982], *Interaction and Coevolution*, Wiley, New York.

Veblen, T. [1902], 'Why is economics not an evolutionary science?', *Quarterly Journal of Economics,* Vol. **4**, pp.373–397.

Witt. U. [1991], *Evolutionary Economics,* Cambridge University Press, Cambridge (forthcoming).

Organization and Information in the Evolution of Economic Systems

NORMAN CLARK
Science Policy Research Unit, University of Sussex, UK

1. INTRODUCTION

This chapter is a follow-up to the paper I presented to the 'Manchester' workshop at its meeting in 1988.[1] In that paper I laid out a number of arguments why in my view the complex open systems metaphor has a number of conceptual advantages over the mechanistic approach which underlies the bulk of mainstream economics – that is when we are dealing with the long run, the evolution of economic systems over medium to long time periods. Specifically these advantages lie in (1) how one defines technology; (2) emancipation from equilibrium states; (3) handling of information and uncertainty; (4) capacity to deal with complex overlapping organizational structures and (5) understanding the role of science and related institutions in bringing about technological change.

There is nothing terribly new about many of these propositions. Some were being suggested around the turn of the century by economists such as Veblen[2],

[1] Paper presented at Workshop on Evolutionary Theories of Economic and Technological Change. Department of Economics, University of Manchester 21/22 March, 1989.

while there has been periodic flirtation with biological concepts since then (witness the debate between Alchian and Penrose in the early 1950s[3], Georgescu-Roegen's exploration of economic change as entropic[4] and also of course the voluminous writings of Boulding). Nor is 'systems thinking' a recent phenomenon. For example Lotka and von Bertalanffy were developing an open systems perspective on all science in the 1920s and 1930s, while more recently Emery, Beer, Ashby and others have done a great deal of valuable work in applying systemic ideas to a wide range of disciplines[6].

What I think is somewhat new is the growing number of economists who are beginning to see the weaknesses of the mechanical metaphor when applied to specific problem areas, and who have therefore been drawn in a systems direction. My own case is an example of an applied industrial economist with interests in the broad developmental relations between science and economic production, who felt he was receiving little guidance from economic theory in the mainstream. Others, like Geoff Hodgson[7] and John Foster[8] in the post-Keynesian tradition, have been driven to an evolutionary, systemic position by the increasing narrowness (and arguably also ideological bias) of neo-classical theory. Many of the neo-Schumpeterian school are now beginning to explore systemic ideas, particularly with respect to 'national systems' of innovation[9], while over the past decade or so the theory of the firm has been opened up in similar ways[10]. The interesting point is that many economists coming from quite different backgrounds and with quite different interests are now beginning to converge on the complex systems metaphor.

Why should this be? I can think of three broad reasons. The first, and in a way the most obvious, is the growing narrowness of the mainstream tradition as its own logic drives it into intellectual positions which it is very difficult for the intelligent layperson to take seriously – for example, the implicit mainstream position that microeconomic behaviour is not influenced by the wider socio-economic environment or Becker's concept of the meta preference function which not only rules the behaviour of all consumers but is also inherently non-identifiable.[11] Developments of these kinds make it difficult *inter alia* to relate experiment to theory and therefore stultify scientific progress.

A second reason may be new developments in a number of natural science fields which appear to emphasize systemic behaviour. Probably the best known example is Prigogine and his colleagues at Brussels who have extended thermodynamics to non-linear, irreversible processes in chemistry and thence more generally to other types of system[12]. In physics it is now well understood that the behaviour of sub-atomic particles cannot be explained in terms of Newtonian theory but requires instead a systemic framework. Developments in quantum field theory, for example, have a holistic bias which is quite alien to the nostrums of classical physics. And similar ideas are now being explored in biology and psychology[13]. What is common to all these critical positions is a sustained attack on the relevance of the mechanistic metaphor as a satisfactory and complete conceptual organizer of recent (and in the case of physics, not so recent) empirical evidence. It would be surprising, therefore, if economics, which has its own problems, were immune from these developments.

However, probably more important than either of these is an intensely practical factor – namely the growing interrelatedness of the international economic system and accompanying competitive imperatives. What this means is that competition (and therefore the technology required for competitive advantage) is a function of unstable and shifting *pre-competitive* alliances which cannot be understood purely in terms of traded goods and services but have also an important informational content. Strategic alliances like those proposed recently between GEC, Plessey and Siemens[14] are very much concerned with technological *competence*, which may be defined as the person-embodied capabilities and institutional capacities needed to enter meaningfully into an economic market. They are therefore a means of acquiring access to long-term informational assets which will lead (hopefully) to economic advantage on the part of the economic actors involved, but because of the increasingly complex and science-related nature of these assets it is very difficult for the state (and indeed for the actors themselves) to understand precisely what is taking place and take appropriate decisions. The problem is that conventional economics provides no conceptual apparatus for evaluating the complex regulatory and informational issues involved and hence there is a 'felt need' on the part of policy-makers for something new.

This paper is a contribution in that direction. It represents a further exploration in the use of the organic metaphor to explain the evolutionary behaviour of economic systems. Its basic assumption is that evolution (and therefore creativity) is a common property of all types of system (physical, natural and social) and therefore its characteristics are *homologous* in the formal sense that they stem ultimately from common roots. Of course there is no way of proving this proposition. However, an exploration of broad evolutionary properties shows some remarkable affinities which are at the very least indicative.

Section II summarizes some reasons why the organic metaphor has tended to be overshadowed by the mechanical metaphor in social and economic theory. Nevertheless the latter is becoming increasingly less tenable in many areas of science where there have been widespread developments of evolutionary thinking in recent years. Since economic morphogenesis is closely associated with technological change, and since there are powerful reasons for believing that the latter is an information-intensive phenomenon, section III attempts to define information and explore its properties in an evolutionary sense. An additional factor is that a closer examination of evolutionary information could eventually provide a theoretical simplicity to match the neoclassical economic apparatus. Finally, section IV gives some empirical content to those ideas by describing briefly a number of examples of how pre-competitive research is organized in biotechnology. The conclusion is that there is a prima facie case for believing that long run economic evolution, like that of other systems, is also ultimately creative, and that the more formal apparatus of conventional mainstream economics is best confined to the short and medium term as a tool of analysis. Throughout the text I use the term 'organic' rather than 'biological' since there are many biologists who tend also to believe in life as mechanism.

2. METAPHORS AND MODELS

The first important point to make is that all conceptual endeavour is metaphoric and conventional economics is no exception to this general rule. When a macroeconomist builds a conventional model of an economic system he/she will normally be making the following implicit assumptions:

1. The system is closed or, if it is open, that there are homeostatic mechanisms which will drive it to a balanced position – i.e. exports and imports will ultimately equilibrate through the free operations of the system itself in relation to other systems.
2. That it can be described satisfactorily in terms of a relatively small range of conventional economic quantities or variables – viz. consumption, investment, the rate of interest, average wages, the rate of profit etc.
3. That these variables enter into relations with each other in such a way that the nature of relationships can be first specified and then estimated by conventional statistical techniques – i.e. the *parameters* of the system are both identifiable and capable of enumeration.

In other words the system behaves as a determinate system whose parts and internal interactions are *knowable* (they exist as *states* independent of the observer), *describable* (in terms of a determinate set of mathematical equations and *computable* (in terms of parameters which define optimum performance). The clear analogy is thus with a machine. The parts of the machine are economic actors (or agents) who are assumed to have, in the aggregate, specific goals which they will seek to satisfy in relation to informational signals from other actors, just as laws of natural systems plus integral information flows govern the behaviour of machines. The 'energy' powering the economic system is the goal directed behaviour (or the intentionality) of the actors. And just as a machine takes in raw materials and converts these into products, so the economic 'machine' transforms resources into output. There are even efficiency analogies in terms of how well respectively the machine and the economy do their jobs and corresponding diagnoses relating to what sorts of problems tend to occur which prevent efficient operations.

Now let me be quite clear on this point – that by suggesting that much of economics is ruled implicitly by a mechanical metaphor I am not by any means arguing that such a procedure is always to be criticized. Quite the contrary, there are many examples in economics where the applied economist is profoundly helped in this way. What I am arguing, however, is *first* that we cannot escape taking up a metaphoric position, even if we are unaware of it, and *secondly* that there are occasions when the mechanical metaphor is inappropriate. It is my general belief that one of these occasions is in the analysis of the long term evolution of economic systems. In this case the organic analogy shows considerable advantages. However, despite a long intellectual history it has always tended to remain on the sidelines, and this is so for a number of reasons.

One of the major intellectual obstacles inherent in using an organic analogy is its links with nineteenth century optimism and the inevitability of social 'progress' in its European form. Goonatilake[15], for example, in his recent detailed survey of evolutionary thought shows how this tradition took root with the eighteenth century 'philosophes' such as Cordorcet, and then in the nineteenth century became fully articulated in the hands of writers such as Comte, Spencer and (in an even more extreme form) Frazer. Comte 'viewed history as an evolutionary sequence of developments, which was accompanied by an increase in intellectual complexity. He viewed changes in the social system as largely due to those in the mental sphere brought about essentially by education, resulting ultimately in the perfection of humanity and of human rationality'[16].

Spencer equated social evolution with 'progress' and later attempted to apply Darwin's ideas of natural selection to the development of human society, thereby giving credence to the view that the socio-economic evils of that epoch were the 'natural' by-products of progress. Frazer took the extreme position that European civilization represented the apotheosis of all evolution and as such pointed the way for all simple peoples oppressed by superstition and ignorant of the liberating message of science. Thus the association of 'social Darwinism' with reactionary views of various kinds has reduced significantly the credibility of the organic analogy. And progressively throughout the twentieth century the rival 'mechanical' analogy has held sway[17].

Of course factors of political ideology are not the only suggested reasons behind the mechanical metaphor. An alternative is the simple elegance of Newtonian cosmology which adopted prior descriptions of the behaviour of clockworks and pumps and used them to portray gravitational astronomy. Deutsch puts the point in the following way:

> To the extent that actual machinery resembled such idealized clockwork, isolated from its environment, unmodified by time and interaction, to that extent it was considered to be nearer perfection.'[18]

However, there was nothing in the classical notion of mechanism which *proved* that all reality was mechanical. Indeed, there were some mechanical properties, particularly the notion of reversible time, which we must suppose could never have been observed. And, as Deutsch points out, the more complex modern mechanical devices have become, the more interdependent their parts are and the more susceptible they are to 'temperature, moisture, electrical and other influences'[19], the more removed they are from the 'ideal machine'.

Whitehead's[20] view was that the mechanical metaphor fulfils a deep human need for psychological security, a need which pre-dates both the renaissance and the scientific revolution, and goes back to Greek times. Throughout the medieval period the Roman Church provided cognitive security for European civilization and gave the search for rationality an unambiguous touchstone – that of revealed truth. What the reformation and the scientific revolution did was to sweep all this away, creating a vacuum which could only be filled by something of corresponding authority. The notion of *natural law* provided precisely such security. It promised a set of simple, unchanging discoverable elements which 'act according to simple unchanging laws; from these. . . . the simple unchanging rules of prudent conduct in politics, economics, psychology, morality, religion – or indeed even in writing poetry – [could] be deduced by reasoning and verified by observation. The elements of the system [might] be the foxy princes of Machiavelli or the wolfish commoners of Hobbes; the prudent businessmen of Adam Smith; the abstract increments of pleasure and pain of Jeremy Bentham, or the 'self-evident truths' and 'inalienable rights' of Thomas Jefferson – but whatever they [were], they [were] as unchanging as the heavenly bodies in [Newton's] solar system.'[21]

Probably both ideological and cognitive factors are at work in the case of economics. From the latter standpoint the neo-classical paradigm in the formal hands of writers such as Walras was seen to resemble closely rational physical laws as described by Newton. From the former its 'free market' injunction to confine the role of the state to ensure the contextual conditions necessary for efficient economic performance, gave ideological support to ruling groups. Paradoxically, therefore, whereas the organic metaphor was rejected (partially) on the grounds of (right wing) ideology there is certainly a case to be made for the acceptance of the mechanical metaphor on precisely the same basis, the main difference, perhaps, being the greater ease with which the organic metaphor could be ideologically labelled.

Two final factors might be fashion (nineteenth century physics created the *zeitgeist* for all the sciences because of its capacity to explain the basis for industrialism, and having achieved this position became fashionable) and secondly the fact that economic systems at the turn of the century were much simpler than they are now (and could therefore be reconciled with the simple system metaphor). However, what is less easy to explain is the progressive growth of formalism in postwar economics, enshrined in the mathematics of mainstream journals, since by that time not only had physics itself abandoned the mechanical metaphor as being only approximately true at macroscopic levels, but many other sciences were beginning to readjust their theories in the face of new forms of evidence.

3. INFORMATION

The notion of information, though clearly central to the evolutionary process, is extremely difficult both to define and to comprehend. Nor is it always used in a particularly rigorous way. Traditional information theory as developed by Shannon and Weaver[22] defines information as a flow of messages between states or events of nature which have 'news value'; that is they cause surprise to the recipient. Hence not all possible communications between states have an information content but only those which have some prospective interest. Conventional information theory measures this information content in probabilistic terms – that is to say the total information content of any message in a given 'ensemble' of possible messages is measured by the probability of its occurrence. Using logarithms to convert combinations of discrete probabilities to a summation provides the following general formula

$$M = \sum_{i=1}^{n} P_i \log_2 P_i$$

where M = amount of information
P_i = probability of the ith message being selected
n = no. of possible messages

A logarithmic base of two is chosen conventionally to define a standard binary unit of information (a 'bit') where there are two possible messages which may be communicated the simplest system. Where such messages are equally probable the information content of the system becomes

$$\log_2 2 = 1 \qquad \text{i.e. 1 'bit'}$$

If there were sixteen (2^4) possible messages, the information content of that system would be

$$\log_2 16 = 4 \text{ bits}$$

Singh[23] shows that no information system has this freedom of information content because of practical limitations. The English language, for example, is such that not all of its letters are equally likely to be chosen. It has the property of 'redundancy' and it is this redundancy that gives it intelligibility. The amount of redundancy in any given system using the above notation is measured by the fraction

$$R = 1 - \frac{\sum_{i=1}^{n} P_i \log_2 P_i}{\log_2 n}$$

Hence the information content of any system is defined in a way which is independent of meaning. It is merely 'a measure of one's freedom of choice when one selects a message from the available set many of which may well be devoid of meaning. And yet there is a curious ambiguity here. For how does a system decide what messages are to be chosen and what not? Clearly the system needs some inherent intentionality within itself either derived from its past history *or* programmed into it by some outside agency. In other words although we may define 'information' independently of meaning (or purpose) its use still requires intentionality whether it be from a computer programmer or from the DNA structure which encodes past instructions.

It is largely for this reason that evolutionists tend to reject the Shannon-Weaver view of information and to replace it with a notion which is much more one of biological process. Maruyama, for example, draws attention to what is a basic contradiction in the Shannon-Weaver position, namely that it emphasises improbability. 'The higher the improbability of the structure the higher the amount of information. Structure in Shannon's epistomology has a tendency to decay to more probable non-structures. All Shannon could do was to combat this decay by means of deviation-counteracting feedback mechanisms'[24]. And again Waddington points out that the Shannon view 'has a technical meaning where "information" refers to specific differences within a given universe of variations. Such a world is static. [It] cannot generate information within itself, it can only decay by losing information'[25]. It is doubtful whether Shannon would have objected to these criticisms, however, since his purpose (and that of contemporaries such as Wiener) was mainly that of providing a mathematical basis for communications engineering – i.e. for dealing with such questions as

> 'how does one measure the amount of information, how does one measure the capacity of a communication channel, what are the characteristics of an efficient coding process by which a message is changed to a signal, when is the coding efficient, what is the rate at which the channel can convey information, what are the general characteristics of "noise", how can noise be minimised and what is the difference in transmitting a signal which is continuous (as say in music or speech) as opposed to a discrete symbol (written speech, telegraphy etc.)'[26]

Thus from an *engineering* point of view all Shannon's definition need provide was a measure of the *potential information capacity* in a system. As an external agent the engineer would then 'give meaning' to the system by programming into it the necessary information structure (or organization) for the job in hand (i.e. telephone messages, radar signals or whatever).

However, the evolutionary theorist needs something different. For him evolution is about the dynamics (or deep structures) of change – how species reproduce, differentiate and become more complex, how environmental 'niches' are identified and inhabited, what the detailed nature of symbiosis is, how information is stored naturally and then accessed when required, and so on. The question is then, how do we get from the traditional abstract definition of information to something which helps us to understand more fully the evolution of natural processes? In other words, how do we 're-ify' or give more content to the concept of information as an integral part of evolutionary activity? And it is here that the concept of entropy proves useful since it provides a physical basis for the notion of information.

The second law of thermodynamics states that in any closed system its entropy (or energy unavailable for useful work) will increase to a maximum level at which point the system itself will reach equilibrium. However, if the system is open there is no reason to presume that entropy will increase, provided there is free energy available which can be imported into the system and (in a sense) 'substituted' for entropy which is then exported into the environment. It was Prigogine's great contribution to recognize that both *physical* and *natural* systems are homologous in this sense – that if energy is applied to any system it is often the case that *decreased* entropy (or *increased* negentropy) will result. Up to that time while the fact of biological evolution appeared to confirm this process for natural systems, there had been no general recognition among scientists that physical structures could evolve as well.

What Prigogine showed was that when an open physical system is far from a thermodynamic equilibrium it can transform its internal structure irreversibly provided contextual conditions are appropriate. He called such structures dissipative structures the most frequently cited example of which is the Belonsov-Zhabotinsky reaction involving the oxidation of malonic acid by bromate in a sulphuric acid solution and the presence of catalytic ions (usually cerium).[27] Dissipative structures which are beyond a certain instability threshold do not return to a thermodynamic equilibrium but in effect export their entropy into their environments, import useful energy and/or matter, which is then used in a reproductive way to renew and expand the original system. Such structures are sometimes referred to as 'self-referential' (or subject to 'autopoesis') in the sense that their metabolic behaviour is not brought about by an outside agency, but rather is entirely self-generated.

What this means is that the distinction often made between *physical* and *natural* systems may not be as clear cut as is commonly supposed. Both experience entropic processes, both are capable of self-organization (and therefore creativity), and both arrest the second law of thermodynamics through the import of free energy from an exogenous source. The main difference appears to be that whereas with living systems the negentropic process is clearly and constantly apparent, in the case of physical systems self-organizing behaviour is observable only under special conditions. Nevertheless the fact that it occurs at all is important. For example, it is now recognized that in the formation of the earth itself there was a period (lasting approximately 1.5 billion years) during which time evolution took place within what is now called a 'primeval soup'. According to Birkett what distinguished this period was the gradual creation of organic polymers out of much simpler (inorganic) chemical structures. Evidence supporting this view was produced by a

> 'classic experiment (which) was performed by Stanley Miller in 1953 to investigate the role of electrical discharge as a source of energy in an atmosphere like that of the primitive earth. Miller circulated methane, hydrogen, ammonia and steam past a repeating electric spark. After a week the water was analysed for organic (carbon-containing) compounds. The results exceeded expectation. Four of the amino-acids commonly found in protein were present (along with several which do not naturally occur in protein) as well as fatty-acids and urea, also fundamental to living cells. Other investigators who repeated Miller's work have added sugars, purines and pyrimidines to the list of spontaneously-formed chemicals, and some have achieved synthesis using ultra-violet light as an energy-source. It seems that the fairly complex building-bricks of which living matter is composed could indeed have been synthesised in quantity by purely chemical means.'

Nevertheless the 2nd law of thermodynamics still holds since the existence of localized evolution is still consistent with global entropy increase and since a closed system isolated from its environment will continue to maximize entropy. And it is here that the link with information becomes apparent since a state of maximum entropy is also a state of minimum information. Thus translated into the language of statistical mechanics, the entropy (or the degree of freedom) of a closed molecular system is a measure of the deviation of molecular (or micro) 'states' from the average (or macro) 'state' of the whole system. It is *also*, however, an inverse measure of that system's information content. Singh puts it as follows:

> 'when the state of motions of the molecules in the body is highly disorganized or anarchic, with each molecule in the chaotic whirl of a law unto itself, the number of microstates leading to one and the same macro state is much more numerous so that its thermodynamic probability becomes exceedingly great. This state of great thermodynamic probability obviously yields much less information about the actual structure of the internal motions simply because there are now so many more alternatives to choose from. The thermodynamic probability of a body thus provides us a measure of information about the state of its internal motions even if it does so in a negative way.'[29]

Conversely the less this deviation, the greater the degree of organization within the system, the smaller the entropy – *or* the greater the 'negentropy'. Put another way, the smaller the thermodynamic probability of the system in question, the more coherent the internal structure of that system is. Hence 'structure' and 'information' are inextricably linked, and the processes which give greater coherence to a system are these same processes which provide it with more information.

Is there then a causal link between the two? Brooks and Wiley[30] have attempted recently to demonstrate this and thereby formally to show that the evolution of physical and natural systems are homologous. Rejecting the Shannon-Wiener definition as being merely an instrumental mechanism, they adopt a concept similar to that used by Collier – 'biological array information'[31] which is physically encoded in the structure of an organism. Evolution proceeds because the internal structures of organisms store information which provides the raw material of further development. Information is of two broad kinds – canalyzed information (that contained within the DNA structure itself which provides the broad genetic heuristic) and non-canalyzed information (that contained in the 'messenger' enzymes and proteins which realize ontogenetic potential). Not all such information is expressed (the organism has a creative potential) while new information may be produced by random mutation.

For Brooks and Wiley then, evolution is an historically-bound entropic phenomenon occurring under non-equilibrium conditions. Energy is imported into a system which is predisposed to a certain developmental path through information encoded in its structure. Potential information is converted into more complex internal structure and entropy is exported into the environment. The process is essentially a self-organizing one involving, in Koestler's terminology, fixed rules (derived from historical information embedded in a structure) and flexible strategies (stochastic engagement with environmental signals) which allow selection to take place. Finally all biological systems evolve entropically, including whole populations and symbiotic systems as well as individual members of a species.

What then are the properties of information when conceived in this way? Clearly *organization* is central since as we have seen information needs this before it becomes useful, and in the absence of organization information merely has potential. Thus in an unstructured system the probability of message transmission is small and as a result the system displays no coherence. Conversely the greater the degree of organization the more information the system possesses. *Secondly*, information flows are *relational*. Rather like alternating electric current, they acquire meaning only in terms of the actors (agents or nodes) who exchange the data for purposive reasons. In the absence of agents for whom the available information has meaning, there would be no information flow.

Thirdly information may be *stored* within the component parts of a system's structure, and may subsequently be used creatively. This properly provides also for *hierarchy* since structures can be 'nested' inside each other and be triggered (or not) depending upon information flow from a higher level of organisation. *Fourthly* the information content within any system has the potential for *continuous increase* since the more structured (or organized) the system becomes the greater the number of messages between the component parts and the greater the evolutionary potential. It is for this reason that information-intensive systems tend to grow in *complexity* as they evolve. *Finally*, information flows are normally accompanied by energy flows but as Deutsch has pointed out the two concepts are not identical.

> 'Power engineering transfers amounts of electric energy; communications engineering transfers information. It does not transfer events; it transfers a patterned relationship between events. When a spoken message is transmitted through a sequence of mechanical vibrations of the air and of a membrane; thence through electric impulses in a wire; thence through electric processes in a broadcasting station and through radio waves; thence through electric and mechanical processes in a receiver and recorder to a set of grooves on the surface of a disk; and finally played and made audible to a listener – what has been transferred through this chain of processes, or channel of communication, is not matter, nor any one of the particular processes, nor any significant amount of energy, since relays and electronic tubes make the qualities of the signal independent from a considerable range of energy inputs. Rather it is something that has remained unchanged, invariant, over this whole sequence of processes.'[32]

Clearly we are dealing with an extremely complex concept. Information is not an entity but is rather a relationship between entities (or states). It is not identical with what we understand as 'energy' and yet it acts along with energy in a creative fashion. It may be measured but only in a probabilistic way. It is capable of continuous increase within a system as that system becomes more structured (or experiences greater organization). Finally it is capable of storage for subsequent use. It is therefore a fundamental basis for evolution since clearly life could not proceed without it.

Let us now apply it to the evolution of economic systems. In another publication[33], a colleague and I tried to give greater economic content to these ideas by portraying economic production (the supply side) in terms of informational flows amongst 4 broad sectors – viz.

(i) The N sector (Nature), understanding of which gives us the power to transform resources into saleable goods and services.
(ii) The S sector (Science) which engages in a process of basic research and discovery of 'facts' and 'laws' having general validity.
(iii) The T sector (Technology) which transforms such knowledge into goods and services having economic value.
(iv) The M sector (Market) which buys and consumes goods and services at the 'final demand' stage.

Several points should be noted about this portrayal as outlined in Figure 1.

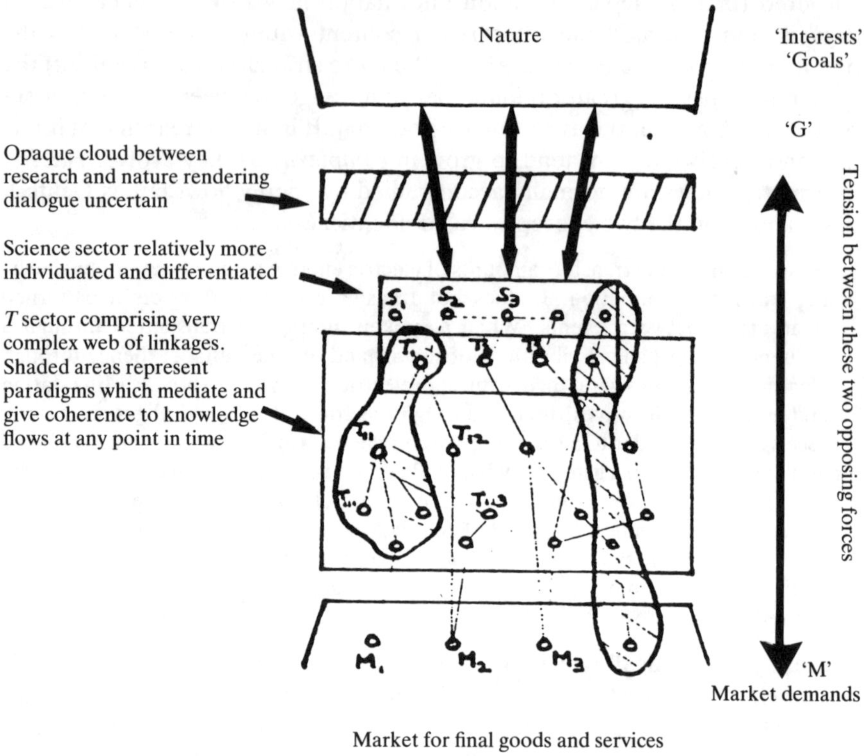

Figure 1 Characteristics: 1) Overlapping activities of T and S sectors (R & D departments of firms do some basic or strategic science, while many basic science labs are to an extent directly concerned with markets). 2) Not all connections are drawn in but one can imagine a multi-dimensional network of very great complexity. 3) Dotted lines represent weak linkages. Solid lines represent strong linkages. 4) Linkages are 'two-way' or 'resonating' flows of information. Often these will be embodied to a degree in economic commodities but of course knowledge flows take many other forms. 5) The 'T' poles are contiguous with productive units and hence may be thought of as akin to junctions in electrical circuitry. Often they act as blockages to the free flow of information (resistors) but they act also as information stores (capacitors) and as information progenitors (amplifiers) where the circumstances are appropriate. 6) The shaded areas represent technological paradigms which mediate and give coherence to information flows at any point in time with respect to any given product type. They should be conceived of also as shifting through time in response to the 'G' and 'M' forces portrayed on the right hand side of the diagram. Technology is thus in a state of continuous flux. There is no equilibrium of the kinds postulated by conventional economic analysis.

a) The relationships are *informational and not economic*. They portray for any economic category of product what flows of information exist in general terms amongst 'nodes' of activity (which may, but need not, be productive in a purely commercial sense). Of course conventional economic relationships often underlie, or shadow, such information flows but the point of the exercise is precisely to emphasise the latter and not the former. Notice that the distinction is not the conventional one of demarcating the 'research' system from the 'productive' system which is quite often used in the economics of technological change. Rather it seeks to point out that in all relationships having an economic, or potential economic, content there is in addition to the transfer of property rights a transfer of information, and it is this second type of transfer which provides economic systems with the capacity for evolutionary change.
b) the relationships are *complex and unstable*. this follows from two factors – first the 'roundabout' nature of activity whereby very many nodes may have economic relevance, and second the fact that every piece of information contains the potential for some form of economic change. Indeed it is certain that the much increased complexity of modern, differentiated economic systems has considerably increased instability due to the very much greater 'potential information' such systems contain.
c) a special feature of these properties therefore is the need for 'potential information' to be *contextualized* in specific circumstances which in turn requires *organization*. Organization is of two types viz. (i) cognitive and (ii) institutional. The first of these corresponds roughly to the technological paradigm but broadened to recognize also its *socially agreed* nature whereby the accepted means of transforming economic resources is enshrined in specific beliefs, designs, codes of conduct, standard models etc. which are used by the practitioners in the field. The second type reflects the constellation of actual institutional mechanisms which provide the necessary *physical* coherence for resource transformation – mechanisms such as joint venture agreements, temporary inter-firm alliances, loose and flexible commercial arrangements such as the Japanese Keirctsu, pre-competitive 'directorates' having significant state involvement, specialist clubs etc. etc. In the diagram these organizational forms are portrayed as amoeba-like structures which in a sense give temporary coherence to unstable information flows and thereby permit economic production (and indeed pre-competitive research) to take place.
d) Paradoxically, however, such organizational forms are continually threatened by the *innate creativity* of component parts of the four systems, whether this be new scientific knowledge or the pressures of economic competition. There is therefore a constant tension between (inter- and intra-systemic) creativity on the one hand and the inertia needed for economic activity to take place at all. In a sense the raw material for this dialectical

relationship is information which, rather like alternating electric current, continuously *resonates* or *fluctuates* between nodes of activity and is understood (or not) within each node depending upon complex behavioural circumstances.

e) Finally, although this is not demonstrated explicitly in the diagram except in the limited sense of overlapping S & T sectors, information flows and their contextualization in organizational form are capable of being arranged as a *hierarchy*. For example cognitive systems spawn sub-systems which themselves differentiate into finer sub-systems and so on. Nested hierarchies thus help to ensure informational stability since they enable *storage* of routines which in turn helps to provide greater inertia than would otherwise be the case.

4. ORGANIZATIONAL FORMS

It would appear then that there is a *prima facie* case for regarding the evolution of economic systems as an entropic phenomenon but with information rather than energy providing the main propagating role. Technological change is viewed as a series of shifting constellations of contextualized information which de-stabilizes established methods of economic production and which takes place as a pre-competitive activity. The efficiency with which any economic system orchestrates this process largely determines economic competitiveness and in turn the main vehicle for such orchestration is the capacity to foster novel organizational forms which enable speedy transition of pre-competitive research to commercial sale. What follows are a number of examples drawn mainly from my own current area of interest – namely the national and international diffusion of biotechnology. These seem to be consistent with the general conceptual ethos of this chapter and have the added advantage that they reflect also the current embryonic, and therefore fluid, state of this exciting but as yet incompletely articulated technology.

Many developing countries (LDC's) now possess expertise in biotechnology but this is confined predominantly to the research end of the spectrum within publicly funded laboratories in universities and research institutes. Outside a few of the richer countries like Brazil, very little manufacture takes place, and what there is is confined to downstream packaging and sale. The major problem for the LDC's is how to formulate viable strategies for absorbing biotechnological expertise which will both complement indigenous technological resources and promote dynamic economic growth in selected areas of relevance. One such area is plant breeding and propagation[34] where the use of modern cell and tissue culture techniques could (and probably will) revolutionize the range, quality, resistence properties and economic costs of most known plant varieties. They will also create new economic opportunities through the discovery and commercialization of new flora and the creation of market opportunities for new cash crops. Another area is the development of monoclonal antibody technology which shows great potential in the production of diagnostics and vaccines to combat endemic diseases. A third example is the use of genetically engineered microbes in waste management and the production of protein supplements for the food industry.

For fairly obvious reasons many LDC's believe that they will only be able to engage sensibly with biotechnology when they themselves have built up a critical minimum level of biotechnological competence. The notion of 'technological capability' which has become increasingly prevalent in recent literature, is designed to capture conceptually precisely this idea of *competence* – of being able to control the ways in which a new technology is deployed for socio-economic ends. And it has been emphasised largely because most technology transfer mechanisms often fail to bridge the 'technology gap' between rich and poor countries.

Bell[35], who has written a great deal in this area, argues the point in terms of 3 distinct types of 'flows' of knowledge and expertise which are associated with the transfer of production facilities: viz.

a) that needed for the transfer and set up of production facilities and all the various services needed to make operational the investment project in question – i.e. feasibility studies, plant commissioning and start-up services, design engineering, training etc.
b) the know-how required to operate and maintain the new system once it has been installed. This is often embodied in people but will also be codified in written forms in manuals, schedules, charts, diagrams etc. People-embodied know-how is often fostered through training and information services, but is also developed through on-the-job learning as well.
c) knowledge and expertise for implementing technical change which involves both the underlying know-why of the technological system itself as well as the various techno-managerial capabilities needed to evaluate and transform existing plant to meet new and innovative operating conditions.

Bell's point is that *genuine* technology transfer requires the eventual flow of all three types of capacity and that a gradual progression of (a) type transfers through to (c) type transfers represents a desirable state of affairs. Very often, however, technology suppliers are unwilling to provide 'core' technology (for strategic reasons), while recipient firms are often quite content with (a) and (b) type transfers where these provide a reasonable level of profit. And although (a) and (b) types transfers usually bring along with them an element of (c) flows there is also a need for government measures to bridge the gap. Usually these will involve identifying missing capabilities and encouraging appropriate organizational means of developing them.

In the case of biotechnology these relate firstly to bioprocessing technology

– i.e. the fermentation, separation and recovery stages – secondly to marketing and sale and thirdly to pre-competitive research where local laboratories may have incomplete capabilities at the 'leading edge'. In order to fill these gaps recent years have seen a whole variety of novel organizational arrangements. These involve small new-technology-based-firms located in the industrialized countries, technology brokerage firms, international aid agencies, non-governmental organizations, research institutions, corporations and government departments. And what is clear is that the permutations involved are very great indeed.

For example, a recent initiative of the World Academy of Art and Science (WAAS) funded by a Swedish Industrial Foundation hopes to establish a Biotechnology Resource Development Corporation (BRDC) which would help to commercialize and stimulate biotechnological innovations of relevance to the Third World. The WAAS has hired a city-based (London) consultancy firm to test the proposal's feasibility, assess start-up financial sources and draw up a business-plan prior to full-scale launch. In the course of its activities this firm is having detailed discussions with corporations, finance houses, universities, LDC governments and international bodies so as to collect and synthesize economic, scientific, technological, financial, marketing and regulatory information of relevance to such a venture.[36]

A second example is a small technology-based consultancy firms operating out of a U.K. university which is currently trying to put together a phytochemical screening programme which involves U.K. and LDC research institutions, pharmaceutical companies, LDC governments and aid agencies. As a secondary activity it is investigating the feasibility of pilot plant antibiotic manufacture in LDC's as a means of developing indigenous technological capabilities.[37] A third example is the Singapore PLANTEK initiative.[38] Here the Singapore government have put up $500,000 to finance a project which has brought together a U.S. high-tech biotechnology firm, two large Asian marketing conglomerates (Indian and Japanese), a local firm and the University of Singapore. The U.S. firm provides the technology and the Indian and Japanese firms provide the marketing know-how, the long term objective being to foster high technology capabilities in Singapore in the area of genetic engineering of tropical cash crops such as papaya and banana. A final example mentioned by Boyson is the 'recent proliferation in OECD countries of state and regional programs for technology development, including public/private research consortia, venture capital funds and 'incubators' for technology based startup companies. In the U.S. at least 43 states have science and technology programs and in 1986 over $700 millions were spent on these initiatives.'[39] Boyson goes on to show how many of these, like the Massachusetts-Puerto Rico program, involve complex tie-ups between high technology sources in the U.S. and commercially-based applications in the South.

What is abundantly clear from these and very many other examples I could mention is the great number and variety of new organizational forms which underlie and buttress the pre-competitive phase of a new generic technology. What is common to all of them is the need to seek out, collect, process and disseminate information relevant to new economic initiatives which have yet to take place – a process which may take many years to complete.

5. CONCLUSIONS

In this paper I have made a broad case for viewing evolutionary economic activity as an organic phenomenon which has close commonalities with the corresponding evolution of natural and physical systems. The metaphor is that of relatively indeterminate open systems (as distinct from that of mechanism) and the raw material, or lowest common denominator, of evolution is information. Put another way, underlying all economic and structural changes are corresponding technological changes which in turn are determined by shifting informational relationships which take the form of changing types of organization – organization of both a cognitive and institutional kind. Organization then becomes the key policy variable for economic change and it is how well new information is put together and contextualized (or organized) which determines the rate and direction of economic evolution.

The examples given of pre-competitive research in biotechnology were chosen by way of illustration but many other areas could equally have been chosen. Thus the recent literature on corporate strategy and intra-corporate structure discussed by Kay[40], Williamson[41], Teece[42] and others is equally amenable to this treatment as is the 'user/supplier' literature of von Hippell[43], Rothwell[44] and Lundvall[45]. Other examples are the 1988 Metcalfe/Gibbons[46] paper which applies a systematic approach to technical change and long-run economic performance and recent research by Fiorenze Belussi[47] on technical change in traditional sectors of the Veneto region in Northern Italy. In all these cases the role of information (in some cases labelled 'knowledge') is central to the relationships among the actors concerned. Hence although it might smack of 'casual reductionism' to give such a primary place to information in this entropic sense it would seem to me that the point is both worth making and capable of further development.

REFERENCES

1. N. Clark, 'Evolution, Complex Systems and Technological Change' published in *Review of Political Economy*. **3** March 1990.
2. T. Veblen, 'Why is Economics not an Evolutionary Science' *Quarterly Journal of Economics* **12**, pp. 374–97.

3. See A. Alchian 'Uncertainty, Evolution and Economic Theory' *Journal of Political Economy LVII*, 1950 pp. 211–21; E. Penrose 'Biological Analogies in the Theory of the Firm' American Economic Review XLI (5) December 1952, pp. 804–19; and then in *AER Communications XLIII* (4) Part I pp. 600–609 (both authors).
4. N. Georgescu-Roegen *The Entropy Law and The Economic Process* (Cambridge, Mass: Harvard U.P., 1971).
5. See for example K. Boulding *Ecodynamics: A New Theory of Societal Evolution* (London: Sage, 1978).
6. See F.E. Emery (ed.) *Systems Thinking* (Harmondsworth, Penguin, 1969) for a useful summary set of readings involving many of these authors. For references to Lotka see N. Georgescu-Roegen op. cit.
7. G. Hodgson 'Economic and Systems Theory' in J. Pheby (ed.) *The General Theory and After: Essays in Post Keynesiansim* special edition of *Journal of Economic Studies* **14** (4) 1987, pp. 65–86.
8. J. Foster *Evolutionary Economics* (London: Pergamon, 1988).
9. G. Dosi et al. (eds.) *Technical Change and Economic Theory* (London: Pinter, 1988).
10. See, for example, N. Kay *The Evolving Firm* (London, Macmillan, 1982).
11. See G. Hodgson *Economics and Institutions* (Cambridge; Polity Press, 1988) for a detailed discussion of these points.
12. I. Prigogine and I. Stengers *Order Out of Chaos* (London: Fontana, 1985).
13. See F. Capra *The Turning Point: Science, Society and the Rising Culture* (Wildwood House, London, 1982).
14. K. Morgan et al. 'The GEC-Siemens Bid for Plessey: The Wider European Issues' PICT Working Paper No. 2, SPRU, University of Sussex, January 1989.
15. S. Goonatilake *Meta Evolution Information Lineages in Gene, Culture and Machine* (mimeo, 1989 and to be published by Pinter Publishers, London, 1990). I have used Goonatilake extensively in what follows.
16. Ibid. Chapter 1 p. 8.
17. See also N. Clark 'Some New Approaches to Evolutionary Economics' *Journal of Economic Issues* **XXII** (2) June 1988, pp. 511–31.
18. K.W. Dueutsch 'Mechanism, Organism and Society: Some Models in Natural and Social Science' *Philosophy of Science* **18** (3) 1951 p. 235.
19. Ibid. p. 234.
20. A. Whitehead *Science and the Modern World* (London: Free Ass. Books, 1985).
21. K.W. Deutsch op. cit. p. 235.
22. C. Shannon and W. Weaver *The Mathematical Theory of Communication* (Urbana Ill., Illinois U.P., 1949) cited in Goonatilake op. cit.
23. J. Singh *Information Theory, Language and Cybernetics* (London, Constable, 1966).
24. M. Maruyana 'Towards Human Futuristics' in M. Maruyana and A. Harkins (eds.) *Culture of the Future* (The Hague, Mouton, 1976), quoted in Goonatilake *op. cit.* chapter 10. p. 7.
25. C. Waddington 'Evolution in the Sub-Human World' in E. Jantsch and C. Waddington (eds.) *Evolution and Consciousness: Human Systems in Transition* (Mass. U.S., Addison-Wesley, 1976) quoted in Goonatilake *op. cit.* chapter 10, p. 7.
26. S. Goonatilake *op. cit.*, chapter 10, p. 3.
27. For a more detailed discussion of this and related points see E. Jantsch *The Self-Organizing Universe* (Oxford: Pergamon, 1980). Also N. Clark and C. Juma *Long-Run Economics* (London: Pinter, 1987), Chapter 4.
28. C. Birkett *Heredity, Development & Evolution* (London; Macmillan, 1979) p. 168.
29. J. Singh *op. cit.* p. 75.
30. D. Brooks and E. Wiley *Evolution as Entropy: Towards a Unified Theory of Biology* (London: Chicago U.P., 1986).
31. Ibid. p. 64.
32. K.W. Deutsch op. cit. p. 241.
33. N. Clark and C. Juma *op. cit.* see Chapter 5.
34. For a detailed account of these technological applications see A. Sasson *Biotechnologies and Development* (Paris; UNESCO, 1988).

35. R.M.N. Bell (ed.) *The Acquisition of Imported Technology for Industrial Development: Problems of Strategy and Management in the Arab Region* United Nations E/ESCWA.NR/85/16, 1986. See Chapter 1. See also M. Fransman and K. King (eds.) *Technological Capability in the Third World* (London: Macmillan, 1984) for a more general account of this area.
36. Private Communication.
37. See R. Thomas 'The Biotic Resource' Guest Editorial *MIRCEN Journal* **3**, 1987, pp. 197–199.
38. S. Boyson 'Commercializing the Spectrum of Biogenetic Resources: Policy Challenges and Opportunities for the South' *Development and South/South Cooperation*, International Center for Cooperation With Developing Countries, Yugoslavia, Spring 1989.
39. Ibid. p. 14.
40. N. Kay 'The R&D Structure: Corporate Strategy and Structure' in G. Dosi et al. (eds.) op. cit. pp. 282–94.
41. O. Williamson *The Economic Institutions of Capitalism* (New York: Free Press, 1985).
42. D. Teece 'Technological Change and The Nature of the Firm' in G. Dosi et al. (eds.) op. cit. pp. 256–81.
43. E. von Hippell *The Sources of Innovation* (Oxford; Oxford U.P. 1988).
44. See for example P. Gardiner and R. Rothwell 'Tough Customers: Good Designs', *Design Studies* **6** (1), January 1985.
45. B.A. Lundvall 'Innovation as an Interactive Process' in G. Dosi et al. (eds.) op. cit. pp. 349–69.
46. J. Metcalfe and M. Gibbons 'Technology, Variety and Organization: A Systematic Perspective on the Competitive Process' in R.S. Rosenbloom and R. Burgelman (eds.) *Research on Technological Innovation, Management and Policy* **4** Jai Press, 1988.
47. F. Belussi *Innovation Diffusion and Innovation Performance in Traditional Sectors: An Empirical Investigation on the Veneto Region of Italy* (SPRU University of Sussex, mimeo, 1989).

Evolution and Intention in Economic Theory

GEOFFREY M. HODGSON[1]
Department of Economics and Government, Newcastle upon Tyne Polytechnic, Newcastle upon Tyne, NE1 8ST, UK

INTRODUCTION

Whilst the relationship between economics and evolutionary theory has been long-standing[2], a number of antinomies and fundamental theoretical problems remain. The first problem can be stated thus: if economic development is determined by some process of natural selection, with something analogous to genetic replication and to random variation or mutation, then what role remains for the notions of intentionality, pruposefulness or choice, which economists from many schools of thought have held so dear? As John Gray (1984, p. 53) remarks: 'The problem with the natural-selection approach is that in accounting for individual character traits, dispositions, and so on by reference to their survival values, it deprives individual choices and purposes of their place at the terminal level of social explanation.'

Thus the preservation of the notion of purposeful behaviour is a problem for those who wish to apply to society the metaphor of evolution in biology. As Gerald Silverberg (1988, p. 539) observes: 'Almost without exception, workers in the field of social evolution acknowledge that human societies are characterised by an emergent property almost totally absent from the bioogical domain – the

[1] Paper originally presented at the University of Manchester Workshop on Evolutionary Theories of Economic and Technological Change, March 21–22, 1989. Second version, June 21, 1989.

presence of conscious goal-seeking behaviour partly guided by mental models of the world which attempt to anticipate the future course of the individual's environment'.

Second, if socio-economic development is evolutionary, what is the analogue to the transmission of information by the gene? 'Whereas the gene has come to be recognised as the fundamental unit of selection in biology, it is still unclear at what level evolutionary selection and innovation operate in socio-economic systems' (Silverberg, 1988, p. 539).

It is the purpose of this article to explore these problems. In the first section the notions of choice and purpose in economics are briefly surveyed. It will be argued that whilst economists have traditionally emphasised purposefulness and choice, these concepts are not well founded in orthodox economic theory, neither is the space in existing ideas for an evolutionary approach well defined.

In the second section the same fundamental issues are raised in regard to biological evolution. It is not the aim here to suggest a slavish imitation of biological theory in social science, but merely to indicate that some underlying issues and controversies are highly illuminating for the matter under discussion.

The third and final section suggests a preliminary theoretical framework of dealing with the problems that have been raised. This features several overlapping and multi-levelled hierarchies, where there are different degrees of purposefulness and a spectrum of notions of causality, with an important place for both habits and institutional routines. Whilst the stress on habit and routine provides a basis for both sustaining the evolutionary analogy in economics, and for building economic models with some tentative capacity to estimate the future, it is nevertheless argued that there must also be a space for some degree of purposefulness and choice, and this must be founded on some notion of objective indeterminacy.

This the main thrust of the paper is to propose a means of combining ideas of purpose and determination, causality and evolution, which previously may have been regarded as difficult to reconcile.

1. EVOLUTION, CAUSALITY AND CHOICE IN ECONOMIC THEORY

1.1 The Rhetorical Background

The idea of individuals with preference is so central to the discourse of orthodox economics that the subject itself has sometimes been defined as 'the science of choice'. The elevation of this concept to this position of supremacy occurred with the 'marginal revolution' in the 1870's. Subsequently, the notion has often been taken for granted. Clearly, to be meaningful, the idea of choice must involve intentional or purposeful behaviour and not a programmed response.

The concept of intentional behaviour is found in a wide variety of economic writings. For instance, when discussing the production process, Karl Marx wrote that 'what distinguishes the worst architect from the best of bees is that the architect builds the cell in his mind before he constructs it in wax' (1976, p. 284). Not only is work a 'purposeful activity' for Marx; in the sphere of exchange the juridical contract is seen as 'a relation between two wills' (p. 178).

Leon Walras, the most formalisitic of the founders of neoclassical theory, saw economic agents as 'endowed with reason and freedom' (Walras, 1954, p. 55). Indeed, for Walras and many other writers, it is the 'exercise of the human will' (p. 61) which distinguishes the social from the natural world.

The American institutionalists also had an explicit notion of purposeful, individual human action. John Commons (1950, p. 36), in particular, argues that the 'science of the human will' acting in both 'individuals and all collective organizations' is the 'twentieth centry foundation' of economic science.

The idea of action as purposeful behaviour is emphasised by the Austrian School, notably von Mises. In his *Human Action*, he forcefully asserts that the analysis of economic and social phenomena requires the premise that human action is purposeful and goal-directed.

The fact that the notion of purposeful behaviour is not deeply rooted in orthodox theory is demonstrated when attempts are made to incorporate Darwinian ideas. In particular Armen Alchian (1950) argued that maximising behaviour by economic agents does not have to be justified in terms of their explicit objectives, but by the 'evolutionary' contention that maximising, and thus 'fit', firms and individuals are the ones more likely to survive and prosper. Such an argument clearly downgrades the theoretical status of the concept of purposeful behaviour, making it irrelevant to the explanation of the competitive process. Given his emphasis on the political desirability of 'choice', it is somewhat paradoxical that a similar argument has been taken up by Milton Friedman (1953) in his famous methodological essay.[3]

Notions of consciousness and purpose cannot be verified empirically, but in that respect they are not unique. All sciences have concepts at their 'hard core' which are not open to testing or falsification, but which help to define the character of the science. However, the principle of the purposeful individual has some basis in real-world institutions. For in practice the principle must be *assumed* by the members system working with individual property rights. The core of the legal system in a market-dominated society is contract law. For this system to operate it must typically be able to attribute consciousness and purposefulness to the parties in any contract. Thus in the modern legal system the assumption of the purposeful individual has its 'material foundation'.[4]

1.2 Determinism, Intentionality and Choice

Nevertheless, despite the rhetoric, it is doubtful that the formalistic models of contemporary neoclassical theory can successfully accommodate such notions as purposeful behaviour and choice. The problem, in brief, is that neoclassical behaviour is essentially predetermined by mechanical functions: the individual is seen as maximising utility with given preferences and objective constraints.

The full consequences of this deterministic view are rarely made explicit. An exception is Vilfredo Pareto, when he wrote: 'The individual can disappear, provided he leaves us this photograph of his tastes' (1971, p. 120). With such a taste-satisfying machine, questions of real will or purpose fade away. As George Shackle (1972, p. 122) puts it: 'if the world is determinist, then it seems idle to speak of choice'.

Furthermore, to be meaningful, choice must take place in conditions of uncertainty. As Brian Loasby (1976, p. 5) has argued: 'If knowledge is perfect and the logic of choice complete and compelling then choice disappears; nothing is left but stimulus and response. If choice is real, the future cannot be certain; if the future is certain, there can be no choice.

Even with the relaxation of the assumption of perfect knowledge in recent neoclassical models, choice is not necessarily re-instated. A Bayesian (or other similar) model with a random element does not necessarily admit true sovereignty or spontaneity for the individual concerned. Action enslaved by the throw of the dice may not be quite as rigidly determined, but it is no more purposeful or free.

Some useful clarifying distinctions have been made in this area by two systems theorists, Russel Ackoff and Fred Emery (1972). They argue that a distinction between purposeful and goal-directed behaviour lies in the set of possible responses to the structural environment faced by the individual. Simpler goal-seeking devices (such as a thermostat) respond in a unique and predetermined manner to changes in their environment. The most sophisticated type of goal-seeking behaviour is that of a computer or machine that can 'learn' from its mistakes in pursuing goals, and thus can respond in different ways to the same repeated problem. However, in both these cases, the goals are still fixed.

The human, purposeful agent is essentially different in that he or she can change goals, and furthermore this may happen without any stimulus from outside. The capacity to change both behaviour and goals without external stimulus means that humans have a will, and that some of our choices are real ones. As Nicholas Georgescu-Roegen (1971, p. 179) puts it: 'If man can *will* his motives freely, then man is free in spite of the fact that all actions follow with necessity from motives.'

Such changes in goals mean that all human action must contain an element of indeterminacy: an uncaused case. As Frank Knight (1933, p. 221) argued long ago: 'If there is a real indeterminateness . . . there is in a sense an opening of the door to a conception of freedom in conduct.' More recently, John Thorp (1980) has argued, such a measure of real and objective indeterminacy is possible at the neurophysiological level, as well as the more familiar indeterminacy in quantum physics. Consequently, and against neurophysiological determinism, there is a place for real freedom of the will.

The question may appear to hinge on the issue of the existence of (non-probabilistic) indeterminacy in the real world. This matter cannot, of course, be resolved here. But the verdict of philosophers such as Karl Popper is clear: after the deterministic drift of nineteenth century science, now the '"natural" view of the universe seems to be indeterministic'.[5] Furthermore, a small number of post-quantum theory economists, such as Knight, have explicitly endorsed the idea of the objective indeterminacy of the world.[6]

We can never demonstrate the existence of indeterminacy because there is always the possibility of a hidden and unknown causal mechanism at work. However, what we do know from the mathematical theory of chaos (Gleick, 1988, Stewart, 1989) is that even if the world is deterministic it would almost certainly behave in an apparently random, even non-probabilistic, and unpredictable way. The possibility of 'deterministic chaos' is thus established. Consequently, even if the world is deterministic, we should have to treat it as if it were indeterministic and unpredictable. Even if all our choices are caused, many of them will appear spontaneous and free. There is thus established a strange ground upon which determinists and indeterminists can meet.

Typically, orthodox economic theory does not include objective indeterminacy, nor purposeful behaviour in the Ackoff-Emery sense, and its models are of goal-seeking behaviour of the simplest type. Behaviour is regarded as a determinate function of external inputs to given preferences. In recent years there has been attention to models where a kind of learning is involved. But, for the reasons given above, the agent is still not endowed with choice.

1.3 The Limits to Indeterminacy

Notably, it is the Austrian School who have put forward a view of the agent where both purposes and actions are not determined by the external environment. A determinate model of human behaviour is not adopted, and consequently real choice is involved. The emphasis is on expectation and imagination: on action in an uncertain world.

However, by taking an extremely individualistic and subjectivist outlook, the Austrian School neglect the forces that themselves may mould – but not necessarily nor completely predetermine – individual purposes and goals. This is done in one of two alternative ways. The first is by the assumption that individual plans cannot be a response 'to anything pre-existent' (Lachmann, 1969, p. 93) and thus they are wholly spontaneous and undetermined. But it seems untenable to deny any possible external influence on the thought processes and purposes of the individual.

The second stratagem is simply to assert that it is beyond the scope of economic theory to enquire as to how purposes and actions may be influenced. As Friedrich Hayek 91948, P. 67) states: 'If conscious action can be "explained", this is a task for psychology but not for economics . . . or any other social science'. But this view involves an unacceptable and rather blinkered compartmentlisation of scholarship: one which not only demarcates psychology from social science, but also denies that it is appropriate for any social scientist to consider the factors that may influence individuals and the formation of their intentions.

The net result, however, is the same; the individual is taken as a given. He or she descends upon the social world, already-formed. The consequences for Austrian theory are that it is incapable of building a model of the economy where the formation or alteration of some individual purposes and goals is taken into account, to complement the complex portrayal of social institutions as the unintended consequences of interacting individuals. Half the picture is missing.

In consequence, Austrian theory is not only ill-suited to consider matters of long-term economic development and growth, it is actually incapable of building any model of the economy which can generate detailed predictions concerning the future. Contrary to many neoclassical theorists, prediction is not all-important. But to ignore it entirely seems to emasculate the science.

In sum, it is desirable to assert the importance of indeterminacy and spontaneity in human action but also to recognise its limits at the same time. In some ranges or dimensions, action may be indeterminate, but in others it is not. To assert indeterminacy is not to deny its limits: that action is also bounded and moulded by the influences of culture, institutions, and the past.

1.4 Matters of habit

The existence of conscious, purposeful behaviour has also itself to be qualified. Frequently, in economics and elsewhere, a Cartesian and dualistic view is taken; the realm of thoughts and intentions is simply divided from the physical and natural world, and consciousness itself is regarded as undifferentiated. The mind, with its rational ideas, is seen to animate the bodily machine.[7]

Notably, the dualistic philosophy, with its unbridgeable divide between the mental and physical, cannot easily accommodate the concept of habit, in the fullest and most meaningful sense, nor recognise its significance in a complex world. For the dualistic philosophy, action is purposeful, and without purpose there is no action. There are no gradations or blurred edges. Habit, which has been defined as a 'more or less self-actuating disposition or tendency to engage in a previously adopted or acquired form of action' (Camic, 1986, p.1044) is either dismissed, devalued or ignored.

In general, neoclassical theorists assume that individual objectives emanate from a single valued utility function. Whilst doubt has already been cast on the ability of such mechanistic functions to represent purposeful behaviour, it is clear that all actions are, so to speak, on the same level. If habits exist they are deemed to result from utility maximisation just like anything else.[8]

Likewise, when confronted with the possibility of such a grey area between the two, Austrian theorists generally react by placing all actions in the same category, that of 'conscious, active, or creative choice' (Buchanan, 1982, p. 15). Hence, when von Mises (1949, p. 13) encountered the prospect of a spectrum of different degrees of purposeful action in his *Human Action* he wrote: 'The vigorous man industrially striving for the improvement of his condition acts neither more nor less than the lethargic man who sluggishly takes things as they come. For to do nothing and to be idle are also action, they too determine the course of events. . . . Action is not only doing but no less omitting what possibly could be done.'

This argument is very flimsy. Von Mises implicitly re-defines action as that which' determines the course of events', in contrast to its spirited definition two pages earlier as 'purposeful behaviour'. In sum, for von Mises, inaction is action. In its first few pages the content of his *magnum opus* is emptied of the vivid promise of its title.

If the abstention from action is to be regarded as purposeful in the manner suggested, then it would imply that we continuously scrutinise not only ourselves but also our entire environment for opportunities to act: a continuous and omniscient monitoring process governing all potential deeds.

Ironically, the reason why such a monitoring process is not feasible derives from the very sort of consideration that the Austrians have quite rightly brought to the fore in a different context: the impossibility of dealing with and processing the vast quantity of information that is involved in the planning of a modern economy (Hayek, 1935; 1948). But a similar point applies to the human mind as well. Both our physiology and our environment are so complex that the human mind cannot commit all the sensory data it receives to the same level of conscious deliberation. Consequently, as Arthur Koestler (1967, p. 238) argues, consciousness 'is not an all-or-nothing affair but a *matter of degrees*. There is a continuous scale of gradations which extends from the unconscious . . . to up to bright, wide-awake states of arousal'.

Such statements are counter to the behaviourist drift in social theory. Under the influence of behaviorists such as John B. Watson and Burrhus F. Skinner, there has been a reluctance, even outside psychology, to use such 'intangible' concepts such as consciousness and intent: 'Merely to mention these pariah words in scientific discourse is to risk immediate loss of attention and audience' (Matson, 1964, p. 174).

To re-assert the importance of the notion of purposeful action it is necessary to establish habitual, unreflexive behaviour, as its real and conceptual opposition. For without such an irreducible hierarchy of levels of consciousness and intent (as proposed by Koestler, 1964; 1967; 1978; 1980) there is the danger that one level or type of consciousness and purposefulness will dissolve conceptually into another.

An excellent illustration of the perils herein is the story of the relegation of the concept of habit in sociology, as related by Charles Camic (1986). After occupying prime of place in the writings of Emile Durkheim and Max Weber, and in sociology generally around the beginning of the twentieth century, the concept of habit was purposefully excised from the discipline. Arguably, a similar process has occurred within economics as well (Waller, 1988).

In part, this was a defensive response to the conceptual homogenisation of action by the behaviourist psychologists after the First World War. Overreacting to this denial of consciousness, reflexivity and rational deliberation, and concerned to maintain its scientific credentials, social science cut loose from the biological and psychological world abandoned any attempt to relate the two,[9] and proceeded to homogenise action at another level: that of deliberative action and choice.

But whilst this defined some distinctive methodological parameters for the subject, the particular choice of level was not itself most crucial. By collapsing all levels of action into one, social science effectively confounded the substance of purposefulness and choice, by denying the obverse with which it could be contrasted. Likewise, behaviourist psychology, by excising consciousness, neglects the essence of habit as well; all is mere behaviour, or all is undifferentiated action.

The re-instatement of the concept of habit is important in a number of respects. By establishing a non-deliberative category of behaviour it is possible, first, to find the basis for some degree of stability and continuity in social life; second, it enhances the idea of choosing or deliberative behaviour with which it contrasts; and third, it provides a basis, as discussed in Section III below, for a genuinely evolutionary theory in economic and social science. But a precondition of this discussion is a closer examination of evolutionary theory in biology.

2. DETERMINISM, CHOICE AND EVOLUTION IN BIOLOGY

2.1 The evolutionary synthesis in biology

The modern, neo-Darwinian, 'synthesis' in biology explains evolution in terms of random variation and natural selection. The notion of causality is neither mechanical, because changes through time are generally irreversible 10, nor strictly deterministic, because of the role of random mutation. It is, to use Jacques Monod's (1971) phrase, a combination of 'chance and necessity'.

What is notable is the absence, at least within neo-Darwinian orthodoxy, of any explanation in terms of intentional behaviour. The element of chance is confined to mutation, not to purpose or action. Given its environment, once an organism is born its behaviour is regarded as being determined simply by its genes.

In neo-Darwinian theory, as Richard Dawkins (1976, 1982) has popularised, natural selection operates not on species, nor groups, nor even individuals, but on combinations of genes. As a result, attempts to explain the social behaviour of gregarious species in terms of 'group selection', such as that of V. C. Wynne-Edwards (1962), have been criticised by sociobiologists such as Edward Wilson (1975). Sociobiologists explain, for example, 'altruistic' behaviour not by its contribution to the capacity of the group to survive, but by the fact that members of the group are likely to be related by descent from common ancestors and therefore share some common genes. Genes are more liable to survive if they include an attribute which disposes that animal to care for its relatives. Consequently, 'caring' or 'altruistic' behaviour can thus emerge by gene-based natural selection.

The work of Wilson, Dawkins, and others is an impressive example of reductionism, reducing animal behaviour to the elemental unit of the gene. It is paralleled by the efforts of neoclassical economists to base all economic phenomena on 'sound microfoundations'. Such an approach is defended in biology on the basis that the genetic material is relatively immutable. However, in economics such a procedure is open to the objection that the microeconomic 'foundations' may be insecurely affected by the macroeconomic environment itself. But, as we shall see, even in biology the degree of mutability of the ensemble of genes is open to question.

It is a feature of reductionist efforts that gross over-simplifications must be made to achieve results, and as Popper (Popper and Eccles, 1977, p. 18) has remarked, there are no known examples of a complete and successful reduction to elemental units in any science, and given the complexities involved we have reason to doubt that it would ever be possible. The dream of Laplace must remain as such.

Another peculiarity of much neo-Darwinian writing is that whilst it excludes the possibility of sub-human species moulding or interfering with their own evolution, this possibility is not denied to humans. (For example, Dawkins, 1976, ch. 11.) The reasoning behind this is usually that humans are affected by their social culture, and not simply their genes, and cultural entities, like genes, have some capacity for replication through time. But it is unclear why other social species should be disallowed the influence of some kind of culture as well.

Counter to the reductionists, there are modern challenges to the idea that the gene is the exclusive unit of selection. Whilst neo-Darwinian biology has a viable explanation of modifications within species, some questions remain concerning the manner in which new species are formed. In other words, success at the level of microbiological theory is not matched by a similar degree of success in explaining macrobiological phenomena such as the development of new species.

Following his joint work with Stephen Jay Gould, Niles Eldredge (1985) proposes that selection mechanisms are hierarchical, involving culls at higher levels that favour whole species, rather than simply genes or individuals. This idea parallels a growing acceptance of non-reductionist and hierarchical views of selection in biology, redolent of the work of biologists such as Ludwig von Bertalanffy, Conrad Waddington and Paul Weiss.[11] Replication, it is argued, is not simply carried out by the gene; it is promoted by chromosomes, organisms and species as well. At each level of the hierarchy the entitity has the capacity to make more of itself.

What is different at each level is the fidelity of the reproductive process; in general, reproductive accuracy declines as the hierarchy is ascended. Furthermore, the longevity of the reproductive unit increases. Although the information in genes is in a sense 'immortal' individual genes do not outlast organisms, and organisms certainly do not outlast the species to which they belong.

Eldredge suggests that a general theory of evolution must incorporate two interacting hierarchies: the genealogical hierarchy, as sketched above, plus the ecological hierarchy. The latter is the hierarchy of organisation of living things as they exist in the real world, namely: molecules, cells, organisms, populations, communities, and regional biota.

Another problem faced by modern biology is to explain the fact that the creation of new species often occurs with relative rapidity, when compared with the longevity of the ancestral and descendant species. These patterns are described by Eldredge and Gould as 'punctuated equilibra'. They argue that rapid environmental change is the main reason for the origin of species, rather than by the simple accumulation of mutations. After a major change in the environment, evolution is seen to slow down, until the population faces another environmental challenge. The theory of punctuated equilibra is an example, therefore, of the interaction fo the genealogical and the ecological hierarchies.

2.2 The question of Lamarckism

A third controversy has been less prominent, but it is no less significant for the line of enquiry traced here. Before Charles Darwin an earlier theory of evolution was developed by Jean Baptiste Lamarck who believed that acquired characteristics can be inherited. Lamarck assumed that organisms have a built-in drive to perfect themselves, and that their consequent adaptations to their environment are passed on to their offspring.

Although revivals of Lamarckism have often concentrated instead on the mechanisms of acquired character inheritance, real intentionality is a feature of

Lamarck's work. Apart from the forceful analytical objections to Lamarckian theory, the idea of intentional behaviour goes against the neo-Darwinian grain, because if all is pre-programmed by the genes, then there is no role left for choice.

Lamarckian ideas are now generally rejected by biologists because no mechanism has been widely accepted as an explanation of how an acquired character would be passed on to the future progeny.[12] For such inheritance to occur the acquired character would have to become encoded in the genes. In fact there are good reasons why acquired characters are very unlikely to lead to such an alteration. The material in which the genetic information is stored is, and has to be, inert and unreactive. As Waddington (1969, p. 369) argues: 'If it was capable of being changed by all sorts of environmental influences, of the kind which exert natural selection on the organisms, it would soon be reduced to a jibbering nonesense.'

Despite the repeated rebuttal of Lamarckian theory, some of the issues it has elevated refuse to be buried. One of these is in regard to the distinction between the genotype and the phenotype (Mayr, 1980, p. 14). Neo-Darwinian theory can explain variation in the genetic information contained in an organism, i.e. the genotype. However, 'natural selection favors (or discriminates against) phenotypes, not genes or genotypes' (Mayr, 1963, p. 184). The actual characteristics of the organism, including those that have been acquired after its birth, are the substance of selection.

This difficulty remains largely inresolved, except by reductionist and deterministic attempts to settle the issue through the dogmatic assumption that *all* relevant behaviour is determined by the genes. In a famous quip, Mayr described such views as 'beanbag genetics'. However, if this idea that natural selection operates on the penotype rather than the genotype is applied seriously and consistently, 'very radical changes in outlook become inevitable' (Waddington, 1976, p. 13). Indeed, it becomes a principal reason for abandoning the reductionist view.[13]

Notably, the deterministic processes of genetic transmission and selection are themselves inadequate to define life. After all, simple and inert crystal patterns can 'replicate' and thus hand on a kind of 'information' as new layers of crystal are formed. As Waddington (1969, p. 362) points out: 'The purely genetically transmissible aspects are simply not interesting enough to be called living'. He argues that the key quality of life is that it interacts with its environment in such a way as to produce a phenotype.

Waddington also points out that 'the environments in which organisms live, develop, and are selected, are neither wholly dependent nor wholly independent of the genotypes of the organisms involved' (1969, p. 365). For instance, an organism may overcome its discomfort in a certain environment not through its own adaptation, but by moving somewhere else. Such a consideration seems obvious, but it qualifies the 'random variation and selection' model somewhat.

Indeed, it becomes almost Lamarckian in its overtones – although not in terms of the actual mechanisms employed – for the behaviour which an animal will exhibit now 'must have been the evolutionary result of natural selection operating on his ancestors according to how they behaved in earlier periods' (Waddington, 1976, pp. 13–14). Clearly the possibility that the environment may be chosen or even shaped is crucial when attention is directed at social or economic rather than biological evolution.

With these considerations in mind, Waddington proposes an essentially Darwinian mechanism of 'genetic assimilation' which nevertheless involves a partial rehabilitation of some Lamarckian ideas:

> 'Natural selection acts on phenotypes, which may have become adapted during their lifetime by physiological proceses to the particular stresses they have had to meet. The selection will act on a population of organisms which have been subjected to the stress, and there is almost bound to be some genetic variation in the capacity of different individuals to respond appropriately to that stress. Selection will therefore depend to increase the frequency in later populations of genes conferring ability to respond adequately and adaptively. This in itself will merely mean that the capacity to respond adaptively to the stress will increase in later generations' (Waddington, 1976, p. 14).

In other words, natural selection will lead to different capacities to respond by further adaptation to changes in the environment. Given that the environment can change progressively, and also the organism can to some extent choose or affect its surroundings, then there is the possibility of an interacting system with circulatory feedback loops of a cybernetic nature.

However, such an interacting system would be ill-adapted if it were oversensitive to environmental changes. In general, there has to be a degree of resistance to change, with small or negligible responses to minor fluctuations in conditions, and stresses have to be quite severe before new developmental adaptations are triggered. Otherwise the organism could be over-reacting to environmental fluctuations with disastrous results. When stresses have become sufficiently strong to trigger the development of a new adaptation, then this will itself become somewhat stable and difficult to alter, that is until fresh and extraordinary stresses once again arise. Consequently: 'An adaptive change, originally induced by some environmental stress, may, after many generations of selection, form part of the hereditary endowment of a later population' (Waddington, 1976, p. 15).

Emphatically, such quasi-Lamarckian developments do not involve the direct inheritance of acquired characters. If a phenotypic modification in response to stress occurs in one generation of organisms then it will not be passed on automatically to the next generation and will probably disappear if the stress ceases altogether. 'However, the capacity to acquire the character – to respond to the stress in this particular way – is inherited.' (Waddington, 1969, p. 374).

After a long and successive process of selection, the population of organisms will still show the phenotypical alteration even after the stress ceases to be present. By this stage the alteration will be genetically assimilated through progressive natural selection.[14]

The so-called Weismann barrier, protecting the genotype from the influence of the environment and acquired characters, is still intact in regard to each individual organism. But in the long run the barrier seems to be violated, as the gene pool hs been winnowed and changed through natural selection, to build up the proportion of organisms showing appropriate responses to environmental change. What is true for one link in the chain of inheritance, and for one individual organism, is no longer true, over time, for the population as a whole.

Clearly, we are back to hierarchies again. Different models and modes of reasoning are required for species or populations when compared with individuals. Waddington himself develops the idea of a *chreod* (from the Greek 'chre', it is fated or necessary, and 'hodos', a path) to conceptualise the long-term development of a population. A chreod is a relatively stable trajectory of development for the population.

Nevertheless, the chreod is not always stable. Just as a river is stabilised by its channel, its head may work back until it breaks into an adjacent and higher valley and 'captures' another river by creating an alternative path. Likewise, in biology, similar topological 'catastrophes' may occur causing one chreod to branch into two, or to suddenly switch its course of development in a relatively short period of time.[15]

2.3 The hierarchies of life

Moreover, in Waddington's work there is an attempt to introduce some notion of purposeful behaviour. Such an endeavour has been endorsed by Popper amongst others, who has accepted that animal preferences and consciousness could be decisive in evolution (Popper and Eccles, 1977, pp. 12–13). In Waddington's argument the organism can, for example, sometimes choose its environment. Furthermore, purposeful behaviour has different effects at differect levels. Whilst individual organisms can make real and effective choices, wholesale environmental changes are required to divert a chreod from its path. At his macroevolutionary level the choices and actions of each individual organism are to little effect.

In consequence, not only is there a hierarchy of degrees of purpose and habit within the individual, these nest within the greater hierarchies spanning populations and species, much in the manner that Eldredge (1985) has described, with different types of causality and determination at each level. The pattern of overlapping hierarchies, or of net-formation or 'reticulation' (as Koestler, 1967, p. 387, puts it), means that each unit or 'holon' (Koestler, 1967, ch. 3) is subject to the influences not only above and below in its own hierarchy but also through the connections with other hierarchies or systems.

However, the actions of holons in general, and organisms in particular, are not completely determined by their status, connections, nor inputs within this network. Although the rules that programme the behaviour of the organism may be fixed, it is suggested by Waddington's work, for example, and argued by Koestler (e.g. 1980, pp. 455–9), that there is room for different and variable strategies.

Georgescu-Roegen (1971, p. 127) proposes that history is the outcome of two conflicting factors: hysteresis and novelty. In a similar vein, Koestler (1980, p. 455) sees the distinction between fixed codes and variable strategies as 'fundamental to all purposeful behaviour'. It is a manifestation of the 'Janus effect' – the idea that organisms 'display both the autonomous properties of wholes and the dependent properties of parts' (Koestler, '967, p. 383), every organism (holon) is possessed of two opposite polarities: 'an *integrative tendency* to function as part of the larger whole, and a *self-assertive tendency* to preserve its individual autonomy' (Koestler, 1980, p. 465). On this basis Koestler thus attempts a break with the dualism of not only mind and matter, but also of the intentional and the mechanical, and of finalism and deterministic causality. It also involves a break with reductionism which is final and complete, precisely because of the degree of indeterminacy at each level of each hierarchy. In this manner the whole can never be explained simply through the interaction of the parts.

These issues cannot be elaborated further here, nor is it possible to adjudicate on the above mentioned disputes within biology. However, it can be noted that in contrast to the reductionist approach of some neo-Darwinians, the hierarchy-orientated developments of Eldredge, Koestler, Gould, and Waddington suggest the most fruitful type of evolutionary analogues for social science. Remarkably, in such hierarchical theories there is once again a place for intentional, as well as habitual, behaviour.

3. ECONOMIC EVOLUTION: INTENTIONS AND DETERMINACY

We now consider the application of the above discussion of purposefulnes and indeterminacy to the social and economic sphere, and the question of the possible economic analogues to the transmission of instructions by the gene.

3.1 *The Replication of habits and routines*

Within an institutionalist perspective, organisational structures, habits and routines play a similar evolutionary role to that of the gene in the natural world. For Thorstein Veblen in particular, institutions themselves are comprised of 'settled habits of thought common to the generality of men' (Veblen, 1919,

p. 239). Similarly, Knight (1947, p. 224) saw the forces that help to mould society as belonging 'to an intermediate category, between instinct and intelligence. They are a matter of custom, tradition or institutions. Such laws are transmitted in society, and acquired by the individual, through relatively effortless and even unconscious imitation, and conformity with them by any mature individual at any time is a matter of "habit"'.

It has been argued above that habits are essential to deal with the complexity of everyday life; they provide us with a means of retaining a pattern of behaviour without engaging in global rational calculations involving vast amounts of complex information. Agents have acquired habits which effectively relegate particular ongoing actions from continuous rational assessment. The processes of action are organised in a hierarchical manner, facilitating monitoring at different levels and rates, and with different degrees and types of response to incoming data.

The capacity to form habits is indispensable for the acquisition of all sorts of practical and intellectual skills. At first, whilst learning a technique, we must concentrate on every detail of what we are doing. Eventually, however, intellectual and practical habits emerge, and this is the very point at which we regard ourselves as having acquired the skill. Thereafter, analytical or practical rules can be applied without full, conscious reasoning or deliberation.

Work itself involves a degree of practical knowledge or know-how which is both acquired and routinized over time. Indeed, the industrial skill of a nation consists of a set of relevant habits, acquired over a long time, widely dispersed through the employable workforce, reflective of its culture and deeply embedded in its practices (Veblen, 1914; Dyer, 1984).

To some degree, habits have a stable and inert quality and tend to sustain and thus 'pass on' their characteristics through time, and from one institution to another. For example, the skills learned by a worker in a given firm become partially embedded in his or her habits. Thus these act as carriers of information, 'unteachable knowledge', and skills. In this respect they have a quality analogous to the informational fidelity of the gene. However, the reproductive accuracy of the gene is clearly much the greater.

The idea that routines within the firm act as 'genes' to pass on skills and information is adopted by Richard Nelson and Sidney Winter (1982, pp. 134–6). Being concerned to show how technological skills are acquired and passed on within the economy, they argue that habits and routines act as repositories of knowledge and skills, as the 'organizational memory' (p. 99) of the firm. As Nelson and Winter make clear, routines do not act as genes in the neo-Darwinian sense because the inheritance of acquired characteristics is possible. Thus the evolutionary process in society is in this respect Lamarckian.[16]

Routinized behaviour replicates itself through its capacity to establish enduring rules and norms. These are not necessarily inviolable, but their relative stability helps agents to estimate the potential actions of others. 'One individual can choose or plan intelligently in a group of any size only if all others act 'predictably' and if he predicts correctly. . . . Without some procedure for co-ordination, any real activity on the part of an individual, any departure from past routine, must disappoint the expectations and upset the plans of others who count on him to act in a way predicted from his past behaviour.' (Knight and Merriam, 1948, p. 60).

Thus routines and formal institutions, by establishing more or less fixed patterns of human action, supply information to other agents, in particular by indicating what other agents might do. In general, the individual is likely to replicate the same sort of behaviour that led to the establishment of the routine. Either because it is seen to 'work', or even by an unconsciously-driven propensity to imitate, the same routine is followed and further reproduced. 'The situation of today shapes the institutions of tomorrow through a selective, coercive process, by acting upon men's habitual view of things, and so altering or fortifying a point of view or a mental attitude handed down from the past.' (Veblen, 1899, p. 190) Consequently, in a highly complex world, and despite indeterminacy and uncertainty, to some extent regular and predictable behaviour is possible.

Because individual tastes and preferences are malleable and will change or adapt, the objectives and behaviour of agents can be moulded or reinforced by institutions. This is partly because institutions have an important cognitive function (Hargreaves Heap, 1986–87; Hodgson, 1988). The information they provide is not transmitted raw; it is affected by the structures of those institutions themselves. They influence the processes through which information is selected, arranged and perceived by agents.

3.2 *Overlapping hierarchies*

Clearly, within institutions, routines are hierarchically ordered just as habits are so arranged in regard to an individual. However, there is not one hierarchy in regard to institutions but several. Formal institutions have an obvious hierarchy in terms of their legal precedence or position in a chain of command. Thus a legislative assembly may be deemed to have sovereignty; and the head office of a corporation has powers over its branches or plants. Nevertheless, the formal hierarchy never corresponds to the pattern of control in the real world. Overlapping with it is a web of informal institutions and routines, embedded in the social culture. Thus there are compelling informal links between many persons at the pinnacles of power, and likewise strong community bonds between actors with less status or influence.

Additionally, the information that is processed and transmitted by institutions itself has its own internal hierarchy of meaning. Thus, at the apex, there is information about information, over-riding instructions or rules, and so on. Lower-level information is often more detailed and specific.

There is an important contrast here with orthodox neo-Darwinian biology. Whereas in the latter information transmission is largely at the single and primary level of the gene, the hierarchy of knowledge in socio-economic systems involves information transmission at several different levels, and incorporating different types of information with different effects.

Notably, non-codifiable knowledge is more difficult to augment or change, involving relatively long periods of 'learning-by-doing' and so on. In contrast, whilst codifiable knowledge may sometimes be complex and difficult to acquire, it can often be altered or augmented more readily. Furthermore, short-run disturbances in the course of economic development are more likely to emanate from changes in codifiable knowledge, simply because it can be communicated more easily.

Hence the immediate sensitivity of the economy to changes in prices or other indicators, and the greater responsiveness of the modernised sectors of the economy to R&D initiative and technological diffusion, in contrast to the sluggishness of the more traditional industries. In short, informational feedback loops are likely to be more immediate and significant in regard to codifiable rather than non-codifiable knowledge.

Other forms of codifiable knowledge concern the political system, formal institutions and so on. Here the transmission of codifiable information can correspond to a change in the rules, or to signals concerning economic and other policies which lead to shifts in the expectations of entrepreneurs and other agents. Thus, in the manner that John Maynard Keynes described in chapter 12 of *The General Theory*, the state of long-term expectation is vulnerable to changes in 'the news'.

In sum, the closer to the codifiable end of the knowledge hierarchy the greater the degree of adaptability and change in the content of that knowledge. Consequently, changes in technological knowledge, or prices, or information concerning governments and laws, are likely to account for the more dramatic changes to the pace and direction of socio-economic development. In contrast, more gradual and enduring change is likely to emanate from the non-codifiable end of the spectrum: in changes to deeply embedded habits and routines.

As a result, the knowledge hierarchy conforms to Koestler's (1967, p. 388) observation as follows: 'Holons on successively higher levels show increasingly complex, more flexible and less predictive patterns of activity, while on successive lower levels we find increasingly mechanized stereotyped and predictable patterns'.

However, Koestler may be wrong in elevating this to a universal principle. For when it comes to the institutional hierarchies – or even the ecological hierarchies in nature – as opposed to the hierarchies of information and knowledge, it is probably more difficult to make changes, whatever the intentions of the agents involved. Thus, for example, it may be easier for most of the population to change the nature or pattern of their employment than a nation to alter the fundamental thrust of its foreign policy. This is because, in the latter case, change is much more constrained, subject to innumberable counter-balancing wills, acting and blocking each other with unintended consequences, and in regard to more numerous and more enduring signals and assurances of stability. Thus historians may explain the development of national foreign policy or the pace of industrial progress not merely by reference to the accretions of months or years, but of centuries.[18]

Such an argument woud not concur with Koestler's maxim, but with the 'principle of *stratified determinism*' adduced by Weiss (1969), i.e. the 'principle of *determinacy in the gross despite demonstrable indeterminacy in the small*'. At the higher and more complex level, spanning many territories and units, the behaviour of institutions is more stable, and more determined or weighed down by the constraints of its past.

Furthermore, in an evolutionary framework, microscopic randomness or indeterminacy may be *necessary* to establish a degree of stability at the macro-level. Without variations in the force and direction of the current the chreodic river could not scour both its channel and its valley. Microscopic variation is essential to both create and stabilise the evolutionary trend.

For this and aforementioned reasons a socio-economic system combines elements of stability with instability. In regard to aspects of the system which are highly dependent on codifiable information, concerning, for example, prices or other economic indicators, activity will be vulnerable to volatile expectations and waves of speculation. In contrast, its structures and social relations will be much less vulnerable to immediate shocks and short-term changes, and severe social and political disruption may be experienced before fundamental changes occur.

3.3 Institutional evolution

Considering these overlapping hierarchies of formal and informal institutions, it is also likely that the capacity to adapt to change, and in contrast to retain habits and routines, is different at different levels. Habitual actions such as driving a car do not need to be re-learned after a break of several months. If the contrary was true, too much time would be wasted in re-learning each skill anew after a lapse in its use. There are good reasons to presume that this capacity of retain habits for some time is part of our biological evolution, as a degree of durability in this regard would clearly be advantageous for an organism. Furthermore, behaviour patterns of greater flexibility at a more conscious level may also be advantageous from the point of view of biological evolution, such as in response to danger or threat.

However, concerning social institutions, we have to treat the biological analogy with care. 'Institutional mutation' (Harris, 1934) is not the same as genetic mutation, and we cannot automatically assume that a Darwinian process of natural selection is at work. The latter involves the gradual accumulation and selection of small mutations over long periods of time, whereas in the case of institutions adaptation is much more rapid with much less opportunity for any efficiency-breeding selection to work. As Philippe Van Parijs (1981, pp. 88–89) observes: 'if the pressures of selection are negligible compared to counteracting pressures of variation, there can be no evolutionary attractor' in the socio-economic domain.

However, whilst institutional variation and differentation occurs much more rapidly and extensively than mutation in the biological world, the observed inertia of cultural and institutional evolution suggests that there are strong stabilising forces at work. Thus it is more reasonable to conjecture that relatively stable, chreodic-type development is more evocative of institutional evolution than the more traditional Darwinian picture of sharpened adaptation in face of a ceaseless and formidable struggle for survival. Likewise, Norman Clark and Calestous Juma (1987) see the chreod as an analogy for the pattern of technological development in an industrial economy. Given the hierarchical control sequences involved, once a technological 'paradigm' is adopted it predetermines a general direction or path of relatively stable development.

If institutional and industrial developments are typically chreodic then the policy conclusions are quite different from the more prominent attempts to bring the evolutonary analogy into social science in the past. In contrast to the theories of Spencer in the last century, and Alchian, Friedman, Hayek and Williamson in this, an evolutionary paradigm is not a basis for the policy outlook of Dr. Pangloss. As selective processes will not ensure a rigorous drive towards greater efficiency, and chreodic development will exhibit a path of development more determined by its past than by its adaptation to the present, there is no grounds for proclaiming that evolution will produce the best of all possible worlds.

Neither can the manifest stability of the chreod be relied upon. As shown by the work of Waddington and Thom, development along a stable path can sometimes lead to catastrophe. Institutions change, and even gradual change can eventually put such a strain on a system that there can be outbreaks of conflict or crisis, leading to a radical change in actions and attitudes. Thus there is always the possibility of a breakdown or regularity. In any social system there is an interplay between routinized behaviour and the variable or volatile decisions of other agents and their outcomes.

This non-deterministic view stresses both the weight of routine and habit in the formation of behaviour and the importance of some elements of strategic deliberation and their possibly disruptive effects on stability. With these ingredients it is possible to envisage processes whereby for long periods the reigning habits of thought and action are cumulatively reinforced and stabilised. But this very process can lead to sudden and rapid change. The very ossification of society could lead to the decimation of the economic system by more vigorous competition from outside, or there could be an internal reaction leading to a newly modernised order.

Conversely, a recklessly dynamic system may suffer from lack of continuity of skill or outlook, and reach an impasse because in its own breakneck pace its members were left without enduring values or goals. In Veblen's view the economic system is not a 'self-balancing mechanism' but a 'cumulatively unfolding process'. Because of the momentum of technological and social change in modern industrial society, and the clashing new conceptions and traditions thrown up with each innovation in management and technique, the cumulative character of economic development can mean crisis on occasons rather than continuous change or advance.

Fortunately, however, there is sufficient capacity for variation in the system that, with appropriate industrial and macroeconomic policies, economic and technological development can be shifted from a less desirable or crisis-prone path to one more advantageous. Whilst the weight of institutions and tradition should never be underestimated, the ever-present margins of variation and indeterminacy ensure the possibility, even necessity, of an interventionist policy.

NOTES

1. The author is grateful to the participants at the University of Manchester Workshop on Evolutionary Theories of Economic and Technological Change for comments on an earlier version of this paper.
2. At the birth of modern economic science, the emergence of social and economic institutions was analysed in terms which we could now broadly describe as evolutionary. This is particularly the case with the Scottish School. Hume grappled with the manner of evolution of laws and conventions, coming close (Hume, 1886, Vol. 2, p. 429) to the idea of natural selection in social evolution. Smith, of course, with his famous 'invisible hand', saw social outcomes as the result of human action but not of human design (Hamowy, 1987).

   Many early evolutionary ideas were borrowed as much from economics by biology as the other way round. Darwin's theory was inspired by the picture of 'the struggle for existence' in the *Essay on the Principle of Population* by Malthus. Furthermore, recent examinations of Darwin's notebooks (Gruber, 1974; Jones, 1986; Schweber, 1977; Vorizimmer, 1977) 'suggest that his reading of Adam Smith in the crucial year 1838 led Darwin to his decisive breakthrough' (Hayek, 1988, p. 146). Thus it has been concluded that 'Darwin grafted Adam Smith upon nature to establish his theory of natural selection' (Gould, 1978, p. 100).

   It is well known that Marx (1977, p. 525) saw the relevance of Darwin's ideas to social science and offered to dedicate the second volume of *Capital* to the great biologist; Marshall (1920, p. xii) stated that 'the Mecca of the economist lies in economic biology rather than economic dynamics'; Veblen (1898) was so influenced by Darwin that he resolved to transform economics into an 'evolutionary science'; and Schumpter (1976, p. 82) insisted that the 'essential point to grasp is that in dealing with capitalism we are dealing with an evolutionary process . . . a fact . . . long ago emphasized by Karl Marx'.
3. The Alchian-Friedman-Williamson verson of the evolutionary analogy has some of the same pitfalls as 'social Darwinism'. One of these, which Darwin himself was keen to avoid, is the assumption that evolution always means increasing progress and efficiency. However, evolution is not a grand or natural road to perfection; it can reproduce error and fail to discover an obscure or tortuous path to improvement. Close examination of the mechanism of 'natural selection' in modern biology shows that it does not even necessarily lead to survival. There is not necessarily

any mechanism 'by which natural selection tends to favour the survival of the species or the population. The population, in fact, may 'improve itself to extinction' (Elster, 1983, p. 54). Like social Darwinism, as Sidney Winter (1964) makes clear, work in the Alchian-Friedman genre fails to specify any plausible and detailed mechanism for the evolution of societies or firms.

4. It should be noted, however, that the modern legal system admits the possibility of actions which are not fully planned or purposive, and the principle of the purposeful individual is strongly qualified or even excluded in a number of areas.
5. Popper and Eccles (1977, p. 32). Elsewhere Popper (1972, p. 213) notes that it was the American philosopher Charles Sanders Peirce who first broached the notion that we have to assume objective indeterminacy in order to understand the diversity of the universe: 'Peirce conjectured that the world was not only ruled by the strict Newtonian laws, but that it was at the same time ruled by laws of chance, or of randomness, or of disorder.'
6. Knight (1933, p. 221) saw reason to contend that "mind' may in some inscrutable way originate action', i.e. as an 'uncaused cause'. For a discussion see Lawson (1988).
7. For relevant discussions see Capra (1982), Koestler (1967; 1980) and Hodgson (1988).
8. Amongst others, Marshall (1920) and Wicksteed (1910) have insisted that habits must directly or indirectly emanate from deliberate choice. In a few cases, neoclassical theorists have tried to model habitual behaviour, including decision-making and processes at more than one level (Pollack, 1970; Thaler and Shefrin, 1981; Winston, 1980). In these models habitual acts are regulated by a secondary preference function to which habitual preferences gradually adjust through time. Whilst such a two-level approach removed the implication that habits are activated in the same manner as higher-level decisions, low-level choices over habitual acts are still made as if with a full calculation of benefits and costs.
9. Whilst it is possible to admit some autonomy for the physical, biological and social levels, it would be unwarranted to render each as an absolute (Timpanaro, 1975). Two social theorists, Hirst and Wooley (1982, p. 24) have argued forcefully that social relations should not 'be rigidly differentiated from biological or psychological phenomena'. Note also the statement by the biologist Dobzhansky (1955, p. 20): 'Human evolution is wholly intelligible only as an outcome of the interaction of biological and social facts'.
10. The irreversibility of economic processes is emphasised by Georgescu-Roegen (1971, 1978, 1979; see also Dragan and Demetrescu, 1986). According to Dollo's law, evolutionary processes in biology in are likewise irreversible.
11. See Bertalanffy (1952), Waddington 1972) and Weiss (1971). Reviewing the arguments from biology, both Corning (1983) and Sober (1984) endorse the non-reductionist and hierarchical approach.
12. Steele's (1979) neo-Lamarckian conclusions are based on experiments in immunology which, nevertheless, have 'failed the test of replicability'. Notably, 'The collective sigh of relief within orthodox evolutionary biology was almost audible' (Eldredge, 1985, p. 122; see Dawkins, 1982, for a more orthodox evaluation).
13. As Elliot Sober (1984, ch. 9) concludes after an extended discussion of this issue.
14. There are some similarities between Waddington's theory and the anti-mechanistic theory of 'organic selection' proposed by James Baldwin, in which directed habit becomes the guiding force of evolution. However, the Baldwin effect depends upon fortuitous mutation after habits are established, whereas genetic assimilation works through progressive selection of the appropriate capacity to respond to stress. (See Bowler, 1983, pp. 81, 131–2; Hardy, 1965, p. 161–70; Koestler and Smythies, 1969, p. 386; Piaget, 1978, pp. 14–21.)
15. René Thom, the founder of catastrophe theory in mathematical topology, was influenced by Waddington when he extended such ideas to morphogenesis (Thom, 19750). Waddington 'anticipated important parts of Thom's biological thought, and was the first scientist of great stature to acclaim catastrophe theory' (Woodcock and Davis, 1980, p. 21).
16. Similar conclusions are reached in McKelvey's (1982) pioneering study of the evolution of organisations.
17. For a fuller discussion of this point, and its application to a theory of economic growth, see Hodgson (1989).
18. For econometric evidence to support such a long-term contention in regard to differences in productivity growth in 16 OECD countries see Hodgson (1989).

REFERENCES

Ackoff, R.L. and Emery, F.E. (1972) *On Purposeful Systems* (London: Tavistock).

Alchian, A.A. (1950) 'Uncertainty, Evolution and Economic Theory', *Journal of Political Economy,* **58**, June, pp. 211–22.

Bertalanffy, L. von (1952) *Problems of Life: An Evaluation of Modern Biological Thought* (New York: Wiley).

Bowler, P.J. (1983) *The Eclipse of Darwinism: Anti-Darwinian Evolution Theories in the Decades around 1900* (Baltimore: Johns Hopkins University Press).

Buchanan, J.M. (1982) 'The Domain of Subjective Economics: Between Predictive Science and Moral Philosophy', in Kirzner (1982, pp. 7–20).

Camic, C. (1986) 'The Matter of Habit', *American Journal of Sociology,* **91**(5), pp. 1039–87.

Capra, F. (1982) *The Turning Point: Science, Society and the Rising Culture* (London: Wildwood House).

Clark, N.G. and Juma, C. (1987) *Long-Run Economics: An Evolutionary Approach to Economic Growth* (London: Pinter).

Commons, J.R. (1950) *The Economics of Collective Action* (New York: Macmillan).

Corning, P.A. (1983) *The Synergism Hypothesis: A Theory of Progressive Evolution* (New York: McGraw-Hill).

Dawkins, R. (1976) *The Selfish Gene* (Oxford: Oxford University Press).

Dawkins, R. (1982) *The Extended Phenotype: The Gene as the Unit of Selection* (Oxford: Oxford Univeristy Press).

Dobzhansky, T. (1955) *Evolution, Genetics and Man* (London: Wiley).

Dosi, G., Freeman, C., Nelson, R., Silverberg, G. and Soete, L. L. G. (eds.) (1988) *Technical Change and Economic Theory* (London: Pinter).

Dragan, J.C. and Demetrescu, M.C. (1986) *Entropy and Bioeconomics: The New Paradigm of Nicholas Georgescu-Roegen* (Milan: Nagard).

Eldredge, N. (1985) *Unfinished Synthesis: Biological Hierarchies and Modern Evolutionary Thought* (Oxford: Oxford University Press).

Elster, J. (1983) *Explaining Technical Change* (Cambridge: Cambridge University Press).

Friedman, M. (1953) 'The Methodology of Positive Economics'. in *Essays in Positive Economics* (Chicago: University of Chicago Press).

Georgescu-Roegen, N. (1971) *The Entropy Law and the Economic Process* (Cambridge, MA: Harvard University Press).

Georgescu-Roegen, N. (1978) 'Mechanistic Dogma in Economics'. *British Review of Economic Issues,* No. **2**, May, pp. 1–10.

Georgescu-Roegen N. (1979) 'Methods in Economic Science'. *Journal of Economic Issues,* **13**(2), June, pp. 317–28.

Gleick, J. (1988) *Chaos: Making a New Science* (London: Heinemann).

Gould, S.J. (1978) *Ever Since Darwin: Reflections in Natural History* (London: Burnett Books).

Gray, J. (1984) *Hayek on Liberty* (Oxford: Basil Blackwell).

Gruber, H.E. (1974) *Darwin on Man: A Psychological Study of Scientific Creativity, together with Darwin's Early and Unpublished Notebooks*, transcribed and annotated by P.H. Barret (New York: E.P. Dutton and Co.).

Hamowy, R. (1987) *The Scottish Enlightenment and the Theory of Spontaneous Order* (Carbondale, Ill: Souther Illinois Press).

Hardy, A. (1965) *The Living Stream: A Restatement of Evolution Theory and its Relation to the Spirit of Man* (London: Collins)

Hargreaves Heap, S.P. (1986–87) 'Risk and Culture: A Missing Link in the Post Keynesian Tradition', *Journal of Post Keynesian Economics,* **9**(2), Winter, pp. 267–78.

Harris, A.L. (1934) 'Economic Evolution: Dialetical and Darwinian', *Journal of Political Economy,* **42**(1), February, pp. 34–79.

Hayek, F.A. (ed.) (1935) *Collectivist Economic Planning* (London: Routledge and Kegan Paul).

Hayek, F.A. (1948) *Individualism and Economic Order* (Chicago: University of Chicago Press).

Hayek, F.A. (1967) Studies in Philosophy, Politics and Economics (London: Routledge and Kegan Paul).

Hayek, F.A. (1982) Law, Legislation and Liberty, 3-volume combined edn. (London: Routledge and Kegan Paul).
Hayek, F.A. (1988) *The Fatal Conceit: The Errors of Socialism, Collected Works of F.A. Hayek*, Vol. 1 (London: Routledge).
Hirst, P.Q. and Wooley, P. (1982) *Social Relations and Human Attributes* (London: Tavistock).
Hodgson, G.M. (1988) *Economics and Institutions: A Manifesto for a Modern Institutional Economics* (Cambridge: Polity Press).
Hodgson, G.M. (1989) 'Institutional Rigidities and Economic Growth', *Cambridge Journal of Economics,* **13**(1), March, pp. 79–101.
Hume, D. (1886) *Philosophical Works*, 4 Vols., ed. T.H. Green and T.H. Grose (London: Longmans, Green).
Jantsch, E. and Waddington, C.H. (eds.) (1976) *Evolution and Consciousness: Human Systems in Transition* (Reading, Mass: Addison-Wesley).
Jones, L.B. (1986) 'The Institutionalists and 'On the Origin of Species': A Case of Mistaken Identity', *Southern Economic Journal*, **52**(4), April, pp. 1043–55.
Keynes, J.M. (1936) *The General Theory of Employment, Interest and Money* (London: Macmillan).
Kirzner, I.M. (ed.) (1982) *Method, Process and Austrian Economics* (Lexington, MA: Heath).
Knight, F.H. (1933) *Risk, Uncertainty and Profit,* 2nd edn. (London: London School of Economics).
Knight, F.H. (1947) *Freedom and Reform: Essays in Economic and Social Philosophy* (New York: Harper and Brothers).
Knight, F.H. and Merriam, T.W. (1948) *The Economic Order and Religion* (London: Kegan Paul, Trench Trubner).
Koestler, A. (1964) *The Act of Creation* (London: Hutchinson).
Koestler, A. (1967) *The Ghost in the Machine* (London: Hutchinson).
Koestler, A. (1971) *The Case of the Midwife Toad* (London: Hutchinson).
Koestler, A. (1978) *Janus – A Summing Up* (London: Hutchinson).
Koestler, A. (1980) *Bricks to Babel* (London: Hutchinson).
Koestler, A. and Smythies, J.R. (eds.) (1969) *Beyond Reductionism: New Perspectives in the Life Sciences* (London: Hutchinson).
Lachmann, L.M. (1969) 'Methodological Individualism and the Market Economy', in Streissler (1969, pp. 89–103). Reprinted in Lachmann (1977).
Lachmann, L.M. (1977) *Capital, Expectations and the Market Process*, edited with an introduction by W.E. Grinder (Kansas City: Sheed Andrews and McMeel).
Lawson, A. (1988) 'Probability and Uncertainty in Economic Analysis', *Journal of Post Keynesian Economics,* **11**(1), Fall, pp. 38–65.
Lerner, M. (ed.) (1948) *The Portable Veblen* (New York: Viking Press).
Loasby, B.J. (1976) *Choice, Complexity and Ignorance: An Enquiry into Economic Theory and Practice of Decision Making* (Cambridge: Cambridge University Press).
Marshall, A. (1920) *The Principles of Economics*, 8th edn. (London: Macmillan).
Marx, K. (1976) *Capital*, vol. 1, translated by B. Fowkes from the fourth German edition of 1890 (Harmondsworth: Pelican).
Marx, K. (1977) *Karl Marx: Selected Writings*, ed. D. McLellan (Oxford: Oxford University Press).
Matson, F.W. (1964) *The Broken Image* (New York: Doubleday).
Mayr, E. (1963) *Animal Species and Evolution* (Cambridge, MA: Harvard University Press).
Mayr, E. (1980) 'Prologue: Some Thoughts on the History of the Evolutionary Synthesis', in Mayr and Provine (1980, pp. 1–48).
Mayr, E. and Provine, W.B. (eds.) (1980) *The Evolutionary Synthesis: Perspectives on the Unification of Biology* (Cambridge, MA: Harvard University Press).
McKelvey, W. (1982) *Organizational Systematics: Taxonomy, Evolution, Classification* (Berkeley, CA: University of California Press).
Mises, L. von (1949) *Human Action: A Treatise on Economics* (London: William Hodge).
Monod, J. (1971) *Chance and Necessity* (New York: Knopf).
Nelson, R.R. (1980) 'Production Sets, Technological Knowledge and R&D: Fragile and Overworked Constructs for Analysis of Productivity Growth?', *American Economic Review (Papers and Proceedings),* **70**(2), May, pp. 62–67.

Nelson, R.R. and Winter, S.G. (1982) *An Evolutionary Theory of Economic Change* (Cambridge MA: Harvard University Press).

Pareto, V. (1971) *Manual of Political Economy*, translated from the French edition of 1927 by A.S. Schwier, and edited by A.S. Schwier and A.N. Page (New York: Augustus Kelley).

Piaget, J. (1979) *Behaviour and Evolution*, translated from the French edition of 1976 by D. Nicholson-Smith (London: Routledge and Kegan Paul).

Polanyi, M. (1957) *Personal Knowledge: Towards a Post Critical Philosophy* (London: Routledge and Kegan Paul).

Polanyi, M. (1967) *The Tacit Dimension* (London: Routledge and Kegan Paul).

Pollak, R.A. (1970) 'Habit Formation and Dynamic Demand Functions', *Journal of Political Economy,* **78**, July–August, pp. 745–63.

Popper, K.R. (1972) *Objective Knowledge: An Evolutionary Approach* (Oxford: Oxford University Press).

Popper, K.R. and Eccles, J. (1977) *The Self and Its Brain* (Berlin: Springer International).

Schumpeter, J.A. (1976) *Capitalism, Socialism and Democracy*, 5th edn. (London: George Allen and Unwin).

Schweber, S.S. (1977) 'The Origin of the *Origin* Revisited', *Journal of the History of Biology,* **10**, pp. 229–316.

Shackle, G.L.S. (1972) *Epistemics and Economics: A Critique of Economic Doctrines* (Cambridge: Cambridge Univesity Press).

Silverberg, G. (1988) 'Modelling Economic Dynamics and Technical Change: Mathematical Approaches to Self-Organisation and Evolution', in Dosi, Freeman, Nelson, Silverberg and Soete (1988, pp. 531–59).

Sober, E. (1983) *The Nature of Selection: Evolutionary Theory in Philosophical Focus* (Cambridge, MA: MIT Press).

Steele, E.J. (1979) *Somatic Selection and Adaptive Evolution: On the Inheritance of Acquired Characters* (Toronto: Williams-Wallace International).

Stewart, I. (1989) *Does God Play Dice? The Mathematics of Chaos* (Oxford: Basil Blackwell).

Streissler, E. (ed.) (1969) *Roads to Feedom: Essays in Honour of Friedrich A. von Hayek* (London: Routledge and Kegan Paul).

Thaler, R.H. and Shefrin, H.M. (1981) 'An Economic Theory of Self-Control', *Journal of Political Economy,* **89**, 392–406.

Thom, R. (1975) *Structural Stability and Morphogenesis: An Outline of a General Theory of Models* (Reading, MA: Benjamin).

Thorp, J. (1980) *Free Will: A Defence Against Neurophysiological Determinsim* (London: Routledge and Kegan Paul).

Timpanaro, S. (1975) *On Materialism* (London: NLB).

Van Parijs, P. (1981) *Evolutonary Explanations in the Social Sciences: An Emerging Paradigm* (London: Tavistock).

Veblen, T.B. (1898) 'Why is Economics Not an Evolutionary Science', *Quarterly Journal of Economics,* **12**, July, pp. 373–97. Reprinted in Lerner (1948) and Veblen (1919).

Veblen, T.B. (1899) *The Theory of the Leisure Class: An Economic Study of Institutions* (New York: Macmillan).

Veblen, T.B. (1919) *The Place of Science in Modern Civilisation and Other Essays* (New York: Huebsch).

Veblen, T.B. (1914) *The Instinct of Workmanship* (New York: Augustus Kelley).

Vorzimmer, P.J. (1977) *Charles Darwin: The Years of Controversy; The Origin of Species and Its Critics, 1859–1882* (Philadelphia: Temple University Press).

Waddington, C.H. (1957) *The Strategy of the Genes* (London: George Allen and Unwin).

Waddington, C.H. (1969) 'The Theory of Evolution Today' in Koestler and Smythies (1969, pp. 357–74).

Waddington, C.H. (ed.) (1972) *Towards a Theoretical Biology*, 4 vols. (Edinburgh: Edinburgh University Press).

Waddington, C.H. (1976) 'Evolution in the Sub-Human World', in Jantsch and Waddington (1976, pp. 11–15).

Waller Jr, W.J. (1988) 'Habit in Economic Analysis', *Journal of Economic Issues,* **221**(1), March, pp. 113–26.

Walras, L. (1954) *Elements of Pure Economics*, translated from the French edition of 1926 by W. Jaffe (New York: Augustus Kelley)

Weiss, P.A. (1969) 'The Living System: Determinism Stratified', in Koestler and Smythies (1969, pp. 3–42).

Weiss, P.A. et al (1971) *Hierarchically Organised Systems in Theory and Practice* (New York: Hafner).

Wicksteed, P.H. (1910) *The Commonsense of Political Economy*, ed. L. Robbins (London: Routledge).

Williamson, O.E. (1975) *Markets and Hierarchies: Analysis and Anti-Trust Implications: A Study in the Economics of Internal Organization* (New York: Fee Press).

Williamson, O.E. (1985) The Economic Institutions of Capitalism: Firms, Markets, Relational Contracting (London: Macmillan).

Wilson, E.O. (1975) *Sociobiology* (Cambridge, MA: Harvard University Press).

Winston, G.C. (1980) 'Addiction and Backsliding: A Theory of Compulsive Consumption', *Journal of Economic Behaviour and Organization,* **1**, pp. 295–324.

Winter, Jr, S.G. (1964) 'Economic "Natural Selection" and the Theory of the Firm', *Yale Economic Essays,* **4**(1), pp. 225–72.

Woodcock, A. and Davis, M. (1980) *Catastrophe Theory* (New York: Avon).

Wynne-Edwards, V.C. (1962) *Animal Dispersion in Relation to Social Behaviour* (Edinburgh: Oliver and Boyd).

On Some Notions of Irreversibility in Economics

GIOVANNI DOSI
Department of Economics, University of Rome 'La Sapienza' and Visiting Fellow, SPRU, University of Sussex, Brighton, UK.*
J. STANLEY METCALFE
Department of Economics, and PREST, Manchester University, Manchester, UK.

1. INTRODUCTION

In general, the notion of irreversibility, in the economic and social domain, has to do with the possibility that present actions – by individual agents or groups of them – might bear consequences which shape and constrain future decision processes, and/or the future structure of the system and/or its path of change.[1] Such a notion entails powerful analogies with a highly general idea, found also in natural sciences, associating irreversibility with the uniqueness of the direction of time. As a quite recent and growing stream of analysis has shown – e.g. in the fields of thermodynamics, chemistry and biology – there exists a wide classes of systems with respect to which one cannot operate the so-called 'Galileo Transformation', i.e. reverse the sign of the time variable and obtain an identical description of their end-states.[2] Hence, for example, with time-irreversibility one may disturb a certain system with an exogenous shock of Δx in a control variable x, wait for whatever adjustment takes place and, then, disturb it again with a new shock, $-\Delta x$, and find that the second end-state does not correspond to the initial one.

More generally, in natural sciences,irreversibility properties are shown by all systems which are, in the currently accepted definition, 'dissipative', in the sense that they exchange energy with the outside environment, and, under certain conditions, conserve and increase their inner order and complexity over time. Moreover, irreversible changes in 'dissipative' systems generally involve 'bifurcations', phase transitions and changes in the structure of the systems themselves.

If anything, social systems magnify the irreversibility properties found in natural dissipative systems and this is closely related to the concept of exchanging information with the 'outside' environment. Certainly, they are generally characterised by growing complexity and non-linear changes in the relationships between microbehaviours and macro outcomes. Moreover, they generally involve also irreversible modifications in the characteristics and behaviours of micro agents (individuals, firms, etc) as a result of their activities of adapting, learning, exploring in environments which continuously change. In these cases the analogue to 'exchange of energy' is 'exchange of information'. S. Winter (1986) recalls the old Eraclitus' statement on the impossibility of bathing twice in the same river. Earlier, Leontief has warned on the difficulty of 'writing history backward' as a task as difficult as providing predictions on the future. (He quotes Balzac: '. . . Predire les gros evenements de l'avenir n'est pas . . . un tour de force plus extraordinaire que celui de deviner le passe. . . . Si les evenements accomplis ont laisse des traces, il est vraisemblable d'imaginer que les evenements a venir ont leurs racines', *Le Cousin Pous*, cited in Leontief (1963)). In essence, Eraclitus' and Balzac's views, together, hint at irreversible processes which are somewhat 'ordered', and hence can be understood, but still leave an irreducible degree of freedom to individual and/or collective actions or even chance.

That history matters is an old and general assumption of most social science disciplines, notably with the exception of a significant stream of modern economic theory. Moreover, the tension between 'freedom' and 'necessity' has been a central concern of Western culture since its origin: such a tension, in fact, implies the tension between irreversible processes – either at the level of individual decisions or of social processes – and the possibility of 'choosing' for the present in ways at least partly independent from what happened in the past. However, for a long time, the discussion of such crucial aspects of social affairs has been confined to the level of 'appreciative' conjectures, or, more often, that level which philosophers would call the 'influential metaphysics' of various theories. Only recently, have formal models been developed which allow for a more precise analysis of irreversible processes and the representation of tensions which they entail between history-shaped constraints and newly emerging opportunities to change.

The order of discussion in this paper is as follows. In section (ii) we shall briefly review different classes of formal models, including classical systems and those involving different degrees and sources of irreversibility.

Section (iii) will browse briefly through the history of economic thought, trying to identify various notions of irreversible economic processes,focussing on Smith, Marshall and Schumpeter as economists aware of significant irreversibility in economic affairs. Section (iv) and (v) will attempt a taxonomic exercise on the sources of irreversibility present in actual economic systems, both at micro level – i.e. for individual decision makers (section (iv)) and for the system as a whole (section (v)). This distinction between the two levels of analysis, we suggest, is particularly important for the understanding of how history matters in social systems and possibly differentiates them from all other systems where micro units do not exhibit teleological (purposeful) behaviours. In this respect, we shall try to show the different notions of irreversibility that apply to either domain and how they link with each other.

2. DYNAMIC SYSTEMS: ANCIENT AND MODERN

For a considerable number of economists, their principal approach to the study of dynamic systems is that contained in Samuelson's *Foundations of Economic Analysis*. This approach draws heavily on the Newtonian classical dynamics of mechanical systems. Classical dynamic systems are intended to explain order and the persistence of phenomena, with the concept of equilibrium playing a central role in the structure of ideas. A system is identified by a given set of state variables and a law of motion defined across the state space to explain changes in those state variables as a function of time. In economics, models of this kind are familiar from equilibrium growth theory and from the discussion of the stability of market systems. However, a sharp distinction must be drawn between linear and non linear laws of motion. Until recently the theory of dynamic systems was based on methods of near to equilibrium approximation and involved the use of linear methods. Only recently has the analysis of far from equilibrium behaviours led to the analysis of non-linear systems which turn out to have quite remarkable properties, as we shall discuss below.

It will be useful to review the general properties of a classical linear system, which are as follows:

a) a sharp dichotomy between the given and invariant laws of motion and the exogenously determined initial conditions of the state variables. The latter are quite arbitrary;
b) the set of state variables is given, in this sense only quantitative change in the set of measured quantities is permitted;
c) the laws of motion are invariant over the entire state space so that qualitative behaviour is the same everywhere;
d) the rates of change of state variables are a function only of current values of the state variables;

e) the limiting states may define single or multiple equilibrium points which may be stable or unstable. These properties of 'equilibrium' are entirely predictable given the laws of motion of the system. Thus a full characterisation of the terminal states of the system is possible e.g. node, focus, centre, saddle point.[3];
f) the system satisfies the time reversal test, namely that (–t) may replace (+t) with no consequent change in the laws of motion.

Such systems define an 'ahistorical' process for which the laws of motion and the set of state variable is time and state invariant. Such a dynamic system is still compatible with the occurance of complex histories for the time paths of the state variables. However, the crucial point is that an 'ahistorical' process is predictable with respect to its end states while a 'historical' process is not. The former is closed and the latter is open.

The classical dynamic process is fundamentally 'ahistorical'. Within such a process a knowledge of the initial conditions allows the entire sequence of the future states of the system to be derived from the given laws of motion. The history and future of any trajectory is fully specified. Moreover, small changes in initial conditions generate, small non-cumulative differences in the corresponding trajectories of the state variables, so that they remain close together.

A closely related property is that the classical system exhibits strong path independence of the equilibrium states. Each choice of initial conditions defines a unique path towards an equilibrium state (given stability) and many paths lead to the same equilibrium. Consequently any exogenous shocks to the state variables at time t, redefine the initial conditions for the subsequent evolution of the system but they do not alter the equilibrium states.

The recent and growing awareness of the properties of non-linear dynamic systems has resulted in a major revision in our understanding of dynamic processes.[4] Even simple dynamic systems can give rise to extremely complex behaviours, in contrast to linear systems where 'complexity' is consistent with simple behaviour patters, whether oscillatory or non-oscillatory. The mathematics of these systems is complex and a detailed representation of their properties is only known for one dimensional systems and a small number of higher dimensional ones, e.g. Lorenz and Rossler systems.

The first important characteristic of non-linear systems is that arbitrarily small differences in initial conditions can result in cumulatively increasing differences in the resulting trajectories, so that they cease to look alike – the phenomenon of exponential divergence. Thus approximate knowledge of initial conditions does not enable one to predict the future states of the system. Since initial conditions cannot be known with any accuracy in the real world, this raises clear difficulties in attempting to forecast future states of a system.[5] The crucial point is that although the imperfections of measurement can be arbitrarily small, for a system exhibiting chaotic behaviour the resulting trajectories converge and diverge in quite unpredictable ways. An immediate consequence of this is the introduction of a new kind of stability condition, relating not to a point or a limit cycle (the traditional notions of stable positions) but to 'strange' or chaotic attractors. This is defined as a bounded region of state space that stretches and folds any ensemble of steady state trajectories, producing sensitive dependence on initial conditions and long term unpredictability.[6] At most, all one can predict from a knowledge of the equations of motion of the system is the nature of the attractor region, and that trajectories will be contained within this region and ultimately converge to its surface. The consequence of this is that the paths followed by the state variables give the appearance of non-deterministic behaviour, when in fact they are completely determinate. Sudden breaks in the nature of the behaviour of the trajectories can be observed and the exact path followed between any two points in time is never repeated in any subsequent such period. The above are properties of a chaotic non-linear system of given structure.

However, not every non linear system exhibits chaos for all values of its parameters. A second remarkable property of some systems is their potential sensitivity to small changes in structure as represented, for example, by a change in a particular parameter. This relates to the phenomena of bifurcation. As this parameter is changed, a critical value (or set of values) is reached at which the equilibrium behaviour of the system changes drastically, e.g. from a point attractor, to a limit cycle, or to some chaotic attractor., Across the bifurcation point the qualitative behaviour of the system changes fundamentally, new solutions emerge, as it were, in competition with each other. Fundamentally, in the neighbourhood of any bifurcation point there exists an essential indeterminacy concerning which of the new solutions will be chosen, even though the system is deterministic. In the neighbourhood of a bifurcation small events or fluctuations shape the evolution of the system. Even a simple, non-linear, logistic diffusion equation possesses a rich bifurcation behaviour with a successive transition from point equilibrium to chaotic attractors as the system parameter is increased.[7]

These are powerful results since they undermine the time reversability property of classical dynamic systems. The qualitatively different behaviour on either side of a bifurcation means that prior states are no longer realisable. Reversing time simply takes the system to different bifurcation points not the original one. At the same time it must be recognised that the systems which have been investigated in the natural sciences are simple in the extreme in comparison to the complexities of economic and social systems. Bifurcations still leave the underlying laws of motion unchanged. Far more significant is the realisation that at critical points the laws of motion may themselves bifurcate in an irreversible way. But this takes us into the conceptual territory of a truly historical system.

Briefly some remarks must be made concerning stochastic (Markov) systems which appear to present an alternative route to modelling historical processes.[8]. A Markov process is defined by a given set of states in which a system may find itself (generally finite, although infinite state space systems may also be treated) and a law of motion represented by a given transition probability matrix which states the probability in a given time interval of the system making a change from state i to state j. Such systems have a number of attractions, in particular, they generate paths which are historically rich and they are susceptible to rigorous analysis.[9] In particular, it is possible to fully partition the given set of states into transient states – the probability of returning to which tends to zero – and recurrent states – the probability of returning to which tends to unity as time tends to infinity. Thus the state is taken as the unit of analysis and the questions posed concern, for example, the probability of returning to any state given that one starts from it. In systems which satisfy the ergodic theorem one may identify a probability distribution of the limiting recurrent states which is independent of the initial state of the system, but not, of course, independent of the transition probability matrix. Other Markov systems do not satisfy the ergodic theorem, e.g. the simple birth process in which all states are transient in that they are left with probability one and never returned to.

While the characteristic of these processes is their openness to a rich variety of system paths they remain at other levels remarkably 'ahistorical'. The key point here is the assumption of a given transition probability matrix which itself is independent of the historical state of the system. As with the other models discussed above it represents a time and state invariant law of motion. However, recent developments, most notably due to Arthur (1983, 1988, 1989) have shown that, in the presence of increasing returns, non-ergodic economic systems can exhibit strongly irreversible behaviours which, for example, result in the 'lock-in' of particular technologies as a function of chance historical events. The insight this work contains is that increasing returns, or any kind of positive feedback mechanism, creates absorbing states for the Markov process. In the simple systems discussed capture by one of the absorbing states is certain but which one is not predictable in advance. Thus events on the path make a difference and the system does not forget its history. Once locked in, a technology may be immune from threats from competing technologies even though the alternatives might have been notionally superior in conditions where the technologies had equal adoption levels.

3. SMITH, MARSHALL AND SCHUMPETER: MASTERS OF HISTORY

One of the surprising but under-researched aspects of classical economic theory was its essential openness to historical, irreversible change. From Adam Smith to Marx it is clear that economic activity is considered to change the conditions for future economic activity. This is most obvious in Marx's 'bifurcation' theory of the breakdown of coherent economic systems and the irreversible transition to new economic and social structures. If we begin with Adam Smith, we find under the guise of dynamic increasing returns some of the essential features of irreversibility. Crucial to this is the idea that the progressive division of labour results in changes in technical and organisational knowledge which are themselves irreversible. Smith's theory of the division of labour is in all essentials a theory of induced technological change in which new productive forms emerge to dominate previous ones. In a fashion made familiar by the biologists, the study of the morphology of a sequence of such innovations gives time's arrow a direction. Nowhere is the irreversible aspect of Smith's account clearer than when he argues that the production of knowledge becomes subjected to the division of labour.

In neo-classical writing these issues all but disappeared as the emphasis shifted from the study of economic development to the study of the allocation of given quanta of resources. Two obvious exceptions to this trend were Marshall and Schumpeter. In their respective ways, Marshall and Schumpeter were acutely aware of the limitations in prevailing understanding of the dynamics of economic change. Although Marshall was a gradualist and Schumpeter a 'saltationist' (or in modern parlance an exponent of punctuated equilibrium) they, nonetheless, shared a common awareness of path dependence and irreversibility in processes of economic change.

Marshall's treatment is contained in the famous Appendix H to the *Principles of Economics*, suggestively entitled 'Limitations of the Use of Statical Assumptions in Regard to Increasing Returns'. Here Marshall makes two points: that consumption activities result in habit formation and change preferences as a consequence; and that productive activity leads to economies of organisation with the consequence that movements along a cost curve are not reversible. In both cases, current behaviour changes the structure of responses to influence future behaviour, and in both cases the source of the phenomena is experience or learning effects. Needless to add this is only possible if the law of motion of the system contain memory dependent functions. Here one finds an elementary statement of path dependence and irreversibility, the property of which is that the laws of motion, as determined by preferences and cost conditions, change as a result of current and past activity.

Schumpeter in the *Theory of Economic Development* attacks this problem on a broader but less formal scale. His concern is with the analysis of revolutionary change, discontinuities which cannot be considered as incremental developments out of pre-exisiting states. Thus, in his famous example, no matter how many stage coaches are added to the transport system one can never approximate the effect on transport productivity of the development of the railroad. In this way the development of a system changes its equilibrium properties, 'it forever alters and displaces the equilibrium state previously existing'. A clear statement of irreversibility if not path-dependence. Moreover, these discontinuities are generated within the economic system and they are the vital distinguishing face of each particular capitalist mode of production.

4. MICROECONOMIC IRREVERSIBILITIES

4.1 The extreme models of micro-behaviour

What is the relevance of irreversibility for individual economic decision-makers? What are the sources of such irreversibilities?

Of course, the idea of reversible behaviour may be intuitively associated with continuous freedom over time to choose within a certain set of actions, while one may *primae facie* consider as irreversible those decisions '. . . significantly reduce for a long time the variety of choices that would be possible in the future'. (Henry (1974), p. 1006). Moreover, as in Marshall, irreversibility may appear simply as the effects of *past* choices on *current* ones, irrespectively of the dimensionality of the current choice set as compared to past ones. Thus, we shall put under the category of microeconomic irreversibility all those behaviours with respect to which *past choices and past events* bear influences which are independent from the current state-of-the-world (and future expectations). That is, we shall consider under the irreversibility heading all those behaviours wherein *individual and collective history matters*.

By way of an introduction to the issue, let us dramatise the alternative between these two classes of behaviour with reference to two extreme archetypes – call the first one the 'economist's model' and the second one the 'sociological model'. (Of course, they are as such caricatures of much more variegated and subtle theories, but they still help in highlighting the basic difference that we are going to discuss between the economist's paradigm and most other social sciences).

As well known, in the pure 'economist's model', one may clearly separate between states-of-the-world, preferences, actions and outcomes, so that the choice process may be represented as a mapping between them which maximises some objective function, given the preferences. Moreover, time and, hence, uncertainty about the future tend to be, so to speak, 'squeezed' into the present via some assumptions on expectation formation and 'certainty equivalents'. Finally, individual preferences are assumed as independent from each other and from the states-of-the-world (that is, preferences cannot take the form ' . . . I'll go along with what other people like . . .', or '. . . I prefer a rainy day if it rains and a sunny day if the sun is shining . . .'). Time, as such, is not a variable which influences choice-sets and decision rules. If one would take two snapshots of the behaviour of an agent, at say, time t_1 and t_2, there would be no theoretically meaningful sense in which one could say that choices at t_1

have 'caused' or 'influenced' the choices at t_2. The economist's model in its bare bones, as formalised, e.g. in axiomatic decision theory[10] is the archetypical model of reversible behaviours.

Conversely, the 'sociological model' typically looks for causes of particular behaviours in the *history* of the agents themselves and, even more often, in the *environment* in which the agents act. Decision rules and preferences are *chained* to the environment in which they emerge. Pushing it to the extreme, 'motivation' simply becomes what *ex-post* makes meaningful an action to an agent who would have done it anyway.[11] Here, there is a high irreversibility of behaviours in the sense that (a) they are somewhat shaped by environmental and personal history and, thus, b) the events (or, less frequently, choices) which made an agent belong to a certain environment/organisation/institution have thereafter long-lasting effects on behavioural patterns. The 'sociological model' is in many respects symmetric and opposite to the 'economist's one': what are axioms in the latter are often endogenous variables in the former and, vice versa, what are outcomes of choices in the latter are given in the former. In particular, the 'sociological model', unlike the other one, accounts for a strong endogeneity of preferences (an economist could not have written the sixties' pop song urging that '. . . if you cannot be with the one you love, love the one you are with . . . '!) Moreover, 'institutions', whenever they are allowed in economist's model, must be somehow derived from a prior individual rationality of interacting agents,[12] while they can easily be taken as ultimate priors by most other social scientists. Finally, rationality itself is possibly the core, invariant, tenet of the economist's model while quite a few sociologists, anthropologists and psychologists would be prepared to argue that there are, pragmatically, different context-dependent rationalities.[13] Of course, all these foregoing three points have a direct bearing on the reversibility/irreversibility of behaviours: indeed, endogenous preferences, endogenous forms of 'rationality' and patterned (institution-shaped) behaviours imply irreversibility at micro level.

More generally, note that microbehaviours showing some features of the 'sociological model' are *sufficient conditions* for irreversibility, at the very least in the sense that individual and collective behaviour is intrinsic to their explanation. However, this is not a necessary condition: micro irreversibility may emerge even in a world wherein one assumes that agents behave according to the 'economist's model', but one also explicitly introduces time in the decision process.

4.2 Time, uncertainty and sequential decisions.

It is by now a well established result in economic theory that some irreversibility conditions emerge even for otherwise standard (and perfectly rational) agents whenever they face a) less-than-perfect information about the future; b) sequentiality of decision processes; c) less-than-perfect liquidity of their assets (a thorough discussion and modelling is in Llerena (1985)).[14]

For example, consider the case drawn, in a simplified form, from Llerena (1985), where choices have a sequential structure, so that given a set of possible events E and a set of actions A, the sequence of feasible actions over time involves a progressive partition of the A × E space: the set of feasible actions B (t), conditional on the action a_{t-1}, taken at (t – 1), is only a sub-set of B(t – 1) feasible at (t – 1), etc. Here, irreversibility is the outcome of rational choices (with imperfect information about the future) that bear forward consequences in the sense that they restrict future options. Hence, also the history of past choices and not only the sequence of (possibly random) events determine actual sequences of behaviours. In this context, further insights can be gained by adding time-dependent flows of information and partial illiquidity of, say, rent-yielding multi-period assets which can be acquired at different dates. From a decision-theoretic point of view all this implies a close link between the notions of irreversibility and that of *inflexibility*. Agents with imperfect information and (partly) irreversible or *costly* reversible decision sequences, are modelled as putting a positive value on the *reduction* of the degrees of such irreversibility (the 'option value').[15] However, this never avoids the dilemma between undertaking *now* actions that appear profitable on the grounds of current information and, so to speak, 'typing one's own hand' for the future.

4.3 Contract Incompleteness, Transaction Costs and Irreversibility.

In relation to the former decision-theoretic sources of irreversibility, it is important to notice that all forms of forward contracting with incomplete contingency specifications, and, even more so, all forms of substitution of organisation for markets as in transaction cost models (Williamson (1985)), involve irreversibility features. It is straightforward that with market incompleteness (in the sense of Arrow and Hahn (1971)) forward contracts may well imply irreversible information available only at a later date (and the possibility of re-contracting does not eliminate the problem). In that respect, future contracts entail irreversibility features of a nature similar to those examined in our previous point, above. From an empirical point of view, note that most observable contracts, especially in the labour and financial market have such forward characteristics, exacerbated by the asymmetric information which they generally involve (Stiglitz (1984), for an overview).

It has been argued that uncertainty, asymmetric information, incomplete contract specification and possibilities of opportunistic behaviours (all major ingredients of *transaction costs*) economically justify the emergence and persistence of organisations, as opposed to markets, as a form of organising exchanges (again, see, of course, Williamson (1985)). Whenever this is so, it also strengthens the case for the existence of irreversible features of behaviour. Organisations may not only be expected to present those behavioural irreversibilities discessed so far but, even more important for our argument, their very existence is justified in a transaction-cost-type of approach, because they generate *and exploit* such behavioural irreversibilities. For example, they are argued to be more incentive-compatible than, say, a market exchange or than incompletely specified contracts, precisely because they fix to some extent behaviours over time and prevent one of the contracting partners from unilaterally benefiting from newly emerging information. Similarly, corporate norms, with the inevitable stigma of the historical time when they emerged, are argued to be superior, as efficient coordinating devices, to other, more reversible, forms of interaction. It is precisely some bahavioural irreversibility which reduces transaction costs.

4.4 Irreversibilities in Technological Investments.

The literature identifies other, more technology-based irreversible micro phenomena. They include indivisibilities of fixed equipment as in Stigler (1939), specific technologies embodied in particular vintages of rather long-lasting equipment as in Salter (1969), different types of capital goods possibly involving different trade-offs between economies of scale and flexibility to unexpected changes in output varieties and rates of throughput as in Dosi (1988) and Columbo and Mariotti (1985). In all these, empirically quite general circumstances, microeconomic irreversibility is the corollary of the non-malleable nature of that part of technological knowledge which is embodied in durable and specific pieces of equipment. As in the old adage, 'bygones are bygones' each agent is bound to stick also in the future with the benefits and constraints of unavoidable past investments (of course, one rules out here and throughout the general fulfilment of *ex ante* technological expectations). Again in this domain, a related notion of flexibility emerges. Investments are in any case irreversible: they are there to stay and, in general, can be eventually sold only at a fraction of the value which would equate opportunity costs (as they are perceived *ex post*), with second-hand prices and actualised discounted revenues (at that date) from the remaining life of the equipment. However, their intrinsic technological characteristics may involve different degrees of adaptability to unforeseen market conditions. As in the domain of exchange where 'flexibility' is linked to marketability without major delays/ transaction costs, so in the domain of capital-embodied production technology 'flexibility' is linked to their adaptability to unexpected conditions of use. Even if the investment is not reversible, different technological paradigms may involve varying degrees of flexibility: current analyses of microelectronics-based automation as compared to earlier electromechanical ones concern precisely these 'flexibility' features.

Finally, it is worth mentioning under this heading the growing stream of literature on irreversibility in trade patterns, even it it does not directly concern the technological properties of investment but, rather, their spatial location.

Irreversible effects of e.g. exchange rate shocks may emerge, it has been shown, whenever export/import decision involve fixed costs related for example to the set-up or the abandonment of distribution networks, lasting advertising campaigns; etc.[16]

4.5 Learning, search, strategies and the importance of history

A rather general property of a quite wide variety of economic models characterised by incomplete information and some search and learning process is precisely that *history matters.* Agents' behaviours depend, other things being equal, also on their memory of the past (the initial conditions and their 'experience'): hence, they present an irreversibility feature, in the definition proposed earlier. It would be impossible to review the huge literature on the subject, ranging from labour economics to adjustments in sequential competitive markets; industrial organisation; game theory. What one can do here is simply to provide some scattered references and some examples.

First, recall the properties of search models (e.g. in product or labour markets): at least since the work of Spence (1974), it has been known that they yield multiple equilibria which depend on initial conditions. In turn, from the point of view of the actors, their sequence of (rational) behaviours depends on these initial conditions (more generally, on 'hysteresis models' of the labour market, see Summers (1988)).

Second, a somewhat similar story is told by those models with strategic interactions whereby each participant must derive some information on the characteristics of the rivals or the nature of the game from the rivals' actions (for a thorough survey and discussion see Kreps and Spence (1985)).[17] In both cases, if some equilibrium is achieved, those mutually consistent expectations and behaviours on which it is based are contingent on specific 'priors' that the agents initially hold (Hahn (1988)).

There is a quite general point here. Whenever an economic process involves information-acquisition via some interactions amongst the agents, it generally happens that the history of such learning process shapes the outcomes, and the description of particular equilibria does not withhold the inversion of the arrow of time. All agents, with what they know *now* would not go back to *yesterday's* beliefs and actions even under *yesterday's circumstances*. Irreversibility is thus a general feature of learning: actions change the basis for future actions.

4.6 Rationality, uncertainty and routinized bahaviours.

We have so far examined several circumstances under which irreversibility emerges as a property of behaviours which nonetheless are modelled within, what we called earlier, the 'economist's archetype'. Indeed, these circumstances are so general that one can safely claim that the timelessness of the purest 'rational' decision-theoretic models represents only an extreme case: some history-dependence of behaviours is the general corollary of uncertainty, sequential decisions and learning-through experience (whether Bayesian or not). However, may these conditions lead to behaviours that are also 'routinized' – that is, guided by relatively fixed rules – and, thus, show some kinds of irreversibility more akin to the 'sociological archetype' discussed above?

Let us mention here two analytical approaches, so to speak, derived from an 'economist's model' of rational decision, some behavioural implications of which are partly consistent with the 'sociological' microdescription.

First, we have already briefly mentioned Williamson's analysis of organisations as coordinating devices alternative to markets which are argued to imply a relatively higher efficiency whenever the cost of market transactions or contracts is particularly severe (in terms of possibility of opportunistic behaviours, asset specificities, incomplete specification of contracts, etc). For the purpose of this work, note that, in turn, organisations imply relative *fixity*, inertia and *event-independence* of norms which are at the core of the 'sociological model'.

Second, from a different angle, Heiner (1983) and (1988) has argued that whenever decision does not imply only imperfect information but also a *competence gap* of the agents in mapping the available information to the 'true' states of the world, routinized behaviours turn out to be more efficient than pure optimizing strategies.

Both Williamson's and Heiner's approaches, starting from models of micro decision, not far from the 'economist's view', attempt to demonstrate the efficiency of departing from the standard ('rational' and reversible) decision model whenever imperfect information is associated with problems of incentive-compatability or information-processing. In that, one derives a sort of 'rational reconstruction' of irreversible institutionalized behaviours.

4.7 Innovation and irreversible bahaviours

The discussion so far has mostly concerned the irreversibility features of microbehaviours which occur in environments that are often modelled to be *stationary*, that is, the so-called 'fundamentals' (technology and preferences) are assumed as given. In such a context, irreversibility is essentially a by-product of imperfect information and sequentiality of learning about a *world which*, however, *is given*. To put it in a somewhat extreme way, irreversibility is represented to be not entirely different from that involved in the temporal sequence between gambling in a lottery and knowing the outcome of the draw, or that involved in two-armed-bandit problems. In most of the approaches discussed so far an agent with perfect knowledge of 'how the world is' at time zero would not need to know anything more at some future date. But what happens if one allows for *innovation*, i.e. for new events and new possible behavioural strategies? As an illustration, take a stochastic game-theoretical set-up such that the decision problem concerns the maximisation of a function $V_i = V_i (a_i, \ldots, a_i, \ldots, a_n; e_i, \ldots, e_i, \ldots, e_m)$ where the a's are the strategies of the i-agents ($i = 1 \ldots n$) (with $a \epsilon A$, the space of feasible strategies) and the e_j are the j-events ($j = 1 \ldots m$). Now suppose that, first, *agents can discover new strategies*. Second, suppose that *actions and events are not independent because the former affect the latter*. After all, an innovation is precisely that: what one agent discovers by means of one's own strategy is a *new and largely unpredictable event*, at least for all other agents. In our example here, innovation introduces a permanent increase of the dimensionality of the A-and E-spaces and a (partial) endogeneity of the events themselves.

Indeed, such a non-stationary world is much more similar to the Eraclitus' metaphor recalled in the introduction: irreversibility is not only the result of imperfect information and sequentiality of decisions but is due to the fact that the world genuinely changes and it changes as a consequence of the very actions of the agents. It is an open world characterised by increasing complexity whereby, as Prigogine emphasises, the future cannot be entirely known by any agent with infinite knowledge of initial conditions and 'laws of motion' of the system. To illustrate all this with a caricatural example, the standard irreversibility problem concerns, say, the decision of planting wheat today not knowing whether tomorrow will rain or not. What we are concerned with, here, however, is a world where agents must still decide whether to plant wheat today, but are also able to 'invent rain' tomorrow, or, for that matter, to invent huge 'umbrellas' that miraculously cover the fields if someone introduces unwanted rain, and also introduce unexpected rivers of milk and honey, as in the old Psalm. . . .

Of course, under such circumstances, whatever agents do and whatever they know, irreversibility is always there, because their very actions make Eraclitus' river flow . . . Relatedly, an irreversible sequence of decision-processes links 'what is required (and what can be done) today (and, to a certain extent, what will be required and what will be possible to do tomorrow) and what happened yesterday'. (Amendola and Gaffard (1988), p.48).

4.8 Freedom to choose vs. power of influencing the world

Let us consider some of the features that are specific to irreversible behaviours in complex non-stationary environments. First, note that in a closed (given) world, 'irreversibility' is largely synonymous with 'maintaining the freedom of choice', within some notional choice set that is initially given. In a non-stationary world, with endogenous innovation, the agents' freedom is essentially linked with the *power of influencing the world*, that is of 'creating new events' and opening up new options that were not previously available: 'a process of new options, associated with the transformation of the environment is fuelled by the learning process that sets in as a result of innovative choice and of the carrying on of innovative processes of production' (Amendola and Gaffard (1988), p.59).

Hence, irreversibility in the world we are describing is not only a constraint on the freedom to choose but also a *continuous source of opportunities*. The fundamental point relates not to the mechanics of optimization but to the way in which agents construct their choice sets from which options follow.

4.9 New events and uncertainty.

Second, the uncertainty that arises in an innovative environment resembles much more closely the notion described by Knight (1921), Keynes (1973) and Shackle (1963) rather than the more familiar notion of risk utilized by decision theory with imperfect information. Indeed, the number of unknown events that can emerge in the future is in principle, infinite, non-enumerable, and, therefore, not subject to probabilistic calculation. Moreover, it is not at all clear how the increase of information about the present, *per se*, should decrease the uncertainty about the future. Continuing our Greek metaphor, increases in current information augment future predictability if they can samples on a closed lake but not necessarily on a flowing river. Of course, it can, if agents hold models of the world and its pattern of change but these models themselves cannot be derived automatically from environmental information (Dosi and Egidi((1987)). Even froms of Bayesian learning could be considered, in this context, as the most naive structure for medelling the world. Agents that are intelligent but finite in their knowledge and power are bound to build models of 'how the world really works', but are bound also to be surprised, at least every now and then, by the unexpected.

4.10 Routinization and innovation

Third, behaviours continuously entail tensions and dilemmas between routinization and innovation. It is precisely the complexity of the problem-solving tasks which makes routinization as the general characteristics of most of individual and, even more so, organizational behaviours (Nelson and Winter (1982)). Unlike a rather widespread opinion according to which routinized behaviour are 'less intelligent' than literally optimizing strategies, it can be shown that relatively robust routines which apply to whole classes of problems are the outcome of 'discoveries' by intelligent actors operating in continuously changing environments (Dosi and Egidi (1987)).

There is a powerful link here between evolution and the 'sociological archetype' of behaviours. Precisely because the world is changing and can also be changed by the actors themselves, the latter can be reasonably represented as developing specific competences and rules which can be repeatedly applied today and also tomorrow even if the world tomorrow will not be the same as today. In a sense, behavioural routines, with all the history-dependence and context-dependence of their emergence, are attempts to 'stationarize' a non-stationary environment. Of course, one can conceive routines with different levels of generality, some of which concern also the generation of innovations (firms' rules on allocation of resources to R&D activities belong precisely to this domain; cf Nelson and Winter (1982)). And it is also plausible to conceive a hierarchical order of various rules within some cognitive structure with 'meta rules' (higher level rules) establishing criteria for change in the lower level ones. Crucial here are the rules by which an organization senses its environment, and the rules by which it learns how to learn. Still, it holds the routinized behaviour is there to control the flow of Eraclitus' river, even ignoring all the information about the characteristics of the water at any particular moment.

Analogously, one can reasonably expect that routinized behaviours will govern a good deal of interaction amongst the agents, set the boundaries of 'what is possible to do', guide expectation formation, etc. In this respect, the threats stemming from complex largely unknown environments – as H. Simon has argued – may well provide an evolutionary foundation for obedience and institution-guided behaviours. Irrespectively of any general anthropological foundation, history-dependent preferences, rules and decision routines imply, of course, the existence of those rather powerful irreversibilities that one generally observes in individual behaviours and social institutions.[18]

Here, we want to emphasise the intrinsic dilemmas that routinized behaviours entail. On the one hand, their apparent mindless repetition hides the functional 'intelligence' of their origin: in fact, they allow some governance and intelligeability of a continuously flowing river. There is no need to recall the long-term/cross-cultural anthropological evidence on harvesting rites, bargaining rules, etc: it should be enough to mention the inertial robustness and specificities of corporate cultures, the bureaucratic procedures of contemporary public and private administrations etc. On the other hand, routinized behaviours of whatever form imply, almost by definition, *a permanent gap between the actual pay-offs that agents reap and notional opportunities*. Loosely speaking, different rule-governed behaviours could be placed somewhere along a notional continuum ranging from an intrinsically mindless but very powerful computer which attempts to optimize some algorithm on event-contingent information and an infinitely powerful, simple, event-independent, rule of an omnipotent entity which can control and shape the universe irrespectively of contingent information. Obvious finiteness of both event-contingent information and event-generating power makes human routines largely 'sub-optimal' and continuously subject to the possibility of discovery and innovation. The dimensionality of the space of opportunities which actual routines have explored, as compared to what is notionally possible, is extremely low. There is always a potentially enormous room for the emergence of 'deviant' behaviours, which, if successful, irreversibly establish new problem-solving criteria and interaction rules. Innovation behaviours are the (dangerous) windows of opportunity within the frame of established routines and also the generating mechanism of what may eventually turn out to be new dominant modes of behaviour.

4.11 Individual uniqueness and social coherence.

Fourth, as a consequence of all that, we suggest that in non-stationary, opportunity-ridden, highly institutionalized environments two somewhat complementary mechanisms are always at work. On the one hand, specific histories of competence-building and expectation-formation, cognitive structure, organisational norms, together, yield relatively unique (hence, heterogenous) micro agents. On the other hand, broader mechanisms of social validation, market selection, institutional coordination tend to reduce such a diversity amongst agents and bring about relative consistency of behaviours. Indeed, various evolutionary models of economic behaviour have attempted to conceptualise irreversible microeconomic changes in terms of varying balances between *learning and selection process* (cf. Nelson and Winter (1982); Dosi and Orsenigo (1988); Metcalfe (1989); Gibbons and Metcalfe (1989); Silverberg, Dosi and Orsenigo (1988); Winter (1971)).

4.12 Micro irreversibilities and systems properties.

The argument so far leads to the conclusion that various sorts of irreversibility in microeconomic behaviours are indeed the general case. Individual and collective history should thus be considered an essential part of the 'explanation' not only of what the agents 'desire', but also, given somehow what they desire, of what they actually do to pursue their aims and of how they modify objectives and actions over time. Moreover, we have argued, in open (innovation-ridden) environments, irreversibility in decision processes cannot be considered only as a progressive fall in the degrees of freedom for choice. It is also intimately linked with the endogenous generation of new opportunities. Certainly, if an agent considers the world as given, irreversibility progressively leads to a 'black hole' where past history completely determine present and future and whose alternative is only the apparent 'liquidity' of waiting and inaction. One has all that in human and, more specifically economic, affairs, but one has also the continuous opportunity of, so to speak, 'breaking out', not by re-writing the past which is indeed unalterable but by turning the irreversibility of the future in one's own favour and creating another (although still irreversible) new world. The Schumpeterian entrepreneur is clearly the archetypical illustration of this possibility of changing rather than accepting environmental events.

The next step of this analysis leads to the investigation of the implication of such microbehaviours for the dynamics of the whole system. (Here and throughout we shall use a rather loose definition of 'system', meaning the environment where the agents interact, be it an industry, a market or the whole economy).

Note, however, that such history-independence, and, hence system-level reversibility depends on the assumptions of (i) one-dimensionality of the system, (ii) linearity in its dynamics, (iii) stationarity – at the very least the dynamics in any *a* parameter must be slower than the adjustment process in the state variable(s) *x* (on all this, cf. Silverberg (1988)).

4.13 Micro reversibility and macro irreversibility

Let us now consider an opposite example whereby micro behaviours are reversible in the sense that agents are perfect maximizers, their history does not limit their 'freedom to choose', and they are perfectly informed (so that one is not bound to account for their learning history), but the system is irreversible at least in the sense that its dynamics depends on initial conditions and, hence, its 'history'. A thorough review and discussion of several examples can be found in Kelsey (1988), from which the following illustration is drawn.

It is by now well known that even first-order non linear difference equations of the form $x_{t+1} = f_r(x_t)$ with $f_r = rx(1 - x)$ may generate a chaotic dynamics for certain values of the parameter r. In turn, this aperiodic motion implies *sensitive dependence* on initial conditions: an ϵ-difference in initial conditions is progressively amplified. Note also that dependence on initial conditions is not restricted to chaotic motions but is a more general property of non-linear systems. Such systems exhibit, in general, those properties of irreversibility (with history-dependence, bifurcations, etc) examined in section (ii). In that respect *sensitive* dependence is only the extreme case.

There is a growing stream of literature showing the possibility of chaotic system behaviour arising out of a microfoundation based on highly orthodox 'rational' agents (see, among others, Benhabib and Day (1981), Boldrin and Montrucchio (1986), Day (1982), Grandmont (1984)). More generally: irreversible system properties are going to emerge whenever one allows non-linear system dynamics and irrespectively of the assumption that one makes on the reversibility of microbehaviours (for general discussions and models, cf. Day (1983) and Goodwin (1989)). Indeed, there is no theoretical or empirical reason to believe that economic adjustments should take the (special) linear form. If anything, the contrary is true: non-linearity and complex dynamics is shown to emerge in utterly simple and empirically highly unrealistic models.

4.14 Externalities, increasing returns and system irreversibility

There is another extremely general class of path-dependent system dynamics emerging, again, irrespectively of whether microbehaviours present some irreversibility features, as defined earlier, or not. This is so whenever the system's behaviour depends on particular sequences of decisions in space or over time. Most forms of externalities, dynamic increasing returns, collective formation of preferences belong to this class of models.

Let us consider here three examples. First, consider the case of alternative technologies presenting increasing returns whereby the incentive to adopt any one of them depends on the number of agents who have already done it. Arthur (2983) and Arthur, Ermoliev and Kaniovski (1987) have modelled the general properties of those non-erogic processes where the probability of choice of the i-alternative (i = 1, ... k) is a function of the vector x, giving the proportion of choices already made in each of the k-categories: that is, at each time the probabilities $p_1(x)$, . . . $p_i(x)$ . . . $p_k(x)$ depend on the whole history of past choices. The process, it is shown, converges with probability one to one of the k-absorbing states, but *which one* of them depends on history and small initial (possibly random) event which get, so to speak 'amplified' by the increasing returns dynamics.

Second, consider the case, presented by David (1989), with 'interdependence of choices made sequentially by agents pursing their private interest within contexts where substantial benefits arise from coordinated or synergetic actions, and where individual decision-agents each are limited (for various reasons) to considering only the actions of the members of a particular reference group [or spatial 'neighbours'] (David (1989), p.23). This is a set of problems to which the 'snow-shoveling metaphor' applies: if it snows over a row of shops, access to the shops by customers depends on whether the outside pavement is cleared and each shop-keeper decides whether to shovel the snow or not depending on the behaviour of the immediate neighbour. What will happen to the whole pavement in front of the shops? Will it be cleared? Will it end up covered with snow? Or will some combination between the two states emerge as the asymptotic state? Well, the answer, under some restrictions, is that the two extreme states (all cleared, all snowed-in) are absorbing states and the probability of each of them (in a one-and two-diminsional spaces) depends on the initial distribution of shoveling/non-shoveling agents (David, (1989)).

Third, consider the case where preferences of any one agent depend on the choices that other agents have made: for example, whenever people want to sit in the second *occupied* row in the lecture room (Schelling (1978) or whenever one's own political preferences are influenced by other people's preferences (Kuran (1987) and (1988)). Indeed, this third case is, in principle, similar to either of the first two, depending on the endogeneity/exogeneity of transition probabilities and on whether the feedback from the state of the system is 'global' or 'local' (that is, whether individual decisions are influenced by the whole universe of other agents or only by the immediate neighbours).

Other cases that are somewhat analogous in spirit concern models of 'local learning' (Atkinson and Stiglitz (1969)), network externalities (Katz and Shapiro (1986)), economies of standardization (David (1985) and (1987)). In general, all of them involve both an influence of microbehaviours on macrostates and a positive feedback from the macrostate to microdecision (Schelling (1978), David (1989)). The latter may well be 'rational' and notionally perfectly reversible, but the collective effect of a multitude of individual choices changes some system parameters and thus also the relative incentive provided by the various options.

4.15 Some empirical examples.

The class of empirical circumstances showing some elements of non-linearity, some externalities, some dynamic increasing returns is very wide: it is, with little doubt, the general case. Indeed, we would find it extremely difficult to present some economically relevant example where none of these conditions occur. It is much easier to list observable phenomena supporting the argument on system-level irreversibilities.

Here we mention only a few (the list contains also the examples from David (1989).

(i) Development of network technologies. All technologies requiring or providing infrastructures, and requiring compatibility with other technologies or among different users show elements of history-dependence and of 'lock-in'. From telecommunications to railways, AC vs. DC electrical systems, video recorders, etc, there is a wide set of technologies whose development and adoption requires the compatibility of different pieces of equipment and whose incentive to adopt depends also on the number of other people who have done so (after all you would like to have a telephone in order to talk to other people!

In turn, this kind of positive feed-back generates history-dependent phenomena of lock-in to technologies that might even be inferior to other notional ones which to for historical contingencies have not taken off.[19]

(ii) Development of specific technological paradigms and trajectories. It has been argued at greater length elsewhere that particular combinations of scientific principles, technological competences and somewhat 'tacit' experience often yield relatively ordered and cumulative patterns of technological learning (a survey is in Dosi (1988)). Relatedly, particular bodies of knowledge grow as distinctive assets of industries, firms, communities of scientists,

engineers, technicians. Of course, in all that, history shapes the selection of particular 'paradigms' and learning trajectories.

(iii) Firms as carriers of specific knowledge. If the accumulation of technological and organisational knowledge is, to some degrees, specific, 'idiosynchratic', cumulative,then business organizations may also be viewed as the development amongst collection of individuals of mutually consistent norms of (relatively) incentive-compatible behaviours and learning patterns.[20]

(iv) Bandwagon-effects on investment. One can also allow for the possibility of imitation affects in investment behaviours and innovation diffusion. Moreover, account for the possibility of endogenous changes in the incentive to adopt new (capital-embodied) technologies, due, for example, to learning externalities, developments of specific skills of the labour force, incremental improvements in the technologies themselves. Then, it is easy to understand how investment patterns can generate some 'bandwagon' and 'swarming' effects, as emphasised by Schumpeter.[21]

(v) Emergence of social customs. '[I]mportant historical processes of institutionalisation of non-obligatory contractual and organisational forms governing long-term trading and employment relations' (David (1989), p.21) are likely to result from the endogenous evolution of rewards, threats and penalties and the influence that such collective phenomena exert on individual behaviours.[22] Hence they can be considered as the result of the collective experience arising from multiple interactions of individual agents.

(vi) Endogenous evolution of consumer tastes. Imitation of the behaviours of others, 'snobbery effects' (and, thus, efforts to differntiate the patterns of consumption), 'conspicuous consumption' a la Veblen, all are phenomena which entail positive feedbacks from collective interactions and microbehaviours.

4.16 The relationships between macro and micro irreversibilities

In the discussion so far we have purposefully proceeded on a rather agnostic assumption on the nature of micro behaviours. Indeed, the foregoing argument on system-level irreversibility is strengthened by the fact that they may well emerge even under those circumstances when microbehaviours are modelled to be perfectly reversible – in the sense of section (iv) and thus history does not place any constraint on the notional 'freedom to choose' of individual agents. However, the characteristics of the environments are not irrelevant for the nature of microbehaviours. For a start, note that, in general, individual learning processes 'will affect the behaviour of economic variables but this will feed back on the learning process itself' (Kelsey (1988), p.22). Obviously, non-linearities in the dynamics of the environment, of whatever origin, will enormously increase the demands that 'rational forecasting' would place on individual calculations. In that respect, a chaotic macrodynamics is an extreme case of the point: in our view, current economic literature largely underestimates the disruptive implications that the possibility of such aperiodic macro movements have for the 'rationality assumption' (and for the very assumptions of correct foresights on which many of the macro models are based).[23]

In general, the more the environment exhibits irreversibility properties – related e.g. to non-linear dynamics, externalities, dynamic increasing returns, etc – the more this strengthens the case for irreversible micro behaviours, based on 'robust routines', specific cognitive frames of knowledge, local experience-based learning. Various sorts of macro irreversibilities, in this sense, represent the *macrofoundation* of institutionalised microbehaviours.[25] It is precisely the gap, in a Shakespearean metaphor, between the dimensionality of the set of all whole possible events (between heaven and earth) and the cognitive power of any one philosophy which underlies institutionalised (hence, in our definition, irreversible) microbehaviours. Moreover, the joint properties of micro and macro irreversibilities further amplify the divergence between notional opportunities and their actual exploitation by empirical agents.

In our view, a distinctive feature of that approach which goes under the heading of 'evolutionary/self-organisation models' is precisely this emphasis on:

a) the permanent possibility of novelty as micro ruptures of behavioural routines;
b) the evolutionary foundations of the routines themselves;
c) the mutually re-enforcing possibility of macro non-linearities/bifurcations and micro innovations;
d) the essential role of micro 'mistakes', disequilibria, diversity, experimentation in macro dynamics.

Of course, all the four points strengthen the argument for the general existence of irreversible processes of economic change.

5. CONCLUSIONS

We have sought in this paper to provide first, a brief exploration of some modelling approaches to irreversible processes and the interpretations of such types of dynamics in the history of economic thought, and, second, an introduction to a few empirical classes of micro and system-level changes that present irreversibility features.

If anything, our conclusions further emphasise the importance of irreversibility in the social (more specifically, economic) domain already highlighted in rapidly expanding theories of natural (physical, chemical and biological) phenomena. It is not only a matter of non-linearity in system dynamics: in social and economic interactions, plausibly general non-linear patterns of change are dynamically coupled with history-dependent and context-dependent processes of learning and socialisation (most economists would call this norms-formation and expectation-formation). In this respect, our whole argument implies that there is a much closer relationship between economic analysis (traditionally based on rather axiomatic and history-free representations of economic behaviours) and other (more institution-based) social disciplines. However, it implies also the representation of a permanent tension between the 'power of the past', in the form of history-dependence of current behaviours, and the 'power to shape the future', as the permanent opportunities of influencing future events by the will of individual actors (or groups of them). Further, it emphasises the permanent possibility of counter-intentional outcomes of individual behaviours (after all, this is a straightforward implication of non-linear system dynamics). Finally, it highlights the permanent possibility of novelty, that is, in a fundamental sense, of new and 'deviant' behaviours which explore yet unknown opportunities.

In all that, we have argued, irreversibility is the necessary implication of individual and collective behaviours which are certainly forged by the history from which they emerge but also always entail the potential of shaping the history yet to come.

In that tension rests, in our view, an essential feature of the analysis of economic coordination and change, of which we have tried to provide a few heuristically introductive examples.

FOOTNOTE

The support to one of the authors (G.D.) by the Designated Research Centre of the ESRC at the science Policy Research Unit (SPRU), University of Sussex, Brighton and by the Italian Ministry of Education ('MPI Priogetti 40%), is gratefully acknowledged. JSM gratefully acknowledges support of ESRC through its continuing programmes on industrial competitiveness.

1. Irreversibility arises as a general class of phenomena whenever the making of a decision changes the decision maker's perceptions of that same and related decision.
2. For thorough discussion, e.g. Prigogine and Stenger (1979) and Nicholis and Prigogine (1987), Prigogine (1980).
3. Sanchez (1968).
4. Prigogine (1980), Prigogine and Stengers (1984), Thompson and Stewart (1986).
5. Baumol and Benhabib (1989).
6. Thompson and Stewart, (1986), p.96, Lorenz (1989).
7. Kelsey (1988), Baumol and Benhabib (1989).
8. David (1989a, 1989b).
9. Feller (1957), Kemeny, and Snell (1960).
10. For a locus classicus, c.f. Arrow (1970).
11. See Luhmann (1975).
12. For example, on the relationship between rationality-grounded games and institutions, c.f. Harsanyi (1977).

13. c.f. Schwartz and Thompson (1985), Douglas and Wildavsky (1982).
14. Who, in turn, originally develops upon, among others, Arrow (1970) A.C. Hart (1949), Kreps (1979).
15. Within a rather wide literature, cf. Llerena (1985) and the extensive bibliography cited there, Dasgupta and Heal (1979), Vercelli (1988).
16. On these issues, c.f. among others, Baldwin (1988), Baldwin and Krugman (1986), Biasco (1988).
17. On this point see also Kreps and Wilson (1982), Milgrom and Roberts (1982), and, for some evidence from experimental economics, Roth and Schoumaker (1981).
18. Some more detailed comments on the relations between institutions and market behaviours by one of the authors is in Dosi (1988a).
19. Discussions of several historical phenomena which broadly belong to this category are in Rosenberg (1976) and (1982), David (1975) and (1987), Freeman (1982) and Freeman and Perez (1988).
20. More on this in Nelson and Winter (1982), Pavitt (1986), Teece (1988).
21. A more detailed argument along these lines is in Freeman and Perez (1988).
22. Interesting cases of 'spontaneous' development of norms governing various sorts of externalities are discussed in E. Ostrom (1988).
23. As Kelsey remarks perfect foresight must rely on agents observing and learning about their environment. With aperiodic motion there is no scope for them to observe regularities in the environment. Moreover, if they understand the sensitive dependence on initial conditions they will realise that it is impossible for them to make forecasts about anything but the immediate future. (Kelsey (1988), pp 22–23). On this point see also Coricelli and Dosi (1988), Coricelli, Dosi and Orsenigo (1989).
24. This is also the original interpretation that Egidi (1981) suggests of Schumpeter's micro analysis.

REFERENCES

A.A. Alchian (1951), 'Uncertainty, Evolution and Economic Theory' *Journal of Political Economy*.

P. Allen (1988), 'Evolution, Innovation and Econmics', in Dosi *et al*, (1988).

R. Amendola and J.L. Gaffard (1988), *The Innovative Choice*, Oxford, Basil Blackwell.

K. Arrow (1970), *Essays in the Theory of Risk Bearing*, Amsterdam, North Holland.

K. Arrow and F. Hahn (1971), *General Competitive Analysis*, San Francisco, Holden-Day.

B. Arthur (1983), *Competing Technologies and Lock-in by Historical Events: the dynamics of allocation under increasing returns*, Laxenburg, Austria, IIASA WP8390.

B. Arthur, Y. M. Ermoliev and Y. M. Kaniovski (1987), 'Path Dependant Processes and the Emergence of Macrostructures', *European Journal of Operational Research.*

B. Arthur (1988), 'Competing Technologies', in G. Dosi *et al.* (1988).

B. Arthur (1989), 'Competing Technologies, Increasing Returns, and Lock-in by Historical Events', *Economic Journal*, vol.99.

A.B. Atkinson and J. Stiglitz (1969), 'A new view of technological change', *Economic Journal.*

Austin (1989), *International Symposium on Evolutionary Dynamics and Non-Linear Economics*, IC2 Institute, Austin, Texas, April 16–19, 1989.

R. Baldwin (1988), 'Hysteresis in Import Prices: The Beachhead Effect', *American Economic Review*.

R. Baldwin and P. Krugman (1986), *Persistent Trade Effects of Large Exchange Rate Shocks*, Cambridge (Mass.), NBER Working Paper no. 2017.

W. J. Baumol and J. Benhabib (1989), 'Chaos: Significance, Mechanism and Economic Applications', *Journal of Economic Perspectives*, vol.6.

J. Benhabib and R. Day (1981), 'Rational Choice and Erratic Behaviour', *Review of Economic Studies*.

S. Biasco (1988), 'Dynamic and Incapsulating Processes in the Generation of World Demand', *Banca Nazionale del Lavoro Quaterly Review*.

R. Boldrin and L. Montrucchio (1986), 'On the Indeterminacy of Capital Accumulation Paths', *Journal of Economic Theory*.
M.G. Colombo and S. Mariotti (1985), 'Note economiche sull automazione flessiblile', *Economia e Politica Industriale*.
F. Corricelli and G. Dosi (1988), 'Coordination and Order in Economic Change and the Interpretation Power of Economic Theory', in Dosi *et al.* (1988).
F. Corricelli, G. Dosi and L. Orsenigo (1989), *Microeconomic Dynamics and Macroregularities: An 'Evolutionary' Approach to Technological and Institutional Change*, Paris, OECD, Working Paper DSTI/SPR/89.7.
N. Cross (ed) (1988), *Unemployment, Hysterises and the Natural Rate Hypothesis*, Oxford, Basil Blackwell.
P.S. Dasgupta and G.R. Heal (1979), *Economic Theory and Exhaustible Resources*, Cambridge, Cambridge University Press.
P. Dasgupta and P. Stoneman (eds) (1987), *Economic Policy and Technology Performance*, Cambridge University Press.
P. David (1975), *Technical Choice, Innovation and Growth*, Cambridge University Press.
P. David (1985), 'Clio and the Economics of Qwerty', *American Economic Review*.
P. David (1987), 'Some new standards for the economics of standardization in the information age', in Dasgupta and Stoneman (1987).
P. David (1989a), 'The future path-dependent equilibrium economics', in Austin (1989).
P. David (1989b), 'Path Dependance: Putting the Past into the Future of Economics', University of Stanford, *mimeo*.
R. Day (1982), 'Growth Cycles', *American Economic Review*.
R. Day (1989), 'Dynamical Systems, Adaptation and Economic Evolutions', in Austin (1989).
G. Dosi and M. Egidi (1987), *Substantive and Procedural Uncertainty: An Exploration of behaviours in changing environments*. Brighton, SPRU, University of Sussex, DRC Discussion Paper.
G. Dosi, C. Freeman, R. Nelson, G. Silverberg and L. Soete (eds.) (1988), *Technical Change and Economic Theory, London, Francis Pinter and New York, Columbia University Press*.
G. Dosi (1988), 'Sources, Procedures and Microeconomic Effects of Innovation', *Journal of Economic Literature*.
G. Dosi (1988a), 'Institutions and Markets in a Dynamic World', *The Manchester School*.
M. Douglas and A. Wildavsky (1982), *Risk and Culture*, Berkeley, University of California Press.
M. Egidi (1981), *Forme della Razionalita e Macrofondamenti della Microeconomia nell analisi Schumpeteriana*, Turin, Laboratory of Political Economy.
G. Eliasson (184), 'Microheterogeneity of firms and the stability of industrial growth'. Journal of *Economic Behaviour and Organisation*.
G.R. Feiwel (ed) (1985), *Issues in Contemporary Microeconomics and Welfare*, London, Macmillan.
W. Feller (1957). *Introduction to Probability Theory and its Applications*, Wiley.
C. Freeman (1982). *The Economics of Industrial Innovation*, London, Francis Pinter, 2nd edn.
C. Freeman and C. Perez (1988). Structural Crisis of Adjustment, Business Cycles and Investment Behaviours', in Dosi *et al* (1988).
M. Friedman (1953). *Essays in Positive Economics*, Chicago, University of Chicago Press.
M. Gibbons and J.S. Metcalfe (1989). 'Technology Variety and Organisation, in R. Rosenbloom (ed), *Research in Technology Management and Innovation*, vol. 4.
R.R. Goodwin (1989). 'Economic Evolution and the Evolution of Economics', in Austin (1989).
J.R. Grandont (1984), *Periodic and Aperiodic Behaviour in Discrete One-dimensional Dynamical Systems*, Stanford, Stanford University, Economic Series Technical Report no. 446.
F. Hahn (1988), 'Information, Dynamics and Equilibrium', *Scottish Journal of Economics*.
L. Harsanyi (1977), *Rational Behaviour and Bargaining Equilibrium in Games and Social Situations*, Cambridge (Mass.), MIT Press.
A.G. Hart (1949), 'Risk, uncertainty and the unprofitability of compounding probabilities', in Lange *et al* (eds), Studies in Mathematical Economics and Econometrics, Chicago, Chicago University Press.
R. Heiner (1983), 'The origin of predictable behaviour' *American Economic Review*.

R. Heiner (1988), 'Imperfect decisions and routinzed production: implication for evolutionary modelling and inertial technical change', in Dosi *et al.* (1988).
C. Henry (1974), '*Uncertainty in microeconomics*, London, Martin Robinson.
R.L. Katz and C. Shapiro (1986), 'Technology adoption in the presence of network externalities', *Journal of Political Economy*.
D. Kelsey (1988), 'The Economics of Chaos or the Chaos of Economics', *Oxford Economic Papers*.
J.G. Kemeny and J.L. Snell, (1960), *Finite Markov Chains*, Van Nostrand.
J.M. Keynes (1973), *A treatise on probability*, London, Macmillan, (vol. vii of the Collected Writings).
F.H. Knight (1921), *Risk, Uncertainty and Profit*. New York, Kelley.
D.R. Kreps (1979), 'A representation theorem for the "preference for flexibility"', *Econometrica*.
D.R. Kreps and R. SPence (1985), 'Modelling the role of history in industrial organisation and competition', in Feiwel (1985).
D.R. Kreps and R. Wilson (1982), 'Sequential Equilibria', *Econometrica*.
T. Kuran (1987), 'Preference falsification, policy continuity and collective conservatism', *Economic Journal*.
T. Kuran (1988), 'The Tenacious Past: Theories of personal and collective conservatism', *Journal of Economic Behaviour and Organisation*.
W. Leontief (1963), 'When should history be written backwards?' *Economic History Review*.
P. Llerena (1985), *Decisions avec incertitude er irreversibilie: doudaments de la valuez d'option et application auz investissement productifs*. Strasbourg, Universite Louis Pasteur, Doctoral Thesis.
H.W. Lorentz (1989), Spiral-Type Chaotic Attractors in Low-Dimensional Continuous-Time Business Cycle Models', in Austin (1989).
N. Luhmann (1975), *Macht*, Stuttgart, Ferdinand Duke Verlag.
J.S. Metcalfe (1989), Evolution and Economic Change in A. Silberston (ed), *Technology and Economic Progress*, Macmillan.
P. Milgrom and J. Roberts 91982), 'Limit Pricing and entry deterence Under Incomplete Information', *Econometrica*.
R. Nelson and S. Winter (1982), *An Evolutionary Theory of Economic Change*, Cambridge (Mass.), The Belknap Press of Harvard University Press.
G. Nicolis and I. Prigogine (1977), *Self-organisation in non-equilibrium systems*, New York, Wiley.
E. Ostrom (1988), 'Institutional Arrangements and the Commons Dilemma', in Vo. Ostrom, D. Feeny and H. Picht (eds.), *Rethinking Institutional Analysis and Development: Issues, Alternatives and Choices*, San Francisco, ICS Press.
K. Pavitt (1986), 'Chips and Trajectories', in R. Macleod (ed.), *Economics and the Human Prospect*, London, Francis Pinter.
I. Prigogine and I. Stengers (1979), *La nouvelle Alliance*, Paris, Gallimard. Also published as I. Prigogine and I. Stengers (1984), *Order Out of Chaos*, Heinemann.
I. Prigogine (1980), *From Being to Becoming*, Freeman.
R. Rosenberg (1976), *Perspectives on Technology*, Cambridge, Cambridge University Press.
R. Rosenberg (1982), *Inside the Black Box*, Cambridge, Cambridge University Press.
A. Roth and F. Schoumaker (1981), 'Expectations and Reputations in Bargaining: An Experimental Study', *American Economic Review*.
W.E.G. Salter (1969), *Productivity and Technical Change*, Cambridge, Cambridge University Press, 2nd edn.
D.A. Sanchez (1968), *Ordinary Differential Equations and Stability Theory*, Freeman.
G.L. Shackle (1968), *Decision, Order and Time in Human Affairs*, Cambridge, Cambridge University Press.
T. Schelling (1978), *Micromotives and Macrobehaviour*, New York, Norton.
P. Schwartz and R. Thompson (1985), 'Beyond the Politics of Interest', in M. Beckmann and W. Knelle (eds.), *Plural Rationality and Interactive Decision Processes*, Berlin/New York/ Springer Verlag.
G. Silverberg (1988), 'Modelling economic dynamics and technical change: mathematical approaches to self-organisation and evolution', in Dosi *et al.* (1988).

G. Silverberg, G. Dosi and L. Orsenigo (1988), 'Innovation, Diversity and Diffusion: A Self-Organisation Model', *Economic Journal.*

H. Simon (1983), *Reason in Human Affairs*, Stanford, Stanford Univeresity Press.

R. Spence (1974), *Market Signalling*, Cambridge (Mass.), Harvard University.

G. Stigler (1939), 'Production and distribution in the short run'. *Journal of Political Economy.*

J. Stiglitz (1984), 'Information and Economic Analysis, San Francisco, *Economic Journal.*

R. Summens (1988), 'Should Keynesian Economics Dispense with the Phillips Curve?' in Cross (1988).

D. Teece (1988), 'Technological change and the nature of the firm', in Dosi *et al* (1988).

J.M.T. Thompson and H.B. Stewart, *1986, Non Linear Dynamics and Chaos,* London, Wiley.

A. Vercelli, (1988), 'Technological Flexibility, Financial Fragility and the Recent Revival of Schumpeterian Entrepreneurship', in R. Goodwin and A. Vercelli (eds.), *Technological and Social Factors in Long Term Fluctuations*, Berlin/New York/Springer Verlag.

O. Williamson (1985), *The Economic Institutions of Capitalism*, New York, Free Press.

S. Winter (1986), 'Comments on Arrow and Lucas', *Journal of Business*, Special Issue.

Evolutionary Human Systems: Learning, Ignorance and Subjectivity

P.M. ALLEN and M. LESSER
International Ecotechnology Research Centre, Cranfield Institute of Technology, Bedford, MK43 0AL, U.K.

1. INTRODUCTION

What is the nature of evolution? Is it about the steady 'improvement' in performance of the individual parts of a system? Or is it about the 'improvement' in functioning of the system as a whole? Does one imply the other, or is 'improvement' defined as 'what happens'? What could 'improvement' mean? Is it a purely subjective judgement, or is evolutionary change characterized by some fundamental, (thermodynamic?) property or measure? Even if it were, then should subjective judgements be subordinated to the 'natural' one?

The basis of scientific understanding has traditionally been the mechanical model[1, 2]. In this view, the behaviour of a system can be understood, and anticipated, by classifying and identifying its *components* and the *causal links, or mechanisms*, that act between them. In physical systems, the fundamental laws of nature *govern* these mechanisms, and determine what must happen.

In isolated or closed systems these restrictions place such limits on the behaviour of the system that we can in fact predict the properties of the final

*Prepared for the Workshop on Evolutionary Economics held in Manchester, 21st March, 1989.

state, *thermodynamic equilibrium*, quite generally for almost any system, however complex. This was such a triumph for classical science, that it was believed (erroneously) that analogous ideas must apply in the domains of biology, ecology the human sciences, and particularly of course, economics. It was argued that the processes of evolution and interaction present in such systems would lead necessarily to an equilibrium state, which was not only predictable, but was in some sense 'optimal'. This was of course a justification for the unfettered working of a free market, and of the 'invisible hand' of Adam Smith, since natural evolution was viewed as leading necessarily to the 'optimal' solution[3, 4].

But such ideas were misguided, being based on belief rather than on any real understanding or scientific proof. In fact systems encountered in ecology and economics are always complex, and any 'understanding', or 'cognitive map' must always involve a reduction of complexity. In order to even think about a situation we must invent words and concepts with which to discuss it, symbols which are simpler than what they represent. This corresponds to discussing some complex biological, ecological or human systems in terms of a few 'variables' and parameters. The particular set of variables chosen to represent the state of the system depends on the questions being considered (e.g. flows of money, goods and services, carbon, nitrogen or energy etc). But what is important to understand is that *none of these* can provide a description of reality which will *determine* the future.

Reality, whatever it is, is something in which all detail exists and each point is unique. But in our minds, we can only carry a reduced description of this, in terms of typical elements of the system and stereotypes, according to the classification scheme that we have decided to apply. Underneath any 'model' or 'understanding' there will always be the greater particularity and diversity of reality.

2. THE EVOLUTIONARY PARADIGM

We can distinguish three types of 'explanation' of complex systems. The first, supposes simply that the system being observed has undergone some unstated evolutionary process, and all the agents inside it have reached equilibrium. Mathematically, it is described by simultaneous equations which suppose that *at each instant* the values of the variables 'explain' each other through sets of parameters, and calibration consists in fitting the parameters to observation. Unfortunately, the parameters change unpredictably and require recalibration over time, blurring the difference between a variable and a parameter.

The behaviour of each agent (and even of the system as a whole) is supposed to be 'optimal' under whatever constraints exist, and *change* is, therefore, something entirely exogenous to the system. For example, technology, demography, culture and society are all supposed to drive the system from outside. If change is explored at all, it is examined by supposing linear perturbations of the existing quantities and flows. Linear programming models merely allocate proposed changes linearly, ignoring completely the non-linear interdependencies that characterize physical and human reality, and assuming essentially that nothing really changes. This must surely be a dangerous basis on which to plan some action.

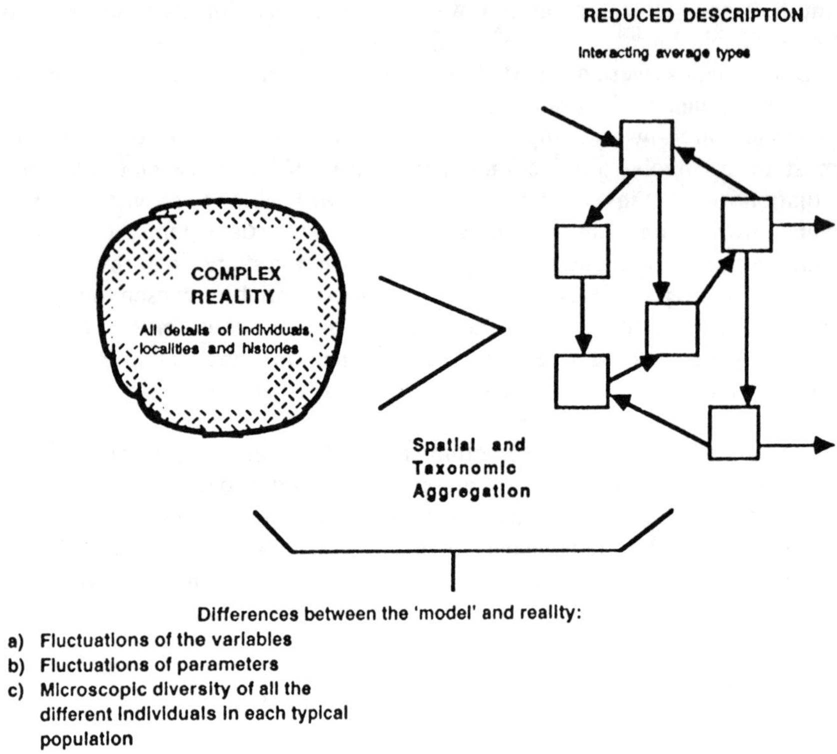

Figure 1. Quantitative studies of human systems used a 'equilibrium' assumption which necessarily viewed change as exogenous.

The second method is that of System Dynamics. In this, equilibrium is not assumed and the future of the system can apparently (but wrongly) be predicted by the simple expedient of considering the behaviour of the *dynamical system* represented by the model. In order to take this step, the only assumption that must be made is that the elements making up the variables (individuals within a species, firms in a sector etc.) are all identically that of the *average* type. In which case, the model reduces to a 'machine' which represents the system in terms of a set of differential (perhaps non-linear) equations which govern its variables.

This is the *Newtonian vision* of the world as a vast and complex clockwork mechanism. In this view, predictions can be made by simply running the model forward in time, and general statements about the future, under given conditions, can be made by studying the types of solutions that are possible for the equations in the long term. Explanation of the world is obtained then in terms only of the internal functioning of the system, and this functioning itself is viewed as a long term solution of the dynamical equations expressing the maximum or minimum of some unseen potential.

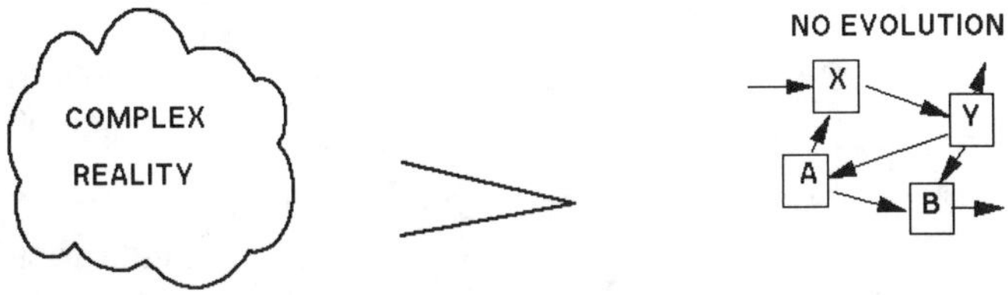

Figure 2. The mechanical paradigm connects components with fixed linkages, and predicts the future on the basis of these. Learning, adaptation and changing responses are treated as exogenous factors.

It was this, of course, that suggested that 'evolutionary superiority' must result from just letting the system run – the 'survival of the fittest', or 'laisser faire' economics and the free market, which supposed that observed behaviour in a 'mature' system must express some optimality. While such a view may offer the convenience of avoiding the need to take difficult decisions, the problem is that it is wrong.

When non-linear mechanisms are present, the system may continue to change indefinitely – either executing a cyclic path of some kind, or possibly even a chaotic movement around a 'strange attractor'. More importantly, its evolution may involve structural changes of spatial and hierarchical organization, in which qualitatively different characteristics emerge. New problems, satisfactions and issues emerge, and hence new factors would be 'turned on' in any value system used to assess the 'optimality' of the situation. How could the reaction to really new aspects be predicted?

Which solution actually occurs will depend on the accidents of history, the contextual and non-average details, which are the source of creativity. This is *not contained* in the *dynamical equations* which appear on the right hand side of figure 2. They are capable of *functioning* but not of *evolving*.

The third approach attempts to tackle the problem of evolution without either assuming equilibrium, or that only average individuals are present.

From figure (2) we see that evolutionary change must result from what has been 'removed' in the reduction to the deterministic description, and if this is so, then clearly, the future of any system will be due to two kinds of terms: changes brought about by the deterministic action of the typical behaviour of its average components, and structural qualitative changes brought about by the presence of *non-average* components and conditions within the system.

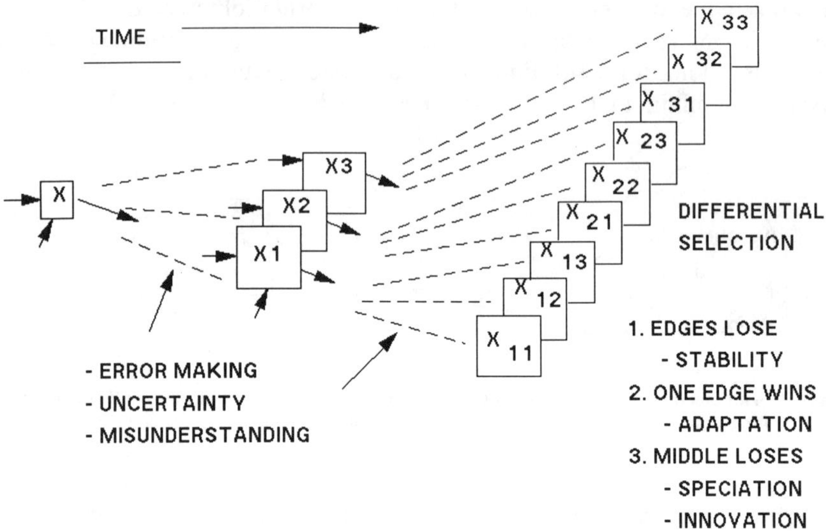

Figure 3. An initially 'pure' population x will explore different behaviours because of imperfect reproduction, bounded rationality, and individual subjectivity. Selection then operates *inside* our variables.

This views evolution as being the result of a dialogue between the 'average dynamics' of the chosen description, and the exploratory, unpredictable 'non-average' peturbations around this that result from entropic processes, imperfect understanding and individual differences[5].

3. THE NATURE OF CHANGE

The consequences of viewing evolution as a dialogue between 'average' processes and 'non-average' detail leads to the concept of 'evolutionary drive'[6, 7]. It shows that evolution selects for populations with the *ability to learn*, rather than for populations with *optimal behaviour*. This corresponds to the selection of 'diversity creating' mechanisms in the behaviour of populations and groups, initially involving genetics, and later cognitive processes.

This work was based on a 2-D 'adaptive landscape' for the most simple population equation possible:

$$dx/dt = bx(1 - x/N) - mx$$

The adaptive landscape considered was one of some 'characters' or strategies' of the population which affected the values of b, the birth rate, and m, the mortality rate.

In this simple system, evolutionary progress corresponds to diagonal movement as both the birth rate and the longevity of the population x increase. The work addressed the issue of exactly how 'improvement' really occurs.

Error making in reproduction was introduced into the population dynamics. It was assumed that parental character that corresponds to a particular value of b and m is not reproduced faithfully enough in an off-spring, which, therefore, appears at a different point in the b/m landscape. In accordance with common sense, we have further supposed that random errors in reproduction of a complex organism give rise much more frequently to a *lower performance than to a higher one*.

The occurrence of errors over successive generations, therefore, results in an initially homogeneous population occupying a single cell in b/m space gradually spreading into a cloud of populations, mostly located lower on the hill than the initial position. However, the populations which are lower down reproduce slower and have a higher mortality rate than those higher up, and so gradually the cloud slides *up the hill* because of the differential performance of the populations present. We may look at this result and interpret it as being due to the forces of *selection*. There is an internal selection process which is operating *inside* the variable 'x', which changes the nature of the 'average type' over time. As we shall see, the internal selection could not only simply be that of competition between different strategies, but complementarities could also emerge. One population could split into several, as strategies were invented which *escaped* from each others competition. Ecologies are the natural result of this evolutionary dynamics, even if the starting point were a single population.

Any particular population always inhabits a system which is populated by others who are also using a similar evolutionary mechanism of 'search' and 'differential reinforcement' to further their own success. In short, for any real population, b and m concern the *relative performance* of the population studied with respect to the others in the ecosystem – *who will also be evolving* – which means that instead of a fixed 'hill' which can be climbed, we have a 'down escalator' on which 'climbing' is necessary just to stay at the present level! This leads to a view very similar to the Red Queen Hypothesis of Van Valen[8], but we have here quantitative models which can be used to examine the question.

The conclusion of this work is that evolution generates diversity, and diversity drives evolution.

An important aspect which was considered concerned the manner in which evolution can itself adjust the degree of error making and variability which is leading to evolutionary change. By supposing that the fidelity of reproduction was an hereditary property which could itself be passed on imperfectly it was shown how the dynamics of such systems gave rise to an apparently 'willfull' reaction wherein a population increased its 'error making' quite violently in order to climb an adaptive slope, and then increased its fidelity when it had succeeded.

A different dimension of evolution was explored by considering the effect of 'error making' on a term of positive feedback which might enhance the reproduction of a particular character or type. In other words, we examined what would happen if a character or strategy existed which would increase the reproduction of that particular character or strategy. Error making was able to move the population a certain way up the adaptive landscape, but then was unable to move it further. The population became 'locked in' to a certain character by its own positive feedback. It could neither increase its own positive feedback, nor climb the slope of improved reproduction. The mechanism of self-reinforcement becomes so dominant that it outweighs the advantages in improved birthrate that deviants may detect.

We have called this phenomenon the Positive Feedback Trap. It suggests that although in many dimensions evolution will lead to error making and adaptability enabling them to adjust to their changing circumstances, some characteristic or strategy could exist involving mutual recognition and cooperativity for example, which would result in positive feedback enhancing this type and trapping it. An example of this from biology is the 'Peacock's tail', where a gene produces the beautiful tail in the male, and makes such a tail attractive to the female. In sexual reproduction, anything which enhances the probability of mating produces a positive feedback on its own population dynamics, and fixes itself. However, it is at the expense of functionality with respect to the external environment. Peacock's tails are not an aid to finding its food better, or escaping predators, but only a characteristic marker of a positive feedback trap.

In human systems, such positive feedback systems abound. Much of culture may well be behaviour which is fixed in this way. In most situations imitative strategies cannot be eliminated by the evolutionary process, and so fashions, styles and indeed cultures rise and decline without necessarily expressing any clear *functional* advantages. Indeed, 'culture' can be viewed as being an expression of ignorance of other ways of doing things. Similarly, academic disciplines such as economics offer almost perfect examples of such autocatalytic systems, where abstract theorems and proofs are rewarded with prestige and medals, even though it is quite clear to outside observers that there is no contact with reality. So much of human preoccupation is with playing a role in such groups where values are generated internally, and the physical world outside is largely irrelevant.

In some further experiments our models were extended to show how 'adaptive landscapes' are really generated by the *mutual interaction* of populations. In character space we suppose that closely similar populations are most in competition with each other, since they feed off the same resources, and suffer from the same predators, but that there is some 'distance' in character space, some level of dissimilarity, at which two populations do not compete with each other.

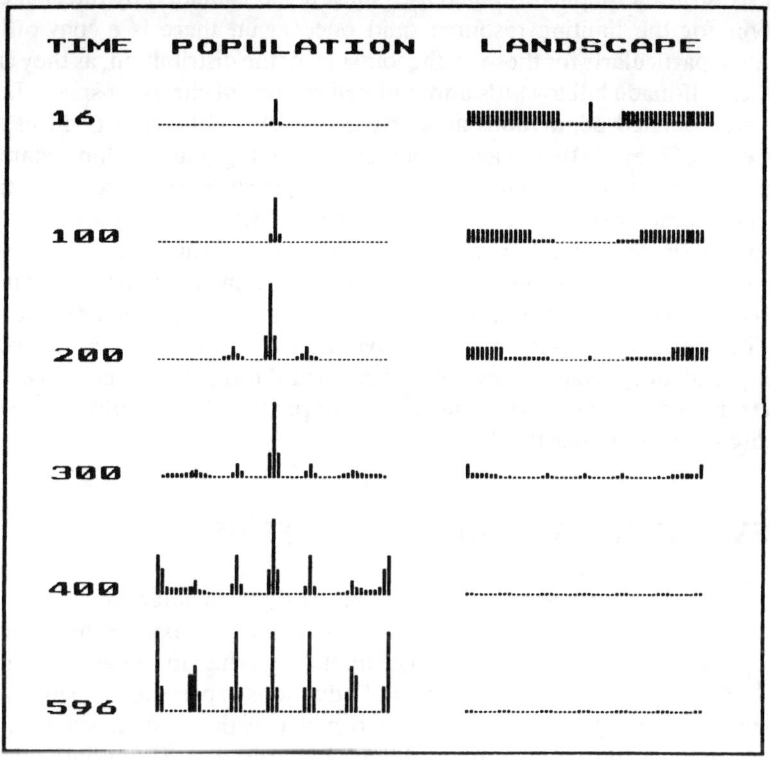

Figure 4. In a 1-dimensional character space, imperfect reproduction creates a simple ecology of populations. We see successive population distributions, and 'self-induced adaptive landscapes'.

Such an idea can be expressed in a 1-D character space by modifying the competition term in the 'logistic' bracket so that it expresses the decay of niche overlap with increasing separation in character space.

We start a simulation with a single, well-defined population. Initially, this population grows until it reaches the limits set by the competition for underlying resources. At this point, there is a 'pay-off' for 'error makers', because of the negative feedback due to competition. We could say that although initially there was no 'hill' to climb, the population effectively digs a valley for itself, until there is a 'hill' to climb on either side of the present character centroid. However, over some distance in this space the populations cannot multiply, because they are still in the 'competitive shadow' of the original population, and so they diffuse in small numbers up the slope away from the original type.

After a certain time, however, small populations arise which are sufficiently different from the original type that they can grow and multiply on the basis of some other resource.

In its turn, the population increases until it too is limited by internal competition for the limiting resource, and once again there is a 'pay-off' for deviants, particularly for those on the 'outside' of the distribution, as they climb another self-made hill towards unpopulated regions of character space. In this way, well defined populations appear successively, and colonists diffuse out from each of them as they reach a competitive limit, gradually filling character space with a set of populations separated approximately by some distance which is characteristic of the resource diversity which can be tapped.

From a single population our model generates a simple ecology, and a dynamic one since the identity of each population is maintained by the balance between a continual diffusion of deviants outwards into character space, and the differential reproduction and survival that is due to the presence of the other populations. Random events which occur during the 'filling' process will affect which populations arise, and so it is not rue that the evolution represents the discovery of pre-existing 'niches'.

4. EVOLUTIONARY HUMAN SYSTEMS

Complex systems dynamics resulting from the dynamic interplay of 'average' processes and 'non-average' detail have been used to explore how we can model and understand the evolution of human systems. In the very first models[9, 10], a non-linear dynamical system of equations expressing the supply and demand of different products was made to evolve by the random occurrence of entrepreneurs at different points and times in the system. The supply side was characterized by a non-convex production function for different economic activities, and consumer demand was assumed to reflect relative prices. The random parachuting of entrepreneurs onto the plain of potential demand resulted in the gradual emergence of a stable market structure and pattern of settlement.

The exact results are affected by the precise details, as well as by information flows affecting the mental maps of consumers. Also, the evolution of structure as a result of changing technology, transportation, resource availability etc. can be explored, as the changing patterns of demand and supply affect each other in a complex dynamic spatial process.

The fundamental basis for the models are the decisions of the different types of individual actors considered, which reflect either their 'cognitive maps', or the particular heuristic rules that they have developed. The spatial dynamics gives rise to very complex patterns of structure and flow, and to a structural emergence and evolution at the collective level. In such systems the microscopic and macroscopic levels are not related in a simple fashion. It is not true that the large structure is simply the small writ large. This is because the emergence of macroscopic structure affects the circumstances of the microscopic parts, as they find themselves playing a 'role' in a larger, collective entity. Each actor is *co-evolving* with the others.

These models have been continuously developed from these early simulations in order to provide realistic tools for understanding regional and urban evolution, in which the patterns of structure and flows are the result of an on-going evolutionary process of self organization[11, 12].

In addition, this approach was used to develop dynamical models for the management of natural resources, and the work connected with the simulation of fishing activities proved to be particularly fruitful[13]. This is because the problem that was considered contains all the essential aspects of the evolutionary problem in a strikingly simple way. This work has been discussed elsewhere in detail, but it is worth summarizing briefly some of the main conclusions.

The essential point that emerged was that success in fishing, as in life, requires two almost contradictory facets of behaviour. First, the ability to organize one's behaviour so as to exploit the information available concerning 'net benefits' (to be rational) which we have called 'Cartesian' behaviour. More suprisingly, however, a second ability is required, that is to be able to ignore present information and to 'explore' beyond present knowledge. We have called these kinds of fishermen 'Stochasts'. The first makes good use of information, but the second generates it! At the root of creativity is always this second type.

In the short term it is always true that the more 'rational' actor must out-perform the less, and therefore that for example taking steps to *maximise present profits* must, by that yardstick, be better than not doing so. Nevertheless, over a longer period the best performance will not come from the most rational but instead from behaviour which is some complex compromise. For example, a fleet of Cartesians which goes where *available information* indicates highest profits will in fact lock into zones for much too long, remaining in ignorance of the existence of other, more profitable zones simply because there is *no information available* concerning 'other zones':

'You don't known what it is you don't know'.

New information can only come from boats which have 'chosen' not to fish in the 'best' zones, or who do not share the consensus values, technology or behaviour, and hence who generate information. They behave like risk takers, but may or may not see themselves as such. They may act as they do through ignorance, or through a belief in some myth or legend. Whatever the reason, or lack of it, they are *vital* to the success of the fishing endeavour as a whole. It is their exploration that probes the value of the existing pattern of fishing effort, and lays the foundations for a new one[14, 15].

The model can be used simply as a 'simulator' providing information either for the management of the entire system or for the benefit of any particular fleet wishing to improve its performance. The parameter values which appear in the mechanisms governing the decision making of fishing boats are calibrated so as to give realistic behaviour[16]. This approach can, however, be completely inverted. Instead of supposing that we know, from observation, the parameters which characterize the fishing strategies of different fleets, we can use the model to *discover robust fishing strategies* for us.

To do this, we can run many fleets simultaneously (our current software will run up to 8) from identical initial conditions. The fleets differ, however, in the values of the parameters governing their fishing strategy, and whether or not they spy on, and copy some other fleets. Our model will reinforce the more effective strategies, and gradually eliminate the others, and by including a stochastic change in strategies for losing fleets, our model will gradually evolve sets of compatible behaviours. Each of these will be effective in the context of the others, not objectively optimal. In addition, such a system can adapt to changed external circumstances, and could be used to show us 'robust heuristics' for the exploitation of renewable natural resource. This is clearly similar in aim to the work on 'learning algorithms'[17, 18].

DISCUSSION

Rather than viewing evolutionary dynamics as the ascent of a population up a predetermined landscape of opportunity, our models show that the landscape itself is produced by populations in interaction and that details of the exploration process may significantly affect the outcome. The evolutionary dynamic is characterized by the emergence of self consistent sets of populations posing and solving the problems and opportunities of their mutual existence.

Innovation occurs because of the initiatives of non-average individuals. When this leads to exploration into an area where positive feedback outweighs negative feedback then population growth will occur. Stochastic processes, ignorance of the future, and differing opinions lead to exploration and discovery which change the system and result in new ignorance of the future and new differing opinions. In human systems ignorance permits learning, but learning creates new ignorance.

More generally non-average behaviour generates adaptation which in its turn generates new non-average behaviour. Because biological and human systems are dominated by internal interactions, it is the endogenously created, dynamical hills and valleys which are most significant.

Within each new situation we retrospectively assign values which we assert are improvements of objective parameters of the system. It seems problematic, given the possibility of backsliding, collapse or emergence of greater levels of complexity caused by structural instability, even to assert that, by the invasion and animation of an increasing fraction of inorganic matter, evolution exhibits any net progress. When we rationalize our behaviour we induce, from the false premise of the preexistence of a hill, an exogenous potential which drove events. Hills and hill-climbing are a rationalization of what has occurred. The landscape of opportunity is a dynamic variable which itself depends on the unfolding of events. The alteration of apparently insignificant details may produce a quite different future.

We can learn from the past. We can, in so far as the future resembles the past, learn about the future. However, there are elements of the evolutionary process, perceptible in the present as discoveries or creations, which cause radical and unpredictable differences between the past and the future. In this situation nothing can be certainly identified as being of future advantage. There may be eager climbers but there is no hill.

REFERENCES

1. I. Prigogine and I. Stengers, 1987, *Order out of Chaos*, Bantam Books, New York.
2. P.M. Allen, 1988, 'Evolution: Why the Whole is greater than the sum of its parts', in *Ecodynamics*, Springer Verlag, Berlin.
3. K. Arrow and G. Debreu, 1954, *Existence of an equilibrium for a competitive economy*, Econometrica.
4. G. Debreu, 1959, *Theory of Value*, John Wiley and Sons.
5. G. Nicolis and I. Prigogine, 1977, *Self-Organization in Non-Equilibrium Systems*, Wiley Interscience, New York.
6. P.M. Allen and J.M. McGlade, 1987, 'Evolutionary Drive: The Effect of Microscopic Diversity, Error Making and Noise', *Foundations of Physics*, Vol. **17**, No. 7, July.
7. P.M. Allen and J.M. McGlade, 1989, 'Optimality, Adequacy and the Evolution of Complexity', in *Structure, Coherence and Chaos in Dynamical Systems*, Eds. Christiansen and Parmentier, Manchester University Press, Manchester.
8. L. Van Valen, 1973, 'A New Evolutionary Law', *Evolutionary Theory*, Vol **1**. p. 1–30.
9. P.M. Allen and M. Sanglier, 1979, 'Dynamic Model of growth in a Central Place System', *Geographical Analysis*, Vol **11**, No. 3.
10. P.M. Allen and M. Sanglier, 1981, *Urban Evolution, Self-Organization and Decision Making*, Environment and Planning A.
11. P.M. Allen, 1985, 'Towards a new Science of Complex Systems', in *The Science and Praxis of Complexity*, United Nations University Press, Tokyo.
12. M. Sanglier and P.M. Allen, 1989, 'Evolutionary Models of Urban Systems: an Application to the Belgian Provinces', *Environment and Planning, A*, Vol. 21. pp. 477–498.
13. J.M. McGlade and P.M. Allen, 1985, The Fishing Industry as a Complex System, *Can. Tech. Fish and Aquat. Sci.* No. **1347**, Fisheries and Oceans, Ottawa.
14. P.M. Allen and J.M. McGlade, 1986, 'Dynamics of Discovery and Exploitation: the Scotian Shelf Fisheries', *Can. J. of Fish. and Aquat. Sci.* Vol. **43**, No. 6.
15. P.M. Allen and J.M. McGlade, 1987, 'Modelling Complex Human Systems: a Fisheries Example', *European Journal of Operations Research*, vol. **30**. pp. 147–167.
16. P.M. Allen and J.M. McGlade, 1987, *Managing Complexity a Fisheries Example*, Report to the United Nations University, Tokyo.
17. J. Holland, 1986, 'Escaping Brittleness: The possibilities of General Purpose Machine Learning Algorithms Applied to Parallel Rule Based Systems, in Michalski et al, *Machine Learning: An Artificial Intelligence Approach*, Vol **2**. Los Altos, California, Kauffmann.
18. J.H. Miller, 1988, *The Evolution of Automata in the Repeated Prisoner's Dilemma*.

The Role of Variety in Economic and Technological Development

P.P. SAVIOTTI
Department of Economics, University of Manchester,
Manchester M13 9PL, U.K.

1. INTRODUCTION

The increasing variety of products and services that we use both in our everyday life and in productive processes is one of the most important aspects of economic development. Both the emergence of completely new products (eg computers, radio, television, motor cars, aircraft, telephone, etc) and the growing diversification and specialisation of existing products increase the variety of services available to users. Simultaneously, some existing products and services disappear. The balance between these two processes, the emergence of new and different products and services on the one hand and the disappearance of existing ones on the other hand determines the net variety of the economic system. Impressionistic observations of the structure of household budgets or of the catalogue of the products offered by different shops seem to indicate that the net variety of the economic system has increased.

The increasing number of products and services available to the consumer has been accompanied by a number of related changes: the number of skills and

[1] This chapter is substantically revised version of a paper published in *Research Policy,* Vol 17 (1989), 89–103.

techniques required to produce this increased variety of final output has increased; the nature of competition has changed, having become less dependent on price and more based on products and associated services. Changes in the nature of competition were recognised in the early 1930s by Joan Robinson (1933) and Edwin Chamberlin (1933), who independently published their theories of imperfect competition. On the other hand Schumpeter (1912, 1934, 1943, 1943 p. 83) had already considered the new consumers' goods as 'one of the fundamental impulses that set and keep the capitalist engine in motion'. All the previous considerations constitute suggestive evidence but no definitive proof that the net variety of the economic system increases in the course of economic development.

The role of variety in technological and economic development can be further analysed by following the approach to economic growth and structural change developed by Pasinetti (1981). Pasinetti's economic system is constituted by vertically integrated sectors, each of which produces internally all the primary and intermediate inputs required to produce final goods. Production takes place by means of labour and capital goals, labour being simultaneously the supplier of inputs to production and the consumer of the final goods produced by the system. The economic system is described by a series of input/output equations and matrices. From these basic equations of the system it follows that an effective demand equation must be satisfied in order for the system to be at equilibrium, which in Pasinetti's definition means full utilisation of all productive resources. Population grows at a constant rate g, technological change in each vertically integrated sector leads to a rate of productivity growth ρ_i and the rate of demand growth in each sector is r_i. The effective demand equation takes the form:

$$1 = (1/\mu\nu) \sum_{1}^{n} a_{ni} (t - \theta) a_{in} (t - \theta) e^{(r_i - s_i)\theta} \qquad (1)$$

$$1 = (1/\mu\nu) \sum_{1}^{n} (g + r_i + 1/T_i) a_{nk_i} (t - \theta) a_{in} (t - \theta) e^{(r_i - s_{k_i})}$$

where

$a_{ni} (t - \theta)$ = movement through time of technical coefficients

$a_{in} (t - \theta)$ = movement through time of demand coefficents

μ = proportion of active to total population $(1 > \mu > 0)$

ν = ratio of working hours to total number of hours $(1 > \nu > 0)$

ρ_{k_i} = rate of productivity growth in making capital goods k_i, required to make goods Ki –

θ = finite length of time during which the percentage rate of change remains constant for each commodity

If all technical coefficients are continuously decreasing in time then the effective demand condition (1) is bound to become under-satisfied, unless the demand crefficients coefficients increase in the same proportion. However, demand coefficients are in general not expected to increase indefinitely because per capital demand for any commodity eventually reaches a saturation level. Therefore condition (1) is likely to generate unemployment as time goes on. The system could not keep growing generating more and more unemployment and the economic growth process would come to a halt.

Technological change an then lead to unemployment but in the meantime it can also counterbalance this tendency. On the one hand technological change by leading to increasing productivity and therefore to the reduction of technical coefficients generates labour displacements and therefore the potential for unemployment. On the other hand it keeps generating new products which are added to the economy. These new products lead to the introduction of new technical and demand coefficients in equation (1), which can then be rewritten as:

$$\sum_{1}^{n(t)-1} a_{ni}(t-\theta)\, a_{in}(t-\theta)\, e^{(r_i-\rho_i)\theta} \tag{2}$$

$$+ \sum_{1}^{n(t)-1} (g + r_i + 1/T_i)\, a_{nk_i}(t-\theta)\, a_{in}(t-\theta)\, e^{(r_i-s_{k_i})\theta}$$

$$= \mu(t)\, \nu(t)$$

Where now the total number of commodities/activities in the economic system is variable and increasing with time. Technical change has a twofold effect on equation (2). On the one hand it brings about a decrease through time of most of the a_{ni} a_{in} (the coefficients of the commodities so far produced), on the other hand it keeps on adding new commodities/activities and new coefficients. It is therefore possible for the second tendency to counterbalance the first and for condition (2) to be satisfied.

This is a somewhat oversimplified description of Pasinetti's ideas. For example, it neglects the effect that decreasing the number of working hours (and therefore ν) or the percentage of active to total population (and therefore μ) can have on employment. Changes in both of these variables can bring condition (2) toward fulfilment. Furthermore, changing patterns of international trade can have similar compensating effects on employment. All these types of changes can counteract the displacement effects of technological change and demand saturation. However, their long term effect is limited and it is unlikely that they can keep the system growing smoothly for ever. For example, it is difficult to envisage an unlimited increase in leisure time without an increase in the activities with which to occupy this leisure time. These results can at least be interpreted as implying that the requirement for the continuous introduction of new commodities/activities is one of the most important conditions for economic growth. Naturally this condition is equivalent to saying that a continuous increase in the variety of the economic system is required to maintain long term economic growth.

The previous way of reasoning could be reversed. It could be argued that only when there is continuous technical progress of the type leading to increasing productivity in the production of existing goods there will be the possibility of introducing new goods. In other words the increasing productivity can free resources to be used in the development and introduction of new commodities. The increasing variety and the increasing efficiency of the economic system are therefore two complementary aspects of the process of economic development. However, it must be recognised that at present economics can deal better with the trend towards increasing efficiency than with that towards increasing variety. While increasing efficiency can be represented and theorised by production functions, the role of new products and services cannot be captured by them, and it has to be introduced ex post into the economic system. In other words, the increasing variety aspects of technological change remains totally exogenous to a neoclassical economic approach. It is one of the great merits of Schumpeter (1912, 1934, 1943) to have pointed out the very important role which can be played by the introduction of new products in the process of economic development. However, his approach remained mainly descriptive and historical. The process of qualitative change which gives rise to economic development [See Chapter 1 and Faber, Proops] is one of the fundamental elements defining an evolutionary approach. Clearly then one must make an attempt to represent and theorise the generation and diffusion of new goods, services and activities better than in the neoclassical approach. The concept of variety can be very important as an aggregate measure of the increase in the number of goods, services and activities.

Pasinetti's model of economic growth provides a very interesting interpretation of the role of the variety of the economic system in determining economic growth. However, in the context of evolutionary theories it must be observed that Pasinetti's model is completely deterministic and therefore it does not take into account the role of fluctuations, irreversibility and path dependency. The role of variety then has to be examined also from other points of view. In this context it is interesting to analyse the implications for variety and economic development of the ideas of Prigogine (1976, 1987; Prigogine Stengers 1985) and of his school.

Prigogine's ideas were developed starting from studies of non equilibrium thermodynamics. This is a relatively new specialty since classical thermodynamics was only concerned with closed systems at or near equilibrium. A closed system, if not initially at equilibrium, will move towards a state of equilibrium characterised by the maximum possible disorder and randomness. This is a universal tendency which is embodied in the second law of thermodynamics. If a system is submitted to external constraints (ie opened up) and gradually moved away from equilibrium it starts behaving differently. In the near to equilibrium region a system submitted to weak external constraints gives linear responses to these constraints. On the other hand a system far from equilibrium and submitted to strong external constraints presents responses that are no longer linear. Accordingly the behaviour of such systems displays a multiplicity of stationary states and a set of new related dynamic characteristics such as bifurcation, hysteresis and so on (Prigogine 1987, p. 22). Nonlinearity and far from equilibrium situations are closely related. In contrast to near-equilibrium situations, where we find only a stable state, they lead to a multiplicity of stable states. This multiplicity can be seen on a bifurcation diagram (Figure 1), in which a descriptor of the system analysed is plotted against some bifurcation or control parameter. For some critical value of the control parameter new solutions emerge (ie new values of the description are allowed). Each of these solutions corresponds to a new branch of the bifurcation diagram. As the system moves further away from equilibrium new bifurcations take place which increase again the multiplicity of the stationary states of the system. Clearly this increasing multiplicity of stationary states of the

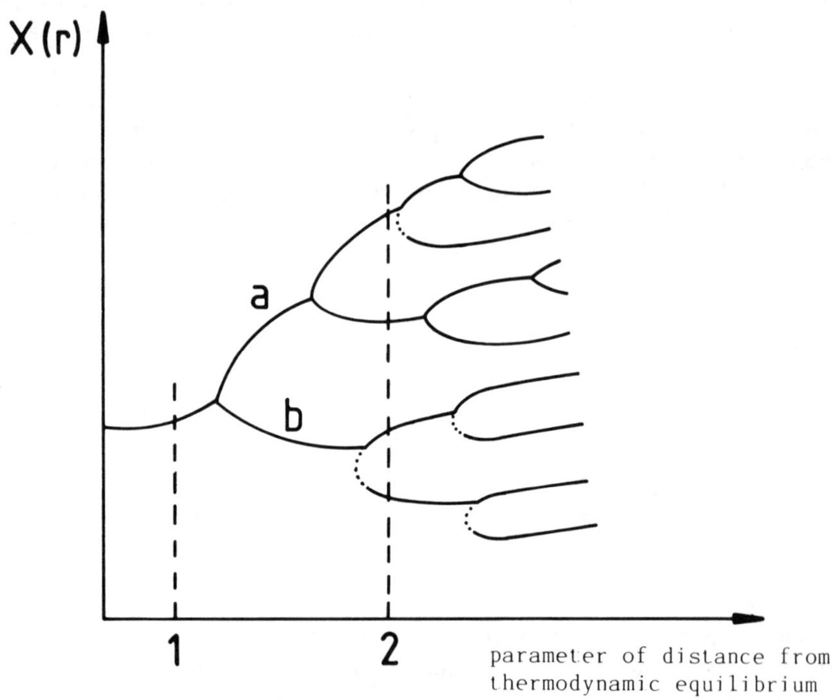

Figure 1. Bifurcation diagram representing the stationary states which are possible for a dissipative structure.

system is equivalent to an increase in variety. An open economic system subject to increasingly strong external constraints is therefore likely to develop an increasing variety.

At this point one could conclude that there is a good, although not yet ideal, degree of support for the proposition that an increasing variety of the economic system is a requirement for the process of economic development. Before turning to the implications of increasing variety for economic development it is important to mention that variety and complexity are two different concepts. In order to appreciate this difference a more rigorous definition of variety has to be introduced (Ashby, 1964, p124). If an economic system is considered a set of elements then the variety of this set will increase when the number of distinguishable elements of the set increases. Consequently one could use various measures of the number of distinguishable elements, the most immediate one being the number itself. In information theory the variety of a set is defined as the logarithm in base 2 of the number of distinguishable elements in the set.

$$V = \log_2 n \tag{3}$$

where V = variety and n = number of distinguishable elements in the set. Given the behaviour of the log function this particular definition implies that the marginal contribution to variety made by new and distinguishable products/activities will decline with an increasing number of products/activities already present in the system. Different definitions of variety will have different implications but they all entail an increase in variety when the number of distinguishable elements of the system increases. The usefulness of this concept of variety in the case of an economic system depends on how easily distinguishable product, services, etc, are. Without getting into a detailed discussion of this problem here we can observe that there are enough cases of distinguishable products, services, etc, to make this treatment worthwhile (for a slightly more extended discussion of this point see Saviotti, 1988a).

Complexity on the other hand can be defined in two different ways (J Nicholis, 1987, p. 163). Structural complexity increases with the number of interacting subunits; the percentage of mutual connectedness among them (dyadic or plural fashion) and the variance of the probability density function of the strengths of interaction between the individual subunits. On a functional level complexity increases with the minimum length of the (most compressed) algorithm from which one can retrieve the full behaviour of the system.

Given the previous definitions the structural complexity of a system could change considerably by changing the interactions between subunits and their distribution function while maintaining the number of subunits, and therefore the variety of the system, constant. Variety is consequently a component of complexity but it is not equivalent to it. Changes in complexity itself can have important implications for economic development but in this paper the discussion will be limited to the effects on variety.

2. INFORMATION, VARIETY AND ORGANISATIONAL STRUCTURES

An increasing variety of the economic system can be considered a requirement for the process of economic growth to proceed. Consumers benefit from this process in two ways: they can afford both larger quantities of existing goods and a greater variety of goods and services. However, this greater variety has also some associated costs. A system characterised by a greater variety of constituent elements is also a system which requires a greater amount of information to be described. The total cost required to describe such a system can be expected to increase with the variety of the system itself, although not linearly with it. Assuming that a growing variety of the economic system leads to positive welfare implications and to the possibility of a continued process of economic growth, this same process will only continue if information costs do not grow as rapidly as the benefits produced by the increasing variety of goods and services supplied to consumers. In other words one can expect that both producers and consumers will tend to reduce the information costs required to generate an economic system of increasing variety. This section is devoted to analysing a number of ways in which producers can try to reduce their information costs.

A number of definitions have to be introduced for the previous purpose. First, new goods and services will in general require new processes. Hence a greater variety of final goods and services will be accompanied by a greater variety of the processes used to produce them. It is thus useful to distinguish between output variety (V_q) and process variety (V_p),where the former is the number (log of) of distinguishable goods and services and the latter is the number (log of) of processes used to produce them. In the course of economic development one could expect producers to tend to decrease the costs that they face to produce a given output variety. These information costs can be reduced in two ways, first by increasing the efficiency with which the information required to described the system is stored and processed, second by changing process organisation in such a way that one has to store and process a smaller quantity of information to achieve the same output variety. For example, let us imagine a given productive process in which information about stock of parts, salaries, etc, is stored and processed manually. If without changing process organisation, and therefore without changing the quantity of information required to describe it, the storage and processing of information is computerised, the information costs of the process are likely to fall. If process organisation is changed in such a way that a smaller quantity of information is required to describe the process a further reduction in information costs will be achieved. In the first case a change in technology would have been introduced without any change in process organisation while the opposite would be true in the second case. Naturally in real life situations changes in organisation and changes in technology are often introduced simultaneously and they are often complementary. The previous considerations were only aimed at separating them conceptually so that their roles and their interactions can be subsequently analysed. The considerations that now follow are mostly aimed at analysing the implications of changes in process organisation for the information costs required to generate a given output variety V_q. In order to do that few more concepts need to be introduced.

2.1 Information, entropy and variety

The content of this section summarises that of a previous paper (Saviotti, 1988a) to which the reader is referred for a more detailed analysis.

A concept related to that of variety but not identical with it is that of entropy. Entropy is generally used to measure the randomness or uncertainty of a given set of elements (Gatlin, 1972, p28). Entropy is higher when the elements are placed randomly with respect to one another and falls when the elements are in an ordered configuration. For example, the entropy of a system constituted by a set of billiard balls of different colours is higher in a random configuration and higher when all the balls of the same colour are grouped together.

Each different configuration of billiard balls represents a microstate of the system. The system can have many different microstates. Most of these microstates are random and very few are ordered, for example, those that have all the billiard balls of he same colour together. Hence the probability of an ordered microstate, given by the number of such microstates, will in general be lower than that of a disordered microstate. If the elements of the system were in continuous motion then the system itself would continuously change between microstates. The system would then spend the majority of its time in disordered/random microstates and an average of the states of the system (macrostate) would resemble more disordered states. Since entropy is supposed to measure the disorder of the system and since disordered states have a higher probability of existence entropy is related to the probability of finding this system in a given state by the formula:

$$S = K \log W \qquad (4)$$

where S is entropy, K is a constant and W is the probability of finding the system in that state, which is given by the corresponding number of microstates.

The previous considerations apply to the case in which the microstates of the system are equiprobable. In this case, and when the microstates are independent, formula (4) can be rewritten as:

$$S = K \log (1/p_i) = -K \log p_i \qquad (5)$$

where p_i is the probability of an individual microstate. However, if the microstates of the system are not equiprobable formula (5) has to be modified and it becomes:

$$s = - K p_i \log p_i \tag{6}$$

This formula was introduced by Shannon and Weaver (1949) in order to deal with the information content of messages. This case is particularly relevant in the case of information since not all the microstates of a message composed of individual symbols (eg letters) are equiprobable, but also in the case of social systems. For example, it is clear that not all the microstates of an organisation, obtained by missing at random the members of the organisation, are equivalent.

It is possible to demonstrate (Gatlin, 1972, p36) that a given set of elements has its maximum entropy when all its microstates are independent and equiprobable. An example of this can be given by an organisation in which initially all the members have the same skills, functions and power. In this case it would be possible to mix at random all the members of the organisation without changing its performance. In this hypothetical case each microstate of the system would be equiprobable. If subsequently all the members of the organisation were retrained and given completely different skills, in such a way that they could now perform different functions and occupy hierarchically different positions, a random mixing of the members of the organisation would not produce equiprobable microstates. In the final state the system is more ordered and structured and has a lower entropy.

The concept of entropy has been introduced here for its relationship to that of information. Information in this context has to be used with the same meaning as in communications theory (Shannon, Weaver, 1949, p8). In this sense information is not measuring knowledge or meaning but refers only to the freedom of choice when one selects a message. For example one unit of information is required to choose between two messages *a* and *b*. When the number of messages increases both the information required to make a choice amongst them and the uncertainty of the system increase. Such a more uncertain system is also more disordered and therefore has a higher entropy. Consequently a system with a higher entropy will require a greater quantity of information to be described.

More specifically, the quantity of information required to choose one message out of a set of n messages increase with the entropy of the set of messages. It is therefore understandable that entropy has been used as a measure of information in communications theory.

Summarising the previous discussion one could say that the greater the uncertainty or randomness of a set of messages (and consequently the greater the entropy of the set) the greater the information required to choose one out of a series of messages. If one moves from the context of communications to that of organisations the analogue o the information required to choose one out of a set of messages becomes the information required by the top management of the organisation to choose one employee or a sub-unit out of the n employees/sub-units which constitute the organisation. In even simpler terms the information which is measured by entropy is the information which is required by top management (or by anyone else who needs it) to 'know' the organisation.

The type of information which has been described above is not the only possible type of information. A communications engineer is interested in the capacity to transmit information. As Shannon and Weaver (1949) put it '. . . this word "information" in communications theory relates not so much to what you do say as to what you could say'. This capacity is called by Gatlin (1972) potential information. Therefore potential information, which increases with messages variety and with freedom of choice, is measured by entropy.

It is not true, however, that as entropy increases information always increases. Let us take a simple example. A library contains information stored in the form of linear symbols ordered according to the constraints of a language. The sequences are contained in books and periodical which are classified and neatly catalogued in shelves. This is obviously a state of very high order. If we were to take each page of each book or periodical, cut them into single letter pieces and mix them at random the entropy of the system would increase but the information stored in the library would decrease to virtually zero. However, starting from this disordered state, many more meaningful combinations of letters could be formed than the one that was destroyed by cutting the books and periodicals into single letter pieces. Hence, while the actual information contained in the library decreased to virtually zero the information that it could contain increased. A difference has therefore to be made between *potential* information and *stored* information. In simpler terms a greater uncertainty or randomness of the system tends to increase its potential information while a grater order and constraint tend to increase its stored information.

Another interesting distinction could be introduced between the *internal* and *external* information required by an organisation. The purpose of an organisation is to transform some kinds of inputs, not necessarily material, into some kinds of outputs. In so doing the organisation must have information about the type of inputs required, the process to transform them into outputs and the users of these outputs or, in general, about the external environment in which it operates. On the other hand the organisation must also have information about its internal structure and the internal resources that it can use to transform inputs into outputs. According to Duncan and Weiss (1979, p. 205):

> 'The objective of the kind of organisation structure that is implemented is twofold: (1) to generate information for decision making that reduces uncertainty, and (2) to generate information that will help to co-ordinate the diverse parts of the organisation.'

where the uncertainty reducing information (1) can be considered mainly information about the external environment. The roles of these two types of information can be better understood by comparing them to what happens in a computer. Particular types of information are stored in a computer at given locations. In order to retrieve this information the address of the location at which it is stored have to be known as well. The addresses are only part of the mechanism by which valuable information is stored. If the capacity of the computer is finite the more memory locations are used for addresses the less space there will be for the storage of valuable information. The uncertainty of the system in an organisation is the analogue of the number of addresses in a computer. In both cases the quantity of valuable external information that can be stored decreases with the uncertainty of the system or the number of addresses.

We can now summarise some of the results of the previous discussion about the effect of the non-independence and non-equivalence of the microstates of the system on its entropy the following statements apply:

a) the larger the number of non-distinguishable elements of the system, and consequently the greater its variety, the larger the amount of information required to describe the system.
b) the greater the degree of non-independence and of non-equiprobability of the microstates of the system, the lower its entropy (Gatlin, 1972, p. 3) and consequently the lower the amount of information required to describe the system.
c) a lower entropy of the system, due to non-independence and non-equiprobability of its microstates, is also associated with a greater capacity of the system to store information.

Before passing to the next section in which some examples will be discussed, three observations are in order. First, the concept of information as used in this paper does not describe all the types of information that an organisation uses in its decision making processes. Types of information which would fall within the definition used in this paper are, for example, that required to choose a particular spare part from a store, to choose the employee who is knowledgeable about a particular topic or to calculate the wages of a group of workers. In other words, it is information of the type which is commonly described as factual. As already indicated, information is not meant to be equivalent to meaning or knowledge. However, the quantity of information of this type that has to be manipulated by organisations is so large that changes in organisational methods and technologies which decrease the costs of processing and transmitting this type of information are bound to have an important economic effect. It could be said that this paper analyses the implications of this type of information only for organisational arrangements or that in other words, it tries to find what would be the preferred organisational structures in a world in which information of the factual type was the only factor determining such structures. Naturally real life organisations are influenced by other factors as well. Whether real organisational structures resemble those which could be predicted on the basis of information alone will depend on the relative importance of information and of these other factors.

However, it is not always true that a lower entropy is an advantage. This can be understood by returning to the previous example of the dictionary. The state in which the dictionary is normally sold has obviously a lower entropy than the state obtained by mixing at random all its letters. Consequently, the information storage capacity of the dictionary in its normal state is much higher than that in its disordered state. However, starting from the disordered state it is possible to re-order the letters in many ways, the original state of the dictionary being only one of them. Many types of meanings can therefore be created starting from the disordered state. In this sense, the disordered state of a library has been described before as having a lower stored information but a higher potential information than the ordered state. The disordered state has a greater flexibility than any of the completely ordered states. In the case of reorganisations this implies that an organisation of lower entropy will be able to store a larger amount of specialised information which allows it to adapt well to a constant environment. On the other hand, an organisation of higher entropy will be more easily adaptable to changes in its external environment which required to store a different type of information. Normally we choose to store particular items of information in order to use them repeatedly. This is an effective strategy if the environment in which the information storing organisation operates is stable. Thus highly structured organisations are likely to be effective when a constant set of routines is appropriate. But if the environment changes quickly highly structured organisations are likely to experience difficulties. This could explain the observation that organisations operating in more dynamic environments tend to have *organic* structures while organisations operating in more static and structured environments are more likely to have *mechanistic* structures (Burns, Stalker, 1961). The information already in them blocks the channels through which new information might come: storing information reduces access to potential information. Alternatively we could say that organic structures have a lower stored information but a higher potential information than mechanistic ones. In a related way one can expect that when the environment changes suddenly and radically the organisation will have to upgrade its external information more rapidly than its internal information. In this sense external information is going to be used relatively more in the formulation of strategies while internal information is likely to be more closely related to organisational structures. If strategy precedes structure (Chandler, 1962, 1977) then it is logical to expect that external information should change in response to environmental changes and ahead of internal information.

3. EXAMPLES

3.1. Example 1 – Division of labour

In this example two different organisational arrangements for the same manufacturing process are going to be compared. In one the process is carried out without division of labour and in the other one the maximum division of labour is used. The process consists of a number of operations which are performed by different people. For simplicity in both cases the number n of operations will be considered equal to the number of people performing them.

Case 1. Each person carries out the n operations in a sequence, therefore producing the final product. Each person works in parallel and independently of any other person involved in the process. In this system, therefore, there is no division of labour. In a formal sense this system does not constitute an organisation, rather each person carrying out the complete process constitute a unit of production and organisation. This productive system constitutes a simplified model of a real system in which individual artisans produce finished products working independently of one another.

The total entropy of this process (formula 4) can be estimated by calculating the probabilities of the individual microstates of the system. We can imagine to derive each microstate of the system by drawing in sequence one person and one operation from two separate boxes and coupling them. A microstate of the system has been generated after n people and n operations have been drawn from separate boxes and coupled. A different microstate can be originated by putting back the people and operations in their separate boxes and drawing in a different sequence n people and n operations and coupling them one to one. In this way all the possible microstates of the system can be originated. Each microstate of the system can be considered as a snapshot in which each person is 'frozen' while performing one of the n operations. In a subsequent snapshot (microstate) each person will be found performing one operation different from the previous microstate.

In each microstate an operation is associated with each person. The following two are examples of microstates:

$$\text{(a)}\quad X_1 0_1, X_2 0_2, \ldots, X_n 0_n \tag{7}$$

$$\text{(b)}\quad X_1 0_3, X_2 0_6, \ldots, X_n 0_n \tag{8}$$

where $X_1, \ldots, X_n$ represent the people and $0_1, \ldots, 0_n$ the operations. The probability of each of them are given by:

$$P_a = P\,[X_1 0_1, X_2 0_2, \ldots, X_n 0_n] \tag{9}$$

$$P_b\, P = [X_1 0_3, X_2 0_6, \ldots, X_n 0_{n22}] \tag{10}$$

Since each person works independently, P_1 can be expressed as the product of the probabilities of each event $X_1 0_1$, $X_2 0_2$, . . .

$$P_a = P[X_1 0_1]*P[X_2 0_2]* . . . *P[X_n 0_n] \quad (11)$$

Given that there is no specialisation and that consequently each person can perform equally well any of the n operations

$$\begin{aligned} P[X_1 0_1] &= P[X_1 0_2] = . . . = P[X_1 0_n] \\ &= P[X_2 0_1] = P[X_2 0_2] = . . . = P[X_2 0_n] \\ &= P[X_n 0_n] \end{aligned} \quad (12)$$

Consequently:

$$P = P_b = \text{probability of any other microstate } P_i. \quad (13)$$

All the microstates of the system are therefore equiprobable. Since the sum of the probabilities of the microstates must be equal to 1 then

$$P_i = 1/n \quad (14)$$

where n is the number of microstates of the system.

Consequently, the entropy of the system will be

$$\begin{aligned} H &= -\Sigma\ 1/n \log 1/n = +\Sigma\ 1/n \log n = n\ 1/n \log n \\ &= \log n. \end{aligned} \quad (15)$$

Here and in what follows formula (4) is used without the constant K because what is important is to estimate the change in entropy between two different situations and not its absolute value.

Since the microstates of the system are independent and equiprobable this type of organisational arrangement corresponds to the maximum possible entropy of the system of n workers and operations.

Case 2. In this case the n operations which constitute the process remain constant but specialisation and division of labour are introduced in their most extreme form. Each person will then be able to perform only one of the n operations. The same microstates of the system can be originated except that now they are no longer equiprobable. In particular, only one microstate, that in which each person is performing the operation that he/she has been trained to perform, has a non-zero probability and this probability is actually equal to 1. Shannon's expression for the entropy of the system is then reduced to a single term:

$$H = 1 \log 1 = 0. \tag{16}$$

In other words, passing from a system in which there was no division of labour and specialisation to one with the maximum possible specialisation and division of labour the entropy of the system has decreased from log n to zero. Since entropy can be considered a measure of the information which is required to describe the system, this example shows that the introduction of specialisation and division of labour decreases the information required to describe the process. Hence, the information that the organisation has to store to know itself and to co-ordinate its own internal processes, what was previously called internal information, decreases with increasing division of labour. Correspondingly, the information that the organisation can store about the environment in which it is operating and about the technologies it is using (external information) increases. This is understandable if we think that information has to be stored or embodied in the human skills of an organisation. In absence of specialisation every person in the organisation would have to store the same types of information. On the other hand with division of labour every member of the organisation has to store only a very limited range of types of information and consequently can store a greater quantity of these types. Collectively the organisation can store more information.

The two cases previously analysed are extreme cases. They are important, however, because any real process is likely to have a degree of division of labour included between these two. The entropy and the information requirements of any real process are therefore going to decrease as the process moves away from the extreme without any division of labour and toward that with the maximum division of labour.

The situation does not change substantially if the number of operation n(0) is different from the number of people n(x). Both for $n(0) > n(x)$ and $n(0) < n(x)$ the entropy of the system will be above its minimum that corresponds to the situation in which each person performs only one operation.

The representation of the division of labour which was given in this section is clearly rather abstract and oversimplified. In a system with a high degree of division of labour the output of the operation performed by each individual worker has to be transferred to the next worker in the sequence of operations. In other words, a transaction occurs between each pair of workers in subsequent operations. A process with n operations will contain (n–1) transactions. To obtain the benefits of the division of labour some co-ordination is required in terms of time, position etc,. This co-ordination requires additional information and therefore causes additional costs. In presence of co-ordination costs actual information requirements would be determined by the balance between operation costs and co-ordination costs. The case previously illustrated is therefore an idealised one in which co-ordination costs can be considered negligible with respect to operation costs.

At a higher level of aggregation, that of an industrial sector or of a national economy, the division of labour in production is equivalent to arms length transactions between independent firms. In this case as well the most extreme form of division of labour and independent transactions would be the organisational structure with the lowest entropy, and therefore with the lowest information requirements, only when transaction costs are negligible with respect to production costs. This situation is similar to that described by Williamson (1975, 1979, 1981) in the transaction cost approach.

3.2 Example 2 – Hierarchical organisations

In this example two different structures of the same organisation will be compared, one with no departmental boundaries and one in which the organisation is constituted by a series of departments. Following Simon's terminology (1981) the first case would be an example of a 'flat' system with only one hierarchical level and with a 'span' of control equal to the number of employees of the organisation. By introducing departments the number of hierarchical levels would have to increase to at least two because now if all the departments but one were at the same hierarchical level the department in charge of overall coordination and planning would have to occupy a higher hierarchical level. Furthermore, the span of control would now be restricted to the employees of each department. Finally, the degree of non-equivalence would increase by introducing departments.

Case 1. The organisation is constituted by n workers performing n operations. The lack of hierarchical structures implies that the operations performed by different people are sufficiently similar that the various workers can be interchanged amongst them. In a first approximation we can assume that these operations are identical and that the workers are completely interchangeable amongst them. This is obviously an oversimplified situation but it is useful to compare this case, in which people can move freely amongst all the operations in the organisation, and case 2, in which people can only be exchanged amongst the operations contained in one department. This situation and the implications for information are clearly different from the effects of the division of labour. The difference can be understood if one imagines to start from n people and n equivalent operations and to generate a real life organisation containing different departments in a number of stages. In the first stage the n people are put together under one institutional roof and assigned in a random fashion to the n equivalent operations. In the second stage the departmental boundaries are introduced and now people can only be interchanged amongst the operations internal to the department. In the third stage the operations in each department are differentiated and division of labour is introduced. The implications of the division of labour for information have already been discussed.

In this example the implications of the introduction of departmental boundaries for information are discussed.

Since in this case people can be interchanged amongst a set of n equivalent operations all the microstates of the organisation are equiprobable. Consequently the entropy of the system has its maximum possible value, given by the logarithm of the number of microstates of the system, which in this case is equal to the factorial of the number of people:

$$H = \log n! \tag{17}$$

When the microstates are equiprobable the entropy of the system is given by Boltzmann's formula (2), which again turns out to be equal to its maximum possible value.

Case 2. When departmental boundaries are introduced only microstates of the system obtained by moving people around within each department are possible. A lower number of microstates can therefore be obtained with respect to the previous case. However, even in this case the microstates of the system are equiprobable, since the operations in department are equivalent. The change in entropy between Case 1 and Case 2 is given by:

$$\begin{aligned} \Delta S(1 \rightarrow 2) &= S(2) - S(1) \\ &= R \log p(2) - R \operatorname{long} p(1) \qquad (18) \\ &= R\log \frac{p(2)}{P(1)} \end{aligned}$$

where p(2) and p(1) represent the numbers of microstates of the organisation in Case 2 and Case 1 respectively. If n(1), n(2), in Case 2 and n the number of people employed in the organisation as a whole then:

$$p(2) = N_1!{*}N_2!{*} \ldots {*}N_k! \tag{19}$$

where $N_1!$, $N_2!$ . . . $N_k!$ are the factorials of n(1), n(2), . . . n(k).

Consequently:

$$\Delta S(1 \rightarrow 2) = R \log \frac{N_1! \;{*}N_2! \ldots {*} \;\; N_k!}{N!} \tag{20}$$

It can be proved that $\Delta S(1 \rightarrow 2)$ is less than zero. If we expand all the factorials:

$$\Delta S(1 \rightarrow 2) - \frac{1*2*3*4* \ldots *n(1)*1*2*3* \ldots *n(2)*1*2*3* \ldots *n(k)}{1*2*3*4* \ldots\ldots\ldots\ldots *n} \qquad (21)$$

If we assume that factorials are ordered in the following way:

$$N_1! < N_2! < N_3! \ldots < N_k! \qquad (22)$$

and if we simplify the factors corresponding to the smallest factorial we find:

$$\Delta S(1 \rightarrow 2) = R \log \frac{1*2*3* \ldots *n(2)* \ldots *1*2*3 \ldots *n(k)}{[n(1)+1]*[n(1)=2]* \ldots *n(2)* \ldots . *n} \qquad (23)$$

To each factor contributing to the second factorial in the numerator will correspond a larger factor in the denominator. Thus, for example, n(1) + 1 will correspond to 1, n(1) + 2 will correspond to 2 etc. The same will occur for each of the subsequent factorials. Consequently the denominator will always be greater than the numerator and ΔS (1→2) will always be smaller than zero.

This will occur as a consequence of the introduction of departments into the organisation. With this change the uncertainty of the system is limited by reducing the freedom of motion of the employees within the organisation. From what was said before it follows that increasing the number of departments leads to a fall in the entropy of the system, and therefore in the quantity of information required to know and co-ordinate the members of the organisation. Intuitively this change can be interpreted as a decrease in the number of information paths possible within the organisation. When there are no departments and all members of the organisation are equivalent and communication may be required between all possible pairs of members. On the other hand communications between members of different departments are much more selective and lower in number.

As it was previously pointed out the information saving effect of the introduction of departmental boundaries is different from that due to the division of labour, discussed in the previous example. The information costs of the organisation would be further reduced if after introducing departments division of labour was introduced as well. In other words, the types of organisational change illustrated in example 1 ad 2 can both lead to reductions in the entropy of a system and these reductions are cumulative.

According to Simon (1981) hierarchical organisations tend to be more stable than unstructured organisations and therefore they have a higher probability of survival in an evolutionary process. The improved chances of survival may be partly due to the fall in information requirements discussed in this example.

3.3 Example 3 – U and M form organisations

This is a further and more detailed application of the case described in Example 2. A schematic representation of the two types of organisation is given in Fig 2. The entropy for the transition from a U to an M form organisation can be calculated in the same way as in Example 2:

$$\Delta S\ (U \rightarrow M) = S(M) - S(U) =$$

$$= R \log p(M) - R \log p(U) = R \log p(M)/p(U) \qquad (24)$$

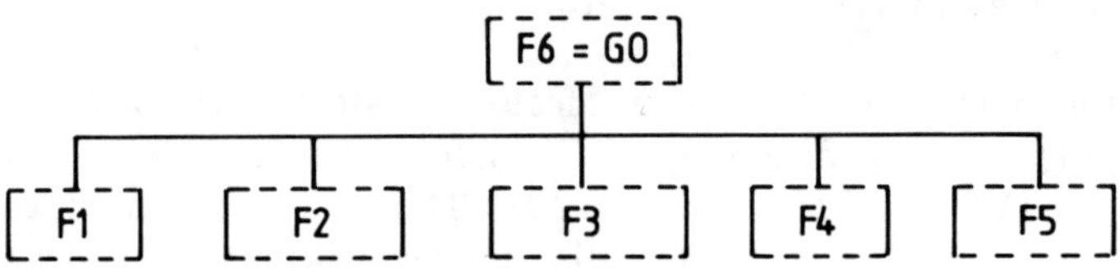

Multifunctional or U form. F6=GO= general office: F1, F2, F3, F4, F5 = functional departments

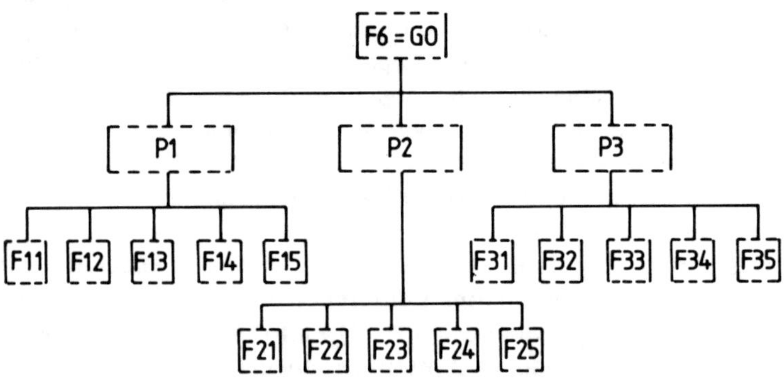

Multidivisional or M form. F6 = GO = general office; P1, P2, P3 = divisional offices; F11, F12, F13.........F34, F35 = functional departments

Figure 2. Schematic representation of the U and M form organisations.

In order to calculate p(M) and p(U) some simplifying assumptions have to be made. First we an assume that total employment N remains constant during the transition U → M. Second we can assume that in both cases employment is evenly distributed amongst different departments, divisions, etc. Therefore if we introduce the following notations:

State U. $n_1, n_2, n_3, n_4, n_5, n_6$ = numbers of people employed in functions F1, F2, F3, F4, F5, and F6 respectively, where F6 coincides with the general office.

State M. $n_{11}, n_{12}, n_{13}, \ldots, n_{21}, n_{22}, \ldots, n_{31}, \ldots, n_{35}$, = numbers of people employed in functions $F_{11}, \ldots$, n_1, n_2, n_3, = numbers of people employed in divisional offices 1, 2 and 3; n_6 = number of people employed in the general office.

The previous assumptions can be stated as:

$$N_T = n_1 + n_2 + n_3 + n_4 + n_5 + n_6 = n_6 + n_1 + n_{11} + n_{12} + \ldots\ldots + n_2 + n_{21} + n_{22} + \ldots\ldots + n_3 + n_{31} + \ldots + n_{35} \quad (25)$$

$$N_T = N_D(U)\ n_U = N_D(M)\ n_M \quad (26)$$

$$n_1 = n_2 = n_3 = n_4 = n_5 = n_6 = n_U = N_T/N_D(U) \quad (27)$$

$$n_6 = n_1 = n_{11} = n_{12} = \ldots\ldots = n_{35} = N_T/N_D(M) \quad (28)$$

where $N_D(U)$ and $N_D(M)$ are the total numbers of subunits (departments, etc) in cases U and M, n_U and n_M are the numbers of people employed in each subunit in ases U and M.

We can then proceed in a way similar to that of example 3.2 to calculate the probabilities of the two states U and M.

$$P_M = n_1! * n_{11}! * \ldots * n_2! * \ldots * n_{21}! * \ldots * n_6! \quad (29)$$

Given the previous assumptions:

$$n_1! = n_{11}! = \ldots n_3! = \ldots\ldots n_6! = n_m \quad (30)$$

Consequently:

$$P_M = N_D(M)^* \ n_M! \quad (31)$$

Similarly:

$$P_U = n_1!^*n_2!^*n_3!^*n_4!^*n_5!^*n_6! \quad (32)$$

but:

$$n_1! = n_2! = n_3! = n_4! = n_5! = n_6! = n_U! \qquad (33)$$

and

$$P_U = N_D(U)^* n_U! \qquad (34)$$

Thus:

$$\Delta S(U \rightarrow M) = R \log \frac{N_D(M)^* n_M!}{N_D(U)\ n_U!} \qquad (35)$$

$$\Delta S(U \rightarrow M) < 0 \text{ if } \frac{N_D(M)^* n_M!}{N_D(U)\ n_U!} \qquad (36)$$

But:

$$\frac{n_U}{n_M} = \frac{\dfrac{N_T}{N_D(U)}}{\dfrac{N_T}{N_D(M)}} = \frac{N_D(M)}{N_D(U)} \qquad (37)$$

and:

$$\frac{N_D(M)}{n_D(U)} = \frac{n_U}{n_M} < \frac{n_U!}{n_M!} \qquad (38)$$

This condition (38) is always satisfied for $n_U < n_M$. Consequently the transition $(U \rightarrow M)$ will always lead to a fall in the entropy of the organisation, and therefore to a fall in information costs per unit of output variety, provided that employment in each subunit of the M state is smaller than employment in each subunit of the U state. This is always going to be the case if total employment in the organisation is kept constant.

This transition leading to a negative change in entropy can be interpreted considering that the organisation is an open system and that the total entropy change for this transition is the sum of the external and internal entropy changes:

$$\Delta S_T = \Delta S_O + \Delta S_E \qquad (39)$$

Where ΔS_T = total entropy change, ΔS_o = entropy change of the organisation and ΔS_E = external entropy change. One can imagine the transition to occur in two states: in the first one product variety increases and in the second one organisational changes are introduced:

$$\Delta S = S(1) \rightarrow S(1,2) \rightarrow S(2) \quad (40)$$

where S(1) is the initial entropy, S(1,2) the entropy after the introduction of product variety and S(2) the entropy after the introduction of new organisational arrangements. Then:

$$\Delta S_E = S(1,2) - S(1) \quad (41)$$

and

$$\Delta S_o = S(2) - S(1,2) \quad (42)$$

Consequently the positive external entropy change due to the increase in product variety is more than compensated by the negative entropy change due to the organisational change (the transition from the U to the M form).

4. PROCESSES OF TECHNOLOGICAL DEVELOPMENT

The previous section dealt with organisational forms and organisational changes that could be used to reduce the information costs that firms face in generating an increase in the variety of products and services that they produce (output variety). This section will be concerned with a different level of aggregation, namely that of a technology or an industry.

Some regularities have been found in the evolution of technologies. For example, Abernathy and Utterback found that a technology progresses from its new, initial phase, characterised by a multiplicity of product designs and by very flexible, unstructured processes, to a matrix phase in which a dominant design has emerged and processes have become structured, efficient and rigid (1975, 1978). In a similar way Nelson and Winter (1977, 1982) have introduced the concept of natural trajectories and of technological regimes, Dosi (1982) that of technological paradigm and Sahal (1981) that of technological guideposts. All these concepts imply that in the process of selection which occurs during the evolution of a technology some constraints are imposed upon the technology itself. Saviotti (1986) has interpreted the emergence of paradigms/trajectories/dominant designs/guideposts as an example of the behaviour of technologies as complex systems and of their tendency to achieve self regulation and invariance with respect to environmental changes.

These processes of technological development have implications for variety and for the information requirements of technologies. For example the emergence of a dominant design is likely to limit the output variety of the technology by reducing the number of distinguishable products. In a similar way according to product life cycle theories of international trade (Vernon 1966; Well 1972) a technology has to mature and become standardised before it diffuses internationally. In a more general way the adoption of dominant design/regime, trajectory/paradigm/guidepost would inhibit wide ranging exploration of a rich field of potential information. By limiting the search in this way the paradigm would create a relatively stable environment in which highly structured organisations which store information efficiently would perform particularly well. Alternatively a radical change in technology or some other important environmental change would invalidate much of the previous stored information and therefore put highly structured organisation at a disadvantage. In these conditions organic structures with their greater flexibility and higher potential information would perform better (Burns, Stalker, 1961). By decreasing input variety and process information requirements one would expect these trends in technology to lead to a fall in process entropy and possibly even in output entropy. These processes function at a level of aggregation higher than that of individual firms/organisations and lower than that of the whole economic system. The problem then arises of the compatibility of these changes in variety, process information requirements and entropy with those at the lower and higher levels of aggregation.

For what concerns individual firms and organisations these trends in technology will impose 'common' constraints which will make firms' technological behaviour quite similar, although not identical in a number of respects. Therefore one could expect that as a paradigm becomes established the variance in ouput quality and (even if perhaps to a lesser extent) in firm behaviour to decrease. Convergence would take place towards common practices and conceptual structures.

From what was said before the output variety of the whole system is expected to grow over sufficiently large periods of time while the output variety of individual technologies and industrial sectors in many cases is expected to fall. If the number of technologies and industrial sectors in the whole economic system were constant this could not happen. However, there are two ways in which these patterns of evolution at different levels of aggregation of the system can be compatible. First, the deceasing output variety of each technology could lead to savings in process information costs and in other dimensions. In a way similar to what followed from Pasinetti's model, savings in individual technologies could allow the accumulation of the resources required to generate new products and services and, therefore, to increase the output variety of the whole system. In this sense paradigms would play a role similar to that of the organisational changes at the level of the firm discussed in the previous sections. Second, it would be possible for a paradigm to define a number of common practices upon which all the firms using the technology would converge but making a 'modular' use of the common practices For example a large variety of models could be produced making use of common parts and or standardised process equipment and practices. In this way it would be possible for process variety and process information costs to fall while simultaneously output variety increased at the level of the individual technology. There are examples of both types of behaviour, with aircraft and helicopters following the former type (Saviotti 1988b, Saviotti, Trickett 1989) and with motor cars and agricultural tractors the second type (Coombs et al 1978; Saviotti et al 1980).

A related point of view about the evolution of technologies and about their gaining economic 'weight' has been developed by Metcalfe an Gibbons (Metcalfe 1984; Gibbons, Metcalfe 1987; Metcalfe, 1987). An important component of their evolutionary approach is the adoption of a population perspective. This implies that it is not only the average (or typical) economic operator, firm, product, technology, etc, which matters but also their dustribution. For each population characteristic such distribution is measured by its variance. They call this variance variety. Their use of the term variety is therefore different from that of this paper. Among others they obtain the interesting result that the rate of change of average population characteristics is related to their variance. This is an economic analogue of Fisker Law (1935), a result that Nelson and Winter (1982, p243) had obtained in a different way for the evolution of unit costs. The significance of these developments in the present context can be gained from the following formula (Metcalfe, Gibbons, 1988, p27):

$$\frac{dh}{dt} = \frac{\delta f}{f+\delta}[V_o\ C(h,\alpha)-V(h)] \tag{43}$$

where h = unit costs, $V(h,\alpha)$ is the covariance of unit cost and product quality α, V(h) is the variance in unit costs, V_o is a constant relating quality adjusted price to product quality, δ and f are parameters describing the environment in which firms operate and the type of competition. In the particularly simple case in which all products are of the same quality (43) reduces to (44):

$$\frac{dh}{dt} = \frac{\delta f}{f+\delta}V(h) \tag{44}$$

which means that average practice unit cost declines at a rate proportional to the variance of unit costs within the technology set (Metcalfe, 1987, p27; Nelson, Winter, 1982, p243). Similar results can be obtained for other characteristics of the population. If one assumes that the variance of unit cost (or of another population characteristic) is higher during the initial phases of the evolution of a technology and declines as the technology matures it follows that the rate of fall of unit cost will fall during the evolution of the technology. This would lead to a pattern of development in which mature technologies would have, for example, a lower rate of productivity growth than new technologies. This type of evolution is compatible with the changes which can be expected to take place within a paradigm. Thus the pattern of convergence on a dominant design is likely to be accompanied by a gradual increase in industrial concentration. Such an increase is particularly fast in the case of industries characterised by high technological opportunity (Nelson, Winter 1982). As the many competing designs and at least some of the competing firms which use them are eliminated by competition, the surviving firms are likely to have not only a higher average productivity but a lower productivity variance. This type of evolution could saw the seeds of its own demise. It could cause a switch to a different technology and a different paradigm and away from one which had been previously dominant due to increasing returns to adoption and to the consequent path dependence (Arthur, 1988). In particular for what concerns the relationship between individual technologies and the whole economic system the decline in variance of population characteristics and consequently in the rate of change of their average values by leading to switches to new paradigms and technologies can significantly affect the output variety of the whole system. This can happen in two ways: first,the increasing efficiency internal to each technology/sector which allows the accumulation of the resources required for the introduction of new technologies/sectors and, second, the slowdown in productivity growth internal to each technology/paradigm as it matures can induce a switch to a new technology/paradigm. In other words, even within a population perspective the trends towards declining output variety and falling rate of productivity grow at a sectoral level can be compatible, and indeed complementary with respect to the trend towards increasing output variety at the level of aggregation of the whole system.

5. ROUTINES

Nelson and Winter (1982) give a sophisticated discussion of the term routine and of its implications for what concerns organisations. They define routines as a general term for all regular and predictable behavioural patterns of firms (p14). Such a definition includes a very large number of routines at very different hierarchical levels within organisations and with very different levels of complexity. Elaborating slightly on their definition one could say that routines are constant patterns of *internal* activities of firms undertaken in response to *environmental stimuli* occurring within a predetermined range.

Some of the negative meanings associated with the term routine and with bureaucrats as routine implementers can be understood in this sense. Thus people (external to organisations) sometimes wonder why their problem, which they consider novel and different, is handled by means of existing routines, designed for different problems (different range of external stimuli).

Routines could therefore be redefined as the union of a set of environmental stimuli with a set of internal organisational responses. The main feature of routines is that the set of internal organisational responses can cope with a large number of external stimuli. This may mean either a relatively constant type of stimulus occurring very frequently or a relatively large range of different types of stimuli. Naturally routines of different flexibility would be required to cope with these two situations. However flexible routines are they tend to have a considerable intertemporal invariance with respect to stimuli coming from their environment, otherwise they would not be routines.

Nelson and Winter (1982) discuss a large number of aspects of routines. Routines can be the memory of an organisation, its skills, a truce among different groups in the organisation or its genes. Also, they discuss the relationship of routines to innovation. The two concepts are not opposites, as it is commonly implied: first, the application of prevailing routines can create puzzles which lead to innovation; second, many innovations are, at least partly, constituted by different combinations of existing routines, in a way that Nelson and Winter find reminiscent of Schumpeter's new combinations of means of production by the entrepreneur (1934, pp. 65–66). Furthermore they interpret heuristics and changes in technological paradigms as the analogue of routines. Routines, heuristics and the technological trends discussed in the previous section can therefore be interpreted as examples of the the stable patterns of behaviour which is followed by organisations.

In a way it may seem surprising that what one could consider as a particular feature of organisations (routines) or of technologies (technological trajectories, paradigms) share some general characteristics. In reality if one considers organisations and technologies as complex systems these regular patterns of behaviour are nothing else than examples of self-regulation and constitute the analogue of what for biological systems is *homeostasis* (Saviotti, 1986). Such patterned behaviour is the logical requirement for the stability of complex systems. A system whose internal structure fluctuated wildly in response to environmental stimuli would have no coherence and no stability.

Thus the concept of routines is of quite general applicability and has many important implications for the behaviour of organisations. For the more restricted purposes of this paper it is important to try and analyse the implications of routines for the information requirements of organisations. In order to do this we start from a definition of heuristics as any principle or device that contributes to the reduction in the average search to solution (Newell, Shaw,

Simon, 1962, p85). In a similar way routines, by sticking to a predetermined course of behaviour in response to a set of environmental stimuli, reduce the search required to 'process' the stimulus. If we imagine an organisation without routines we can realise that the search, and therefore the information required to process a given stimulus, would be much wider in absence of routines. We can therefore argue that one of the functions of routines, although not the only one, is to contribute, together with other organisational devices, to information saving.

Naturally this is only one of the functions of routines. To the extent that they contribute to the observed behaviour of organisations they do so based on all their functions. What can be done in this paper is to isolate the implications of the need to economise the information required in given productive processes. One could therefore say that if to use efficiently information was the only goal of organisations, routines and heuristics would exist anyway. Naturally one could not from this infer that they would have the same form that they have in real life organisations.

In principle one an expect routines to change depending on the functions that they have to perform. For example, we can expect different routines when they function as truce an when they function as information savers. The real routine would be a compromise between the two. However, it is difficult to imagine a prevailing routine as being highly information inefficient whoever are the groups jostling for power within the organisation. If this were to happen it would always be possible for one of the groups to propose a change to a more information efficient routine and to improve its power position in the subsequent truce. This is to say that there is not necessarily a conflict between the best routines which satisfy different functions.

6. ELEMENTARY MECHANISMS IN TECHNOLOGCAL EVOLUTION

It is analytically useful to be able to think of a process as complex and multiform as technological evolution in terms of a common set of concepts that can be used in a 'modular' way to represent it. Technological evolution is essentially a process of transformation (ie qualitative change from something into something else). It would then be useful to have some *elementary* transformations or *elementary* mechanisms such that by means of various combinations of them one could reconstruct any complex process of technological evolution.

A number of these elementary mechanisms exist. For example substitution, the emergence of a completely new product and specialisation are mechanisms which form part of the evolution of most if not all technologies. Therefore, they satisfy the requirements previously outlined. In a similar way one could consider the introduction of new forms (innovation) as opposed to their diffusion.

Most of these mechanisms are generally loosely defined and they have to be reconceptualised to be used in this context. This can be done by means of a twin characteristics approach developed by Saviotti and Metcalfe (1984). In this approach a product is conceptualised as the combination of two sets of characteristics one describing the internal structure of the technology (technical characteristics) and the other one describing the services performed by the product (service characteristics). The two sets of characteristics are linked by a pattern of *correspondence* or *imaging*. This pattern of imaging is very closely related to the concept of *design* because it is related to the transformation of the internal structure of the technology (technical characteristics) into the services performed for its users. Alternatively one can think of technical characteristics as those characteristics which can be *directly* influenced by engineers and designers in order to obtain *indirectly* changes in performance (service characteristics) (Saviotti, 1988b). Finally, demand is generally likely to be demand for service characteristics rather than for technical characteristics. The following is a simplified representation of a product based on the previous conceptual scheme:

$$(X_{ij}) \longleftrightarrow (Y_{ip}) \tag{45}$$

where (X_{ij}) and (Y_{ip}) represent the two vectors of technical and service characteristics of the *ith* product and the arrow represents the pattern of imaging.

On the basis of this conceptual framework the previously mentioned mechanisms can be adequately conceptualised and their implications for variety and entropy can be discussed. Let us imagine a number of examples in each of which two products, a pre-existent one P_1 and a new one P_2, are involved (Saviotti, 1988a). Depending on the relationships between the two products a number of situations can arise.

6.1 Pure substitution

P_2 has a completely different internal structure with respect to P_1 and therefore a qualitatively different set of technical characteristics, but supplies qualitatively identical services and can be represented by the same service characteristics. This situation can be represented as follows:

$$P_2 = [(X_{2j}) \longleftrightarrow (Y_{21})];\ P_1 = [(X_{1j}) \longleftrightarrow (Y_{21})] \tag{46}$$
$$X_{2j} \neq X_{1j}\ ;\ Y_{21} \equiv Y_{1i}$$

If P_2 supplies the common services at a lower cost it is going to be preferred by all users. Therefore P_2 will replace completely P_1. This is an example of *pure substitution*. In this case the output variety V_p of the whole system would not be affected.

6.2 The emergence of a completely new product

In this case P_2 has a completely different internal structure and supplies completely new services. Therefore P_2 needs qualitatively different sets of both technical and service characteristics to be represented:

$$P_2 = [(X_{2j}) \longleftrightarrow (Y_{21})]; [(X_{ij} \longleftrightarrow (Y_{1i})]$$
$$X_{2j} \neq X_{1j} \; ; \; Y_{21} \neq Y_{1i} \qquad (47)$$

If this new product is produced alongside existing products it will obviously lead to an increase in output variety V_Q and consequently in users' information requirements.

6.3 Specialisation or market segmentation

In this case P_2 has a completely different internal structure with respect to P_1 and supplies services which are partly qualitatively different and partly qualitatively identical to those of P_1:

$$P_2 = [(X_{2j}) \longleftrightarrow (Y_{2i})] \; ; \; P_1 = [X_{j1} \longleftrightarrow (Y_{1i})] \qquad (48)$$
$$X_{2j} \neq X_{1j} \; ; \; Y_{21} \neq Y_{11} \; ; \; Y_{22} \neq Y_{12} \; ; \; Y_{23} \neq Y_{13};$$
$$\ldots\ldots Y_{2,i-1} = Y_{1,i-1} \; ; \quad Y_{1,i} \neq Y_{2,i}$$

Users will prefer either P_2 or P_1 depending on the services in which they are most interested. For example, if:

$$Y_{22} >> Y_{12} \qquad (50)$$

Users who have a particular interest in the second service characteristic will prefer P_2. On the other hand, if:

$$Y_{21} << Y_{1l} \qquad (51)$$

users particularly interested in the *lth* service characteristic will prefer P_1.

In this case P_1 and P_2 are likely to survive alongside, each one creating its own niche or segment of the market. The output variety of the system is increased by the emergence of P_2, but not as much as it would have been increased by the emergence of a new product, because some of the service characteristics of P_2 coincide with those of P_1.

6.4 Product evolution or product diversification

In this case P_2 has an internal structure very similar to that of P_1 and supplies the same services, but in addition it supplies some new services:

$$P_2 = [(X_{2i}) \longleftrightarrow (Y_{2m})]; \; P_{21} = [(X_{ij}) \longleftrightarrow (Y_{1i})] \tag{52}$$

$$(\text{i max} > \text{j max}); \; (\text{m max} > 1 \text{ max})$$

$$X_{21} = X_{11}; X_{22} = X_{12} \; ; \; \ldots\ldots \; ; X_{2j} = X_{ij}$$

but no X_1 equal to $X_{2,\,j+1}, \ldots\ldots, X_{2i}$

$$Y_{21} = Y_{11}; Y_{22} = Y_{12} \; ; \; \ldots\ldots \; ; Y_{2i} = Y_{1i}$$

but no Y_1 equal to $Y_{2,i+1}, \ldots\ldots, Y_{2m}$.

This case is similar to specialisation. Depending on how novel and important the new services are and on what trade-offs there are with the old services, P_2 will either become a partial substitute for P_1 or it will specialise in a different niche.

Examples of these four types of mechanisms and of their implications for variety can be found in Table 1.

7. COMPETITION

The generation of new products and services leading to an increasing variety has been considered in this paper one of the most important features of economic development. However, not every new product when launched is successful. Mechanisms of selection and competition screen new products and only some survive. The existence and nature of competition therefore influences the variety of the whole system.

There are different types of competition. One of them is perfect competition. In this type of competition firms are price takers, there is free entry which establishes a position of normal profitability, and all firms produce a homogeneous and equal product. A different type of competition, called *innovation competition*, is described by Metcalfe and Gibbons (Metcalfe, 1984; Gibbons, Metcalfe, 1987; Metcalfe, 1987). This type of competition is driven by technological differences between firms which has as its outcome continuous changes in the relative economic performance of firms (ibid). Clearly the latter type of competition is much more reminiscent that the former, of the Schumpeterian introduction of 'new combinations' by entrepreneurs (Schumpeter 1934, pp. 65–66).

Table 1. Examples of different elementary mechanisms and their contribution to output variety $V_Q - P_2$ = new product, P_1 = existing product

Process	*Example*	ΔV_Q
P_2 completely substitutes P_1	P_2 = digital watch supplying time and date P_1 = mechanical watch supplying time and date P_2 cheaper	0
P_2 almost identical to P_1 except for few new technical and service characteristics	P_2 = digital watch supplying time, date, stopwatch P_1 = digital watch supplying time and date	>0 small
P_2 new internal structure, some services similar to P_1, some new services	P_2 = digital watch supplying time, date, stopwatch P_1 = mechanical watch supplying time, date	>0 larger
P_2 new internal structure, new services	Personal computers, 1976–79	>0 maximum

An additional difference between different types of competition has been pointed out by McNulty (1968). The concept of competition is often used with two different meanings: on the one hand it represents a *force* which by equating prices and marginal costs assures allocative efficiency in the use of resources (analagous to gravitation in the physical/services); on the other hand it is a descriptive term characterising a particular (idealised) situation. In the latter sense competition is not an ordering force but an *assumed state of affairs* (similar to the concept of perfect vacuum in physics). This distinction is relevant because innovation competition is much more a force than an assumed state of affairs.

Perfect and innovation are two idealised types of competition. One can consider them as the two extremes of a range of possible types in which all real life competitive situations can be found. This can be understood if one considers multicharacteristics products as the typical output of industrial firms. If the output of a given set of similar firms can be described by the same n technical and m service characteristics then their representative points in a characteristics space (X space) will occupy close but not identical positions. In other words the population of firms using a given technology will produce a population of output (Revealed technological performance, RTP) characterised by a given dispersion (Saviotti, 1988b). The greater the dispersion, the larger the region of X space in which the outputs of different firms will be found, the greater the average distance of points representing different

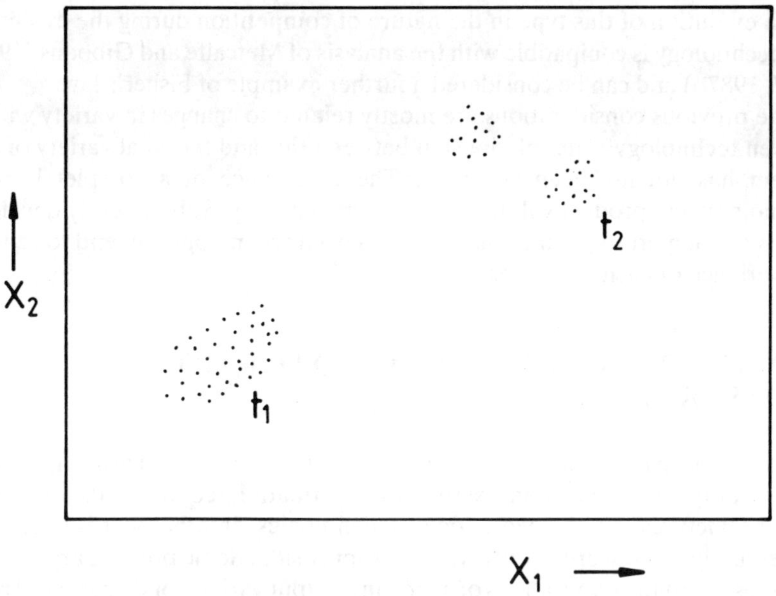

Figure 3. Representation of the changes in a technological population between times t1 and t2 in a bidimensional X space.

products, the lower their similarity. Hence as the population of products becomes more dense and compact in X space the more similar the outputs of different firms become (Figure 3). If such an evolution were to take place in the course of time we would say that competition in the technology considered has become less similar to innovation competition and more similar to perfect competition. If a very different product were introduced by one competitor its representative point in X space would be very far apart from those representing existing products and would therefore raise again the dispersion of the population of technological outputs. (See Fig 3 for a bidimensional representation of this situation).

In another way we would say that innovation and competition acting as a force will tend to destabilise the system by generating new types of output, by giving early innovators an advantage and by creating an adjustment gap (Metcalfe 1981, 1982). An increased dispersion and therefore an increased variance of the population of outputs in X space will result from this. However, in the course of time imitation and diffusion will lead to a fall in the dispersion and therefore in the variance of the points representing the population of technological outputs. Innovation and diffusion therefore act in opposite ways on the variance of the population of technological outputs. In particular diffusion tends to move it towards the extreme of perfect competition. Examples of such changes have been found for the dispersion of the populations of motor car (Saviotti, 1985) and aircraft (Saviotti, 1988b) technologies.

An evolution of this type in the nature of competition during the evolution of a technology is compatible with the analysis of Metcalfe and Gibbons (1984, 1987, 1987b) and can be considered a further example of Fisher's law.

The previous considerations are mostly related to changes in variety within a given technology. The relationship between this and the total variety of the system has already been sketched. The emergence of a completely new technology and product will raise variety and entropy. Subsequently developments in competition and diffusion within each technology will end to reduce the variance of each technology.

8. VARIETY AND THE DEMAND FOR GOODS AND SERVICES

The previous part of the paper dealt with the implications of a growing output variety of the whole economic system for information requirements, organisational structures and the evolution of technologies. In other words so far the paper has been concerned only with the supply side. Some preliminary considerations about the implications of a growing output variety for demand will now be introduced. The purpose of these considerations will be simply to show that the same type of analysis can be used for demand as it has been used for supply. The analysis itself, however, will not be complete by any means.

From the beginning of this paper it has been assumed that a growing output variety would be beneficial to consumers. To be able to use a wider range of products and services can be considered and advantage. However, it is not necessarily an unmixed benefit. To be able to choose only one product limits the satisfaction or utility that one obtains from purchases, but it has associated with it very limited uncertainty and, therefore, very limited information requirements. In other words, a consumer faces information costs (see for example Stigler, 1961) to scan what the market offers, to analyse product characteristics and to compare them to his/her demands. These information costs are likely to rise with output variety, although not necessarily linearly with it (Saviotti, 1988a). One can expect mechanisms to be developed to decrease the information costs required to enjoy the services of a unit of output variety. Better education, specialist publications and advertising supply consumers with an increased flow of information about the goods and services that they are supposed to choose. On the other hand, trade marks, brand names and even the reputation of particular chains of shops and supermarkets function as information savers because they replace the information that consumers would use to evaluate the quality of the goods and services under consideration. A user could simply choose the output of a known trade mark instead of all the goods produced by different manufacturers. In this way the user would not need to have a detailed knowledge of the internal structure of the product or even of its services in order to make a choice. The advertising, trade mark, etc, would function as a source of expert knowledge reducing the quantity of information needed to make a choice while enjoying the services of a greater variety of goods.

9. SUMMARY AND CONCLUSIONS

Variety has been considered in this paper a very important variable characterising economic developments. If qualitative change is one of the most important features of economic and technological evolution then variety, defined as the number of distinguishable products and services in the economic system, is extremely well suited to describe this evolution. It has been argued that in the long term economic and technological evolution leads to a continuously increasing variety of the economic system. Obviously if a process leading to an increasing variety of the economic system takes place then this increasing variety must be beneficial for at least some of the actors in the system itself. Consumers must derive some benefits otherwise, in absence of coercion, they would not buy the increasing variety of output produced. If one then starts from the assumption that consumers welfare is positively affected by the increasing variety of new goods and services a number of problems have to be solved. On the one hand output variety V_q will lead to increasing information requirements and therefore costs. On the other hand consumers themselves will face growing search costs in making a choice of goods and services. One can expect that in the process of economic evolution there will be tendencies to reduce information costs both on the supply and on the demand side. Changes in information costs of these types can be better analysed by using together with the concept of variety that of entropy. On the supply side information costs an be reduced either by changing organisational structures, thus intrinsically reducing the quantity of information to be processed, or by leaving the quantity of information to be processed unchanged but by adopting more efficient technologies to process it. The latter option is clearly the one that is exploited by information technology. In general it is to be expected that the two options will be used simultaneously and interactively. However, it is still useful to separate them for analytical purposes. This paper has been mainly concerned with changes of the former type.

Using the concept of entropy it has been demonstrated thatin absence of coordination problems division of labour will decrease the information requirements of a given process. Similarly, a change in organisational structure in which subunits of some type are introduced into a homogeneous organisation leads to a reduction in information requirements. In particular it was shown that the transition from a U form/multifunctional to an M form/multidivisional organisation leads to a fall in the entropy of the organisation itself.

At a different level of aggregation processes of technological development and the emergence of dominant designs/regimes, trajectories/paradigms/ guideposts were examined for their implications on variety. In a number of cases, although not always, these processes lead to a fall in output variety in the course of time. However the possibility that output variety can fall in particular fields of technology does not imply that it cannot simultaneously increase for the economic system as a whole. A possible combination of these developments at different levels of aggregation can be a fall in the costs (information and general) in each field of technology which allows the accumulation of the resources required to generate new products and services and therefore an increase in variety at the highest level of aggregation.

Amongst elementary mechanisms of technological development some (pure substitution) do not contribute to overall variety while others (specialisation, product diversification, emergence of new products) contribute to different extents. In order for the overall variety of the system to increase variety increasing processes must predominate over substitution.

The general framework described in this paper, based on the concepts of variety and entropy, can lead to analytical models of processes of economic and technological evolution. The concepts of variety and entropy are of very general significance. They are very closely related to what can be considered the central feature of economic and technological evolution, namely the qualitative changes which continuously transform the system and which had already been recognised by Schumpeter as the emergence of new 'forms' from old ones (1934, pp. 65–66). In general this approach is also closely related to modern developments in the non-equilibrium thermodynamics (Prigogine, 1976, 1987; Prigogine Stengers, 1984) which are heavily based on the concept of entropy. According to this approach an open system as it moves away from equilibrium can undergo a series of transitions to a greater and greater multiplicity of stable states, which can be more and more ordered. Some analytical applications have been outlined in this paper but many more can be developed.

Finally, the examples which have been considered show that in many cases the tendency to economise information in given productive processes, technologies and organisational practices, leads to organisational structures and patterns of development which are very similar to those observed in real life examples. In other words, one could say that if the tendency to economise information as defined in this paper was the only factor determining the structure of organisations, such structure would be very similar to the one we can observe in real organisations. This means that either information is one of the most important factors determining organisational structures or that the other factors which could influence them, tend to produce the same type of organisational structure. Naturally it is impossible to give an answer to these questions in this paper because the other factors would have to be taken into account. However, the analysis of one of the factors constitutes a beginning and helps to formulate the problem.

REFERENCES

W.J. Abernathy, J.M. Utterback, A Dynamic Model of Process and Product Innovation in: *Omega,* **3** (6) (1975) 639–656.
W.B. Arthur, 'Competing technologies: an overview' in: G Dosi, et al (eds) *Technical Change and Economic Theory*, (London, Pinter, 1988).
W.R. Ashby, *An Introduction to Cybernetics*, (Methuen, London, 1964).
T. Burns, G.M. Stalker, *The Management of Innovation*, (London, Tavistock, 1961).
E.J. Chamberlin, *The Theory of Monopolistic Competition*, (Harvard University Press, Cambridge, MA, 1933).
A.D. Chandler, *Strategy and Structure*, Cambridge, Mass., MI Press (1962).
A.D. Chandler, *The Visible Hand*, Cambridge, Mass, Harvard University Press (1977).
R. Coombs, et al, *Incremental Innovation in the UK Tractor Industry*, (Manchester University, PREST, 1978).
G. Dosi, Technological Paradigms and Technological Trajectories: a Suggested Interpretation of the Determinants and Directions of Technical Change, *Research Policy,* **11** (1982).
R. Duncan and A. Weiss, Organisational Learning: Implications for Organisational Design in: *Research in Organisational Behaviour,* **1** (1979) 75–123.
R.A. Fisher, *The Genetical Theory of Natural Selection*, (Drew, New York, 1958, originally published 1929).
L.L. Gatlin, *Information Theory and the Living System*, (Columbia University Press, New York, 1972).
J. Marschak, Economics of Inquiring, Communicating, Deciding, *American Economic Review Papers and Proceedings,* **58** (1958) 1–18.
J.S. Metcalfe, On the diffusion of innovation and the evolution of technology in: B.R. Williams, J.A. Bryan-G/Brown (eds), *Knowns and Unknowns in Technical Change*, (London, Technical Change Centre, 1985).
J.S. Metcalfe, 'Impulse and diffusion in the study of technological change', *Futures*, Vol **13** (1981).
R. Nelson and S. Winter, In Search of Useful Theory of Innovation in: *Research Policy*, (1977) 36–76.
R. Nelson and S. Winter, *An Evolutionary Theory of Economic Change*, (Cambridge MA, Harvard University Press, 1982).
A. Newell, J.C. Shaw, H.A. Simon, 'The process of creative thinking' in H.E. Gruber, G. Terrell, M. Wertheimer (Eds) *Contemporary Approaches to Creative Thinking*, New York, Atherton Press (1962).
J.S. Nicolis, 'A study program of chaotic dynamics applied to information processing' in: I. Prigogine, M. Sanglier (Eds), *Laws of Nature and Human Conduct*, (Bruxelles, GORDES, 1987).
L.L. Pasinetti, *Structural Change and Economic Growth*, (London, Macmillan, 1981).
I. Prigogine, Order through fluctuations in Self-Organisation and Social System, in: E. Jantsch and C.H. Waddington, *Evolution and Consciousness: Human Systems in Transition*, (Addison Wesley, New York, 1976).
I. Prigogine, A new rationality? in: I. Prigogine, M. Sanglier (eds), *Laws of Nature and Human Conduct*, (Bruxelles, GORDES, 1987).
I. Prigogine, I. Stengers, *Order out of Chaos*, (London, Fontana, 1985).
J. Robinson, *The Economics of Imperfect Competition*, (Macmillan, London, 1933).
P.P. Saviotti, Information, variety an entropy in technoeconomic development in: *Research Policy*, Vol **17** (1988a) 89–103.
P.P. Saviotti, 'A characteristics approach to technological evolution and competition' presented at the Conference on Recent Developments in the Economics of Technological Change, Manchester, 22 and 23 March 1988.
P.P. Saviotti and J.S. Metcalfe, A Theoretical Approach to the Construction of Technological Output Indictors, in: *Research Policy*, **13** (1984) 141–151.
P.P. Saviotti, R.W. Coombs, M. Gibbons, P.C. Stubbs, *Technology and Competitiveness in the Tractor Industry*, (Manchester University PREST, 1980).
D Sahal, Alternative Conceptions of Technology, *Research Policy,* **10** (1981) 2–24.

J. Schumpeter, *The Theory of Economic Development*, Cambridge, MA, (Harvard University Press, 1934, original edition 1912).

J. Schumpeter, *Capitalism, Socialism and Democracy*, (George Allen an Unwin, 1943, 5th Edition 1976).

C.E. Shannon and W. Weaver, *The Mathematical Theory of Communication*, (University of Illinois Press, Urbana, 1949).

H.A. Simon, The Architecture of Complexity, *Proceedings of the American Philosophical Society,* **106** (1962) 467–482, reprinted in: H.A. Simon, *The Sciences of the Artificial*, (MIT Press, Cambridge, MA,1981).

G.J. Stigler, The Economics of Information in: *Journal of Political Economy* **69** (1961) 213–25. Reprinted in: D.M. Lamberton (ed), *Economics of Information and Knowledge*, (Penguin Books, Harmondsworth, 1971).

R. Vernon, International Investment and International Trade in the Product Cycle in: *Quarterly Journal of Economics*, Vol **80** (1966) 190–207.

O.E. Williamson, *Markets and Hierarchies: Analysis and Antitrust Implications*, New York, Free Press (1975).

O.E. Williamson, 'Transaction costs economics: the governance of contractual relations' *Journal of Law and Economics*, Vol. **22** (1979) 233–261.

O.E. Williamson, 'The modern corporation: origins, evolution and attributes' *Journal of Economic Literature*, Vol. **19** (1981) 1537–1568.

A Computer Simulation of Economic Growth and Technical Progress in a Multi-sectoral Economy

STEPHEN SMITH
Liverpool Polytechnic, U.K.

1. INTRODUCTION

This paper takes as its aim the analysis of the micro foundations of economic growth, focussing particularly on the role of technical progress. The discussion describes a non-equilibrium, evolutionary, approach to the issues, based upon the behavioural theory of the firm and a dynamic theory of competition. These principles are embodied in a microeconomic based computer simulation model, which is used to describe the technological and industrial development of an economy.

The principal facts to be explained by growth theory are generally agreed. At the macro level we see relatively stable behaviour and approximately steady state growth. When we disaggregate to the industry level it is clear that some sectors have developed much more quickly than others, and that sectoral patterns of growth have varied over time. Within any industry some firms prosper whilst others fail. At the micro level disequilibrium obtains.

Neo-classical models, based on rationality and equilibrium, have serious weaknesses and criticism of that approach to economic problems is widespread. Nowhere is this criticism more justified than in the study of economic growth, which focuses, by its very nature, on change rather than equilibrium. Simon (1986) is unequivocal: "existing uncertainties about the correct explanations of economic growth and business cycles cannot be settled by aggregative analysis within the neoclassical framework." p. 21). He argues that "to build an interesting and useful theory of long term economic growth, even for developed countries, we have to go behind the principle of rationality" (p. 28) Kay (1984) sees the problems a also being rooted in system nondecomposability (that the whole cannot just be regarded as the sum of the parts). Pasinetti (1981) argues that neo classical theory emphasises trade, which is essentially about short run considerations, rather than industry and production, which is dynamic and requires a long run perspective. Our simulation model allows investigation of these features.

Nelson and Winter (1982) ascribe the critical problem of neoclassical growth models to a misapplication of the conventional static theory of the firm in a competitive market. Bounded rationality and uncertainty are principal elements of the innovation process, which has a major role in explaining technical change and economic growth. The very idea of innovation implies something previously unknown, and thus not likely to be part of a well defined choice set. Imitation implies that some firms do things better than others, something difficult to incorporate into a world of perfect knowledge. Nelson and Winter (1974) cite studies showing much evidence of the role of insight in the invention process, of differential ability to use knowledge, of considerable differences among firms in the technology they use and in their profitability.

Once firms are differentiated, competition becomes a dynamic process involving struggle and disequilibrium. In a Schumpeterian model of competition, innovation is a method by which firms may seek to gain advantage over others, and so increase their profit. The development of the industry is now dependent on three forces; profit seeking behaviour by firms in their current production decisions, profit seeking by firms by search over uncertain terrain, and from selection of the most profitable firms through their higher rates of growth. It is the continued diversity of behaviour, as a result of innovation and chronic disequilibrium that drives the process of economic growth (Metcalfe, 1984).

2. ELEMENTS OF AN EVOLUTIONARY MODEL

The main building block of an evolutionary model of economic growth is a 'behavioural' theory of the firm. Here a firm consists of boundedly rational people who together use their knowledge and abilities to operate a set of decision rules, which determine the firm's response to a given environment, given its current state. The rules are the outcome of previous decisions, and are the analogue of genes in biology. Day to day rules on production and pricing are rule of thumb, rather than optimising, in nature. They are stable in the short term, but susceptible to change over time as a result of learning, chance and from goal-orientated search; these are economic mutations. The rules governing the firm's search procedure are necesarily qualitatively different from the others. Search, by its very nature implies dissatisfaction with existing rules. Search is a remnant of behaviour motivated by the profit consequences of a contemplated course of action; the innovating remnant (Winter 1971).

The core of evolutionary theory is the dynamic process by which firm behaviour and market outcomes are jointly determined over time. At each point of time firms exist in given states, a legacy of past decisions. The firm's transition to its state in the next production period is determined by two factors, profitability and innovation. Competitive selection has the function of ensuring that the most profitable firms tend to have higher than average growth rates. The selection mechanism consists of a set of rules which translate each firm's performance, as revealed in the market place, into expansion or contraction. This mechanism will also depend on the relative profitability of firms, and the extent to which firms are able to grow. The efficiency of financial markets determines the extent to which each firm's profits can be transformed into new capacity (eg the Soete and Turner 1984 model).

Innovation comes from success in the firm's search for superior modes of behaviour. The nature of search will be partly determined by the economic environment of the firm. Prices determine both the profitability of firms and the direction of search. Search and selection are simultaneous, interacting, aspects of the evolutionary process. Each firm's actions help determine its future. The condition of the industry at any time contains the seeds of its condition in the future.

As technological progress occurs faster in some industries than others, so this will be reflected in prices, growth and a changing structure of industry. Changing relative input prices, via induced innovation will result in firms changing their production processes to reflect the new economic environment. So long as innovation continues the process as a whole does not converge to an equilibrium but continues to evolve.

3. A SIMULATION MODEL

In this section we describe a model which simulates evolutionary growth in a multisectoral economy. The model economy comprises three industries, each of which contains 10 firms. There is one consumption good and two capital goods. Homogeneous labour is in perfectly elastic supply at a fixed wage. All firms use labour and capital of various vintages as their only inputs. The model uses discrete time periods within each of which a complete round of production, sales and investment takes place. In essence the model is a dynamic input-output model. Its constant technology equilibrium properties are analysed by Hicks (1965).

The principal objective of firms is to grow. The first decision the firm must make is to set its price. The firm's price helps determine two variables, its market share, which decrases as price increases, and profits, which increase as price increases. Firms make two decisions concerning output and capacity. First their current level of investment, in either search or new capacity, which is equal to the maximum that can be purchased from current profits, supplemented by borrowing based on the firm's relative rate of return. Thus in seeking to attain its objective the firm sets its price such as to increase market share as fast as possible, commensurate with acquiring sufficient capacity to satisfy demand. We now look at prices, market shares and demand/production in more detail.

Prices. Two types of price are used; firm's and market prices. We model each market as being organised by a marketing board, which buys from firms at the firm's prices and sells to all customers at a single ruling market price. The market price is calculated as a share weighted average of the firm prices, so that the marketing board operates for nothing. Each firm within the industry determines for itself the price at which it will sell to the marketing board, based on a weighted average of a mark up of its costs and the current ruling market price. The weighting in this calculation reflects the degree of consumer loyalty/ monopoly power we give the firm. The price mark up is determined by the extent to which the firm has adapted its capacity to its demand. It is increased if the firm runs low on stocks.

Market Shares. These are adjusted by the marketing board according to firms' prices relative to market price. Adjustment costs are the reason for not allowing share to increase very rapidly for low price, advancing firms; firms are limited in their ability to absorb new capacity and new processes.

Demand and production. At the start of each production period prices are given, and each industry faces a known demand for its product given total wages and profits in the previous period. Each firm within the industry has also already been allocated its share of the demand.

In making its production decisions, the firm operates a percentage of its profitable capacity predetermined by its rules of thumb. The firm then meets its demand from current output with any difference met by an adjustment of stocks. Justification for this behaviour may be found by supposing various communication and organisational difficulties within the firm.

Thus in this model prices and output are set prior to the start of each period, and it is stocks which adjust to match supply to demand. In particular stocks enable the firm to smooth the response to changing patterns of demand.

3.1 The firm

The firm's initial state each period is governed by three main groups of variables as Figure 1 shows. First its external environment gives the ruling market prices, wage rate and demand. Second its economic and physical state given by the quantity and type of capital stock, the stocks of finished product and the funds available for investment. Third, the firm's current decision rules, determined by previous experiences, for deciding output and price.

At the start of a period the first decision is to operate a given percentage of profitable capacity. The output decision rules depend on two parameters: the current utilisation rate; and the desired capacity utilisation, which the firm aims for in the long term. The desired rate is initially set at 90 percent utilisation, the idea being that the firm cannot efficiently use its capital at 100 percent utilisation since this leaves no margin for breakdowns, unexpected surges of demand or whatever.

Output is equal to the current utilisation rate times current capacity. The firm will use its most productive capital at 100% of capacity and may use less productive capital not at all. Having determined its output, the firm fulfills the demand for its product with any shortfall from output being made up from stocks. If the firm cannot meet its demand without its stocks becoming negative the unmet demand is allocated pro rata with market shares to the other firms in the industry. The firm is then penalised by having its market share reduced, and shares for all firms are increased to compensate.

At the end of the production phase, the firm takes delivery of capital ordered at the end of the previous period. The quantity of new capital ordered is that which can be purchased from operating profit plus borrowing and interest (which is earned on all monies carried over from period to period), minus search expenditure. Borrowing increases with the firm's rate of profit relative to the economy average.

The next stage of the program, but conceptually simultaneous with production, is the firm's search for new processes. We distinguish two types of search activity; basic research to discover a fundamentally new 'basic' production process and incremental research to make improvements to existing techniques,

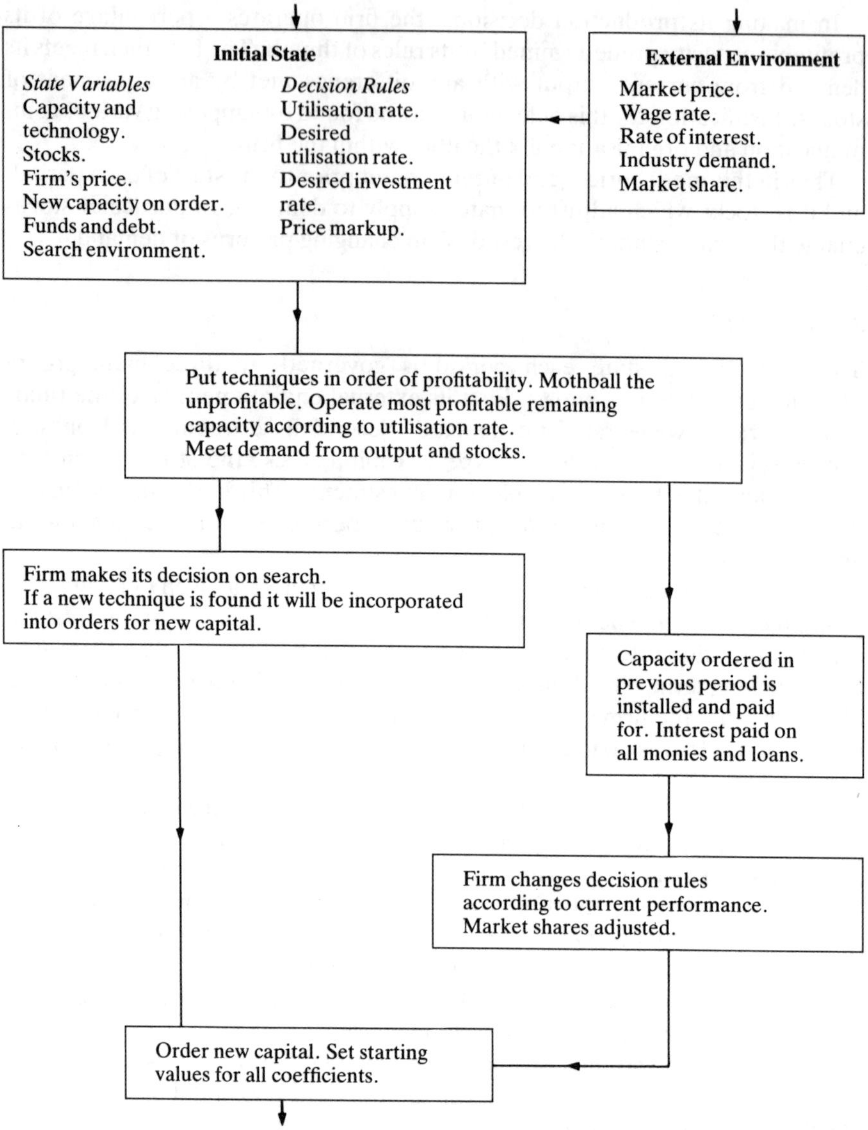

Figure 1. Flow chart of a firm's activities.

(improving on the current basic process). The simulation program explicitly models only incremental search. However, the outcome of that search may be a new basic process if the firm strikes it lucky.

In Figure 2, the firm is currently using basic process (a). Following Binswanger (1974), it has available to it six lines of possible research (of which

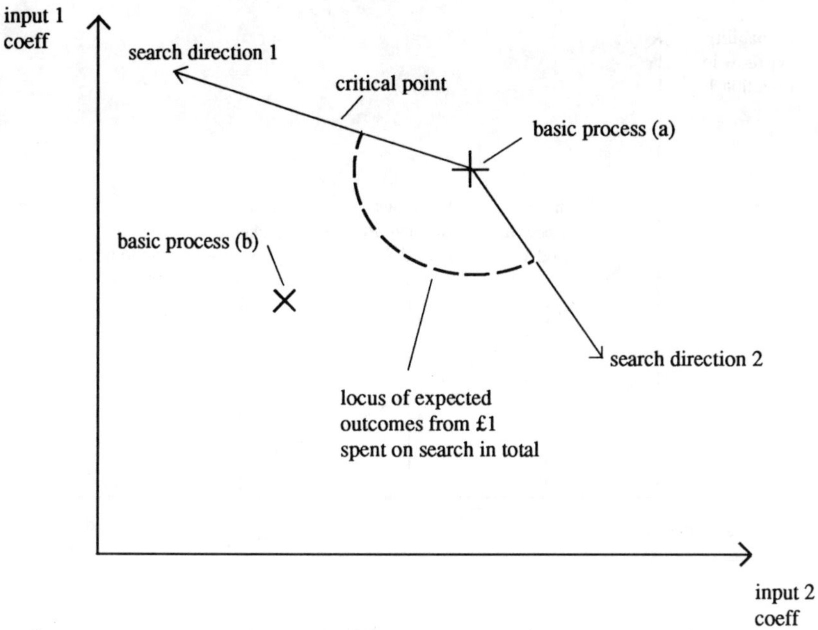

Figure 2. Basic and incremental innovation.

Figure 2 shows two). Each direction reduces one input coefficient at the *expense of increasing another*, the angle in Figure 2 determining the relative weight of changes in each. The firm chooses the intensity of search in each of the possible directions on the basis of the search environment faced and expected prices and output over its time horizon.

The outcome of search is determined according to an exponential distribution, as illustrated by Figure 3. The origin of the probability distribution is the most productive basic process currently in use by the firm, in this case (a). The intensity of search in each period is modelled as the number of draws on the probability distribution, with the search outcome being the best technique reached (Evenson and Kislev, 1976). The firm pays a fixed amount per draw, determined by the type of technology it is investigating.

In the search process, each firm first makes a decision as to what fraction of its investment funds it will devote to new capacity and what fraction it will devote to search. To do this the expected profit from an extra draw in each search direction is first calculated. If the best of these exceeds the current best expected profit, then the number of draws in that direction, for this period, is increased by one, funds for new capacity are reduced by the extra search cost and the sequence of calculations is repeated. Once no additional draws are found to be profitable, the decision on search intensity is complete.

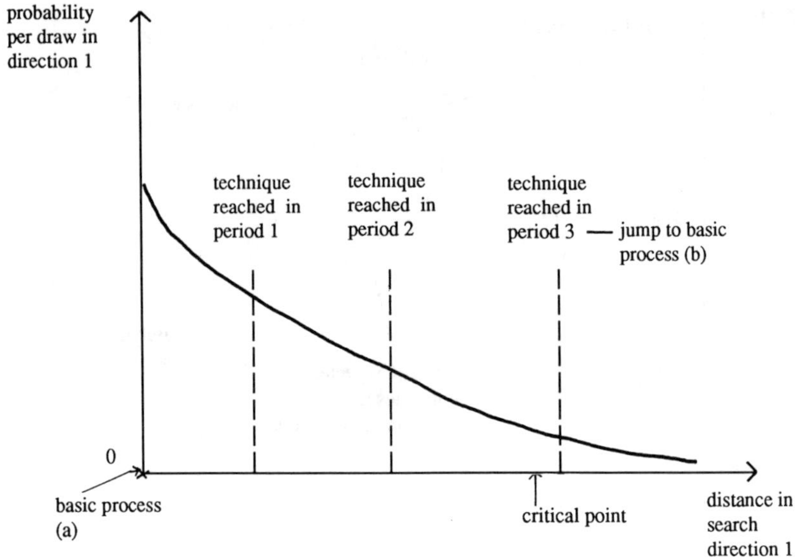

Figure 3. Cross section of Figure 2.: Incremental search in direction 1.

The next stage is to find the actual outcome from search. This involves drawing randomly the chosen number of times, on the appropriate probability distribution. For each search direction, the outcome is the maximum value found from the current sequence of draws. If this exceeds the value already found by the firm in previous periods, in one or more directions, then the firm is deemed to have found a new technique. Finally the new production coefficients are calculated. If only an incremental innovation has been made the new technique has coefficients which are proportional to the previous best technique. At any time all of a firm's new investment will be in its best currently available technique. New techniques do not affect existing capital; technical progress is the embodied type.

The firm perceives the ordinal nature of the probability density function for discovery of a new basic process, that more search increases the probability of success, but is not aware of the cardinal nature of the function. Thus the firm cannot make precise calculations of the optimal amount to invest in search taking this possibility into account. We model this by summarising the potential for fundamental breakthroughs in a single number, the firm's degree of optimism, which increases the perceived variance of the incremental search probability function. Thus an optimistic firm will engage in more search than a pessimistic firm.

The firm is deemed to have attained fundamental breakthrough if it reaches a predetermined distance along one (or more if desired) of the search directions. On reaching such a critical point it then jumps to the next basic process. When a new basic process is discovered its coefficients may be given exogenously, or endogenously as those of the best technique found so far. Having reached such a point the firm is able to engage in further incremental research, starting with a clean slate from a new origin: the production coefficients of the new basic process.

The process of search uses up resources, and these must be accounted for in the simulated economy. A firm expends its investments funds either on search or on new capacity. In the program both activities use the same combination of inputs. In effect search consists of building new plant and testing it to destruction. This may not be very realistic but it is as good a guess as any other as to the resource requirements of research projects.

An important area of policy in the field of research and innovation, is the use of patents. A system of patent rights if included in some simulation runs. Patent rights are awared to the first firm to find any particular basic process. The holder of a patent receives, for a period, a percentage of the market price, as royalty, for each unit of capacity from each subsequent user of that basic process.

The patent system has three important features. First, firms at the current forefront of technology now have the potential of earning royalties, and this encourages them to devote more resources to search. This is modelled by increasing the firm's degree of optimism when making its search decision. Second, in this model a patent encourages the holder to disseminate the knowledge it contains, speeding the diffusion of the innovation through the industry. This is modelled by increasing the effective search variance for those firms not yet using the patent. This means that the diffusion process still takes time. Third, it redistributes funds to patent holders from the other firms in the industry.

Once the production and investment decisions are made and results known, the firm begins to evaluate its rules on production and pricing. Current performance is assessed in two ways: changes in stocks and in market share.

Stock changes are the analogue of responding to delivery lags in a continuous time model. A reduction in stocks increases both price and utilisation ceteris paribus. If the firms were to meet demand conditions in accord with its long run investment plans, so that stocks were always at the desired level, then the price markup will return to zero eventually. (Sufficient rate of return to cover investment for planned growth is included in costs). In the same conditions the utilisation rate will return to the long run desired level.

The simulation model functions adequately with only the single feedback mechanism of destocking. The addition of a second feedback, the change in market share, increases the rate at which technical progress can be absorbed into the simulated economy. It is more forward looking than stock changes.

The feedback from the change in market share works in two ways. A short run effect to improve the stability of the adjustment process; changes in capacity utilisation as a result of changing stocks are reinforced according to the change in market share. Second a long run effect for a firm whose share has increased. Faced with the expectation of being able to grow more quickly, it feels able to take the risk of increasing its desired rate of capacity utilisation; higher growth requiring more planning than faster decline.

3.2 The industry

An industry is characterised by three variables: market price; the total demand for its products; and the vector of market shares for the constituent firms. For the consumer good, demand depends on the total wages carried over from the previous period and the current market price. In this case the market price elasticity of demand is –1. For each capital good industry, demand depends on the total profits carried over and the prices of both capital goods. The individual industry elasticity of demand is thus inelastic, the degree depending on the proportion of capital spending which the industry accounts for. The adjustment mechanisms described above mean that the time path of prices is a second order difference equation. The time paths of output and prices are interdependent, which contributes to the realism of the model.

4. SIMULATION RESULTS

In this section we describe the outcomes from various simulation runs. This will enhance our understanding of technical change in two ways. First, we will see that our model does give plausible results, so that it, and the theoretical understanding is based on, gain in plausibility; simulation is affirmative (McCloskey, 1986). Second, we will have a fuller understanding of the model, and thus to some extent the real system, in that our simulation will suggest one possible way in which economies develop.

4.1 Induced innovation and economic development

The scenario for this section is that all firms engage in search. If and when a firm achieves a certain total distance, summed across all search directions, it is deemed to have discovered a new basic process. The coefficients of the new basic process are those of the best technique so far in use by the firm, thus the technological development of the economy is endogenous, given the search environment. In the absence of technical change the economy would experience steady state growth at a rate of 3%. Industry O produces the consumption good, industries 1 and 2 capital goods 1 and 2 respectively.

Turning to the results themselves we begin at the micro level in a typical industry. Figure 4 shows the time trend of best practice technology as a share weighted average of the best technique in use by each firm. It shows substitution of capital good 2 for labour, with a lesser tendency to substitute capital good 2 for good 1. The final increase in average productivity is 45 percent as six new basic processes are introduced in the industry.

The increase in productivity reflects the major changes in the actual coefficients. The best practice labour coefficient falls by over 50% and the K1 coefficient by 20%. The fluctuations in the trends of the coefficients arise because it is possible that a firm will find different search directions profitable at various stages of the simulation. Each search direction operates to effect a proportionate change in production coefficients. The early search effort is to substitute K2 for labour. As the K2 coefficient increases, it then becomes worthwhile to devote resources to reducing it; induced innovation in practice.

Behind the industry average performance there is the dynamic competition between firms. Figures 5 and 6 show the performance and behaviour of the best and worst firms, classified according to their final market share. Figure 5 shows that by the end of the simulation the best firm is approximately twice the size of the worst. After 150 periods, this is quite a small difference, except when seen in the light of the rapid diffusion paths. Early discovery of a new basic process allows a firm access to finance from both profits and borrowing. Although the best firm is one of the last to find process 2, from then on its one of the first to find each new basic process. The converse is true for the worst firm. The differences between firm prices cannot really be seen from Figure 5, where the scale of the graph is too small to show them. In Figure 6 we see that from period 60 onwards, soon after it gets ahead, the best firm is consistently able to have a higher price markup, whilst at the same time maintaining an increasing market share. The best firm's success is seen to stem from the slightly superior average productivity which it is able to maintain throughout the simulation run, which has a big impact on the financing of new capital. Dynamic competition and economic selection are seen to be operating here.

Figure 6 shows how firms adapt to their changing situation. As firms' relative performance is constantly changing, so market shares adjust. Sharp increases in share coincide with discovery of a new process, small fluctuations being caused by successful innovation in other firms. When the worst firm begins to lose market share rapidly between period 50 to 65, it reduces its price markup and utilisation rate. Most of the change falls on the utilisation rate, which is able to effect a much more speedy response to increasing stocks. The firm is also limited in its ability to cut its price by the need not to lose too high a percentage of its funds through lending.

Figures 7 and 8 show the relative performance of industries and the macroeconomy. We begin by considering the growth in output. Figure 5 shows growth for firms increases rapidly within the first few periods to about 8%

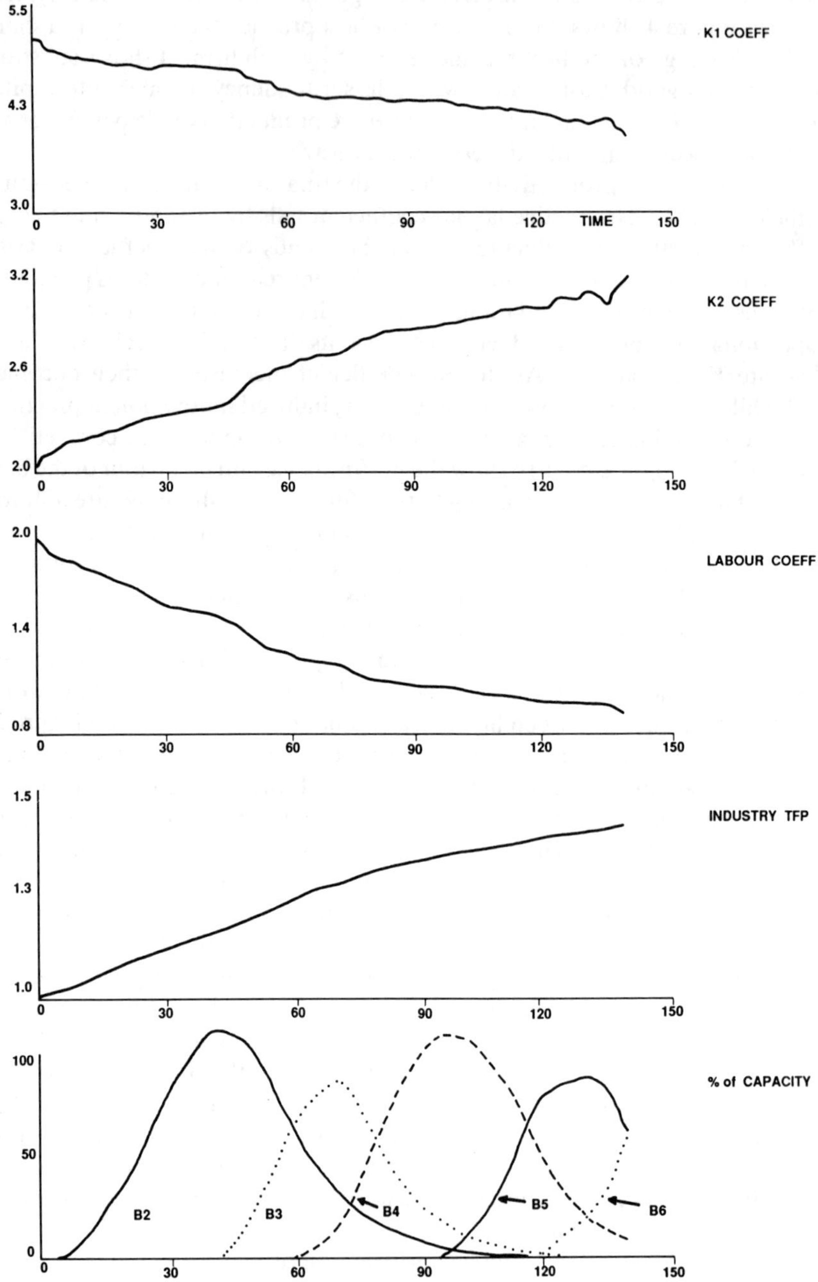

Figure 4. Industry 2 Technology.

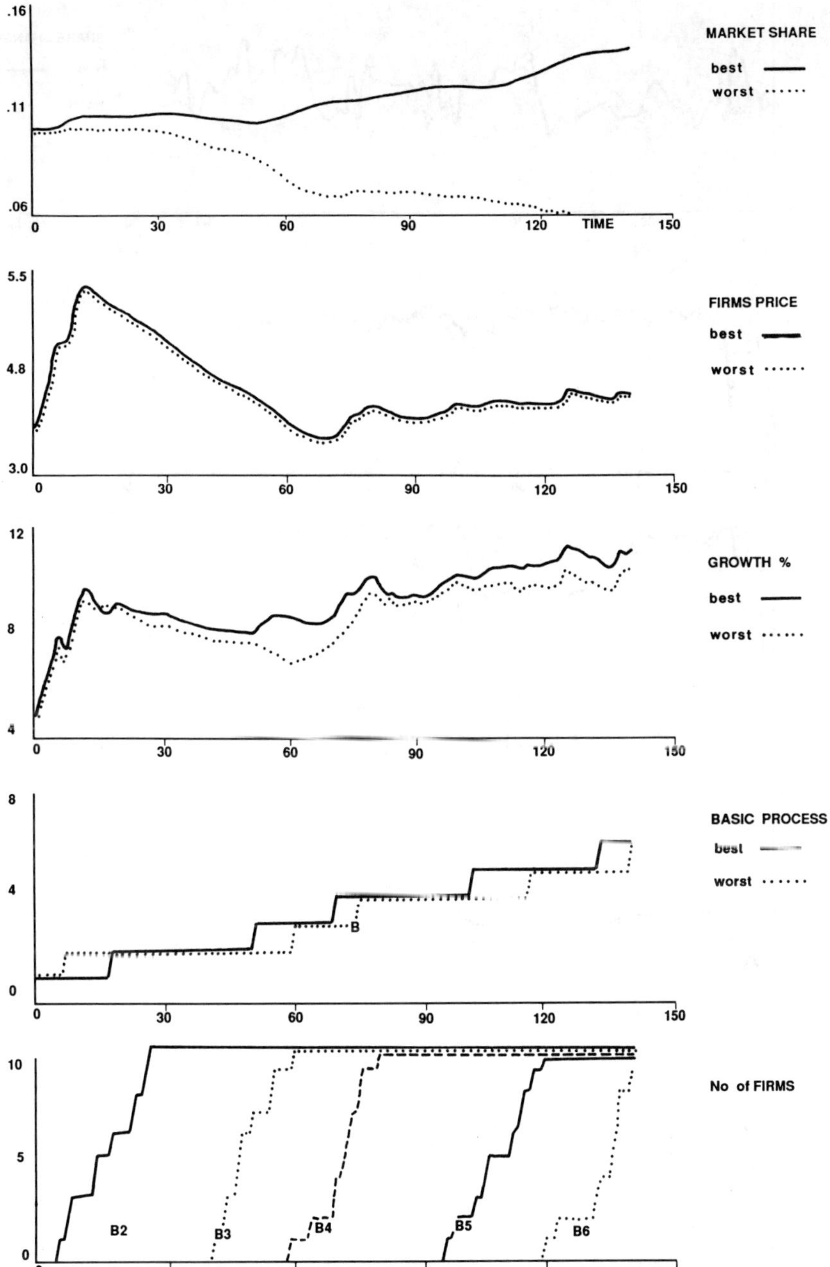

Figure 5. Best and worst performance: Industry 1.

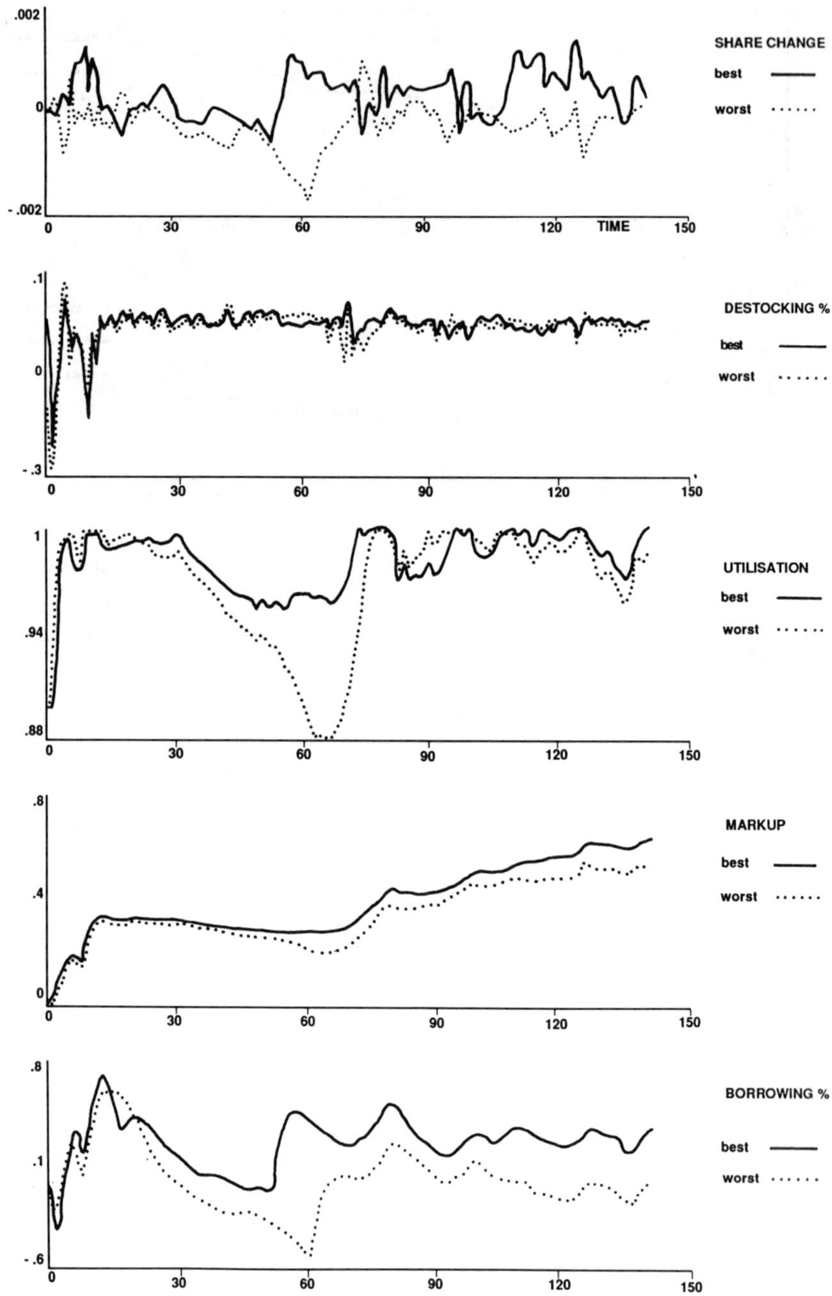

Figure 6. Best and worst behaviour, Industry 2.

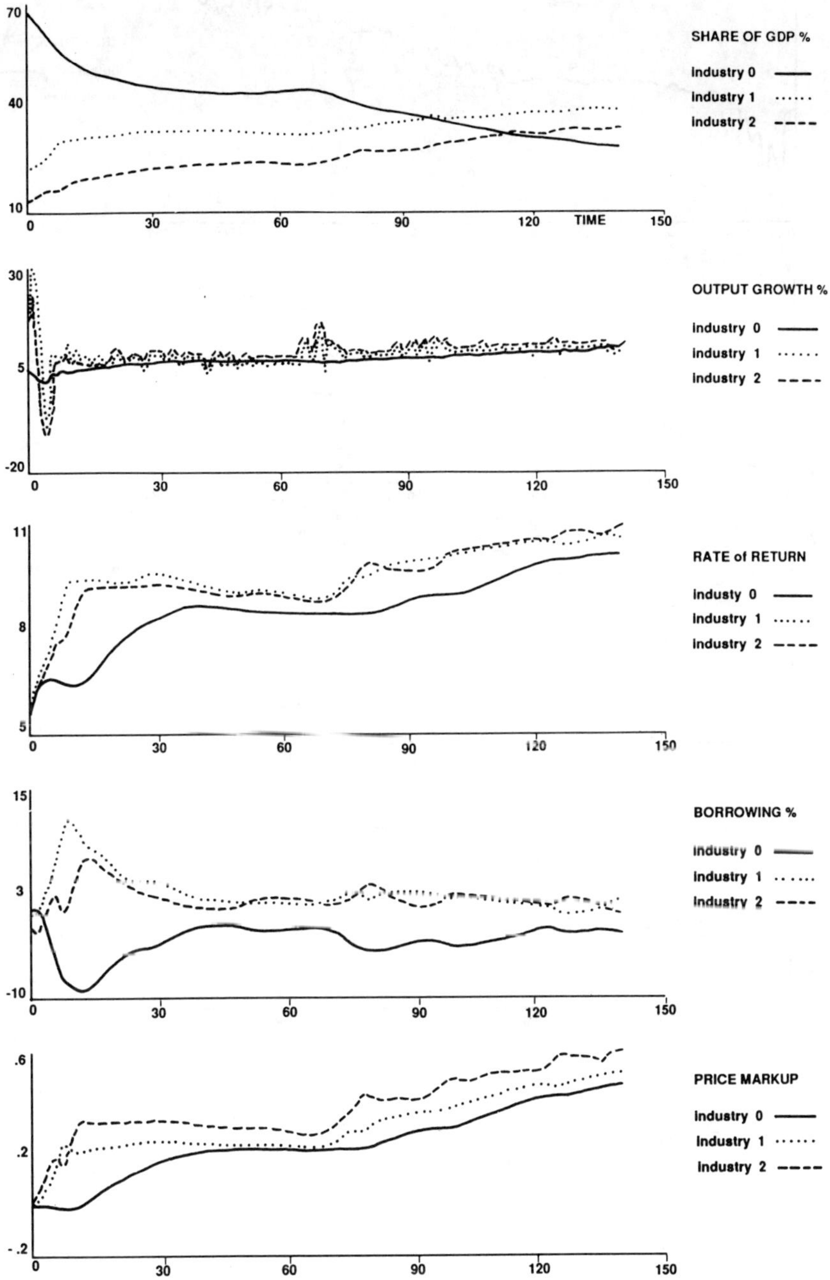

Figure 7. Relative performance of industries.

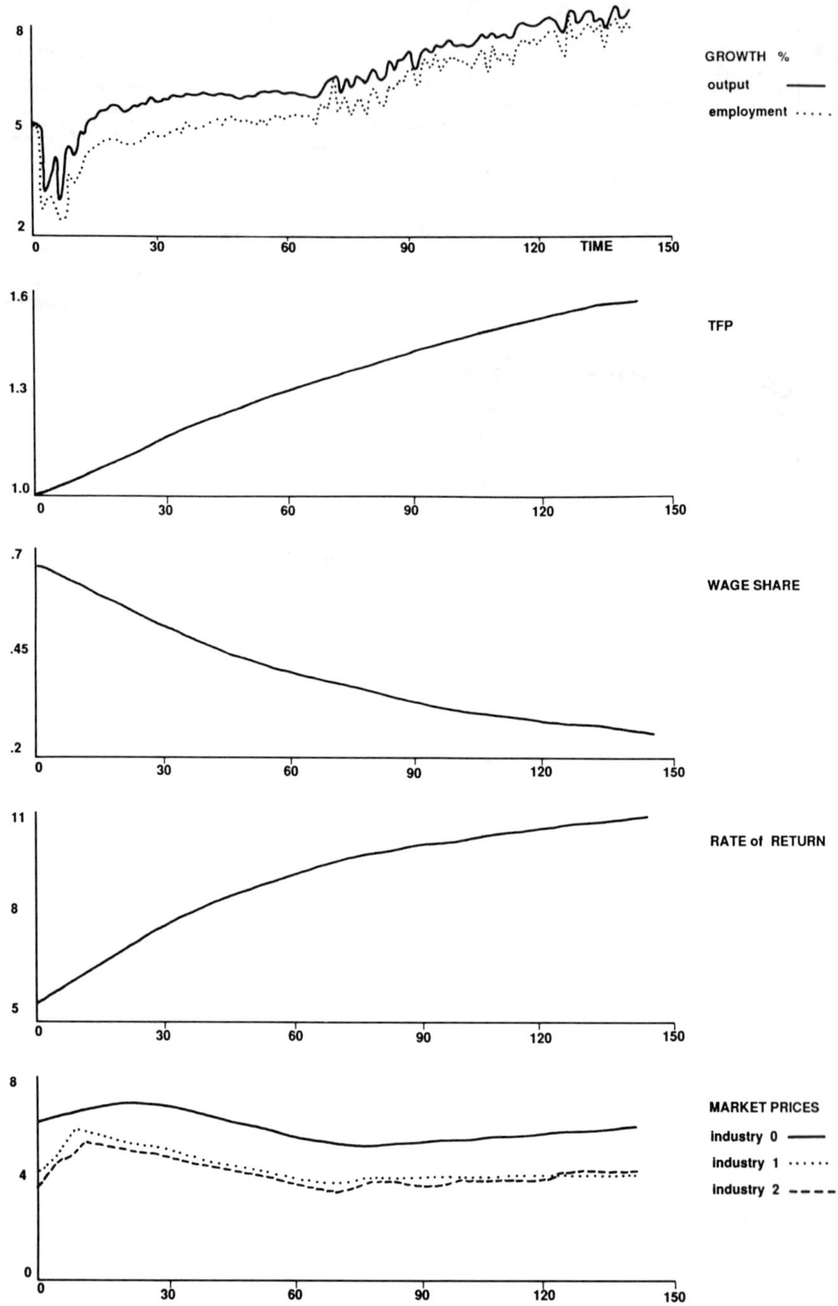

Figure 8. Performance of the economy.

per period, from which level it rises more slowly. Figure 8 shows similar performance for the economy as a whole. The initial cause of this high growth is firms optimism as they increase productivity during the first few periods of the simulation. Each firm acts in isolation, translating its productivity increase into a higher utilisation rate and price markup. This optimism is reinforced as the whole economy behaves in this way. An additional source of growth is the diversion of funds from the consumption good industry to the capital goods industries seen in Figure 7.

Perhaps the most notable feature of Figure 7 is the very high price markups. The effect is that market prices actually increase despite the technical progress. The high price markups are possible because there is no mechanism to slow growth in our model: the economy is closed; there is no monetary constraint; competition is mainly within industries, not between them. The high price markups are an accident of the construction of the model, and do not really affect our story of economic evolution, except in so far as additional induced innovation effects, as a result of endogenously generated input price changes, do not occur. The important point is that the average markup for industry 2 is the highest, reflecting the bias of technical progress towards using K2, whilst the markup of industry O is lowest reflecting the bias way from labour.

The high absolute level of the price markups does have one interesting effect. Technical progress in our model is directed towards reducing the labour input. Thus the growth of employment is below that of GDP, even during the period when prices are falling. Figure 8 shows that the share of GDP accounted for by industry O falls from 68% to 30% during the simulation, in line with the reduction in labour coefficients. Even though the demand for labour is growing more slowly than the economy as a whole, high price markups in the consumption good sector are needed to stop it losing investment funds to the other sectors. This has the effect of reducing the real wage and demand for the consumption good grows less quickly than it otherwise would. The effect of the high price markups is essentially to reduce real wages and divert funds to the accumulation of capacity in the capital goods sectors. If the wage rate is increased, say in line with the increase in total factor productivity (TFP), then the share of wages falls more slowly and the growth rate is reduced.

Impact of a major innovation. Figure 9 shows the impact of an economy wide exogenously determined innovation. The coefficients of the third basic process for each firm are set at 70% of the coefficients of the best technique so far in use by the firm. We might suppose that the third basic process is a superior mode of production, such as production line technology, which is equally applicable in all industries.

Figure 9 shows the effects of the technological revolution. Once the first firms find their third basic process, the growth rate begins to accelerate. The key point from Figure 9 is that the increase in growth rate is not once and for all.

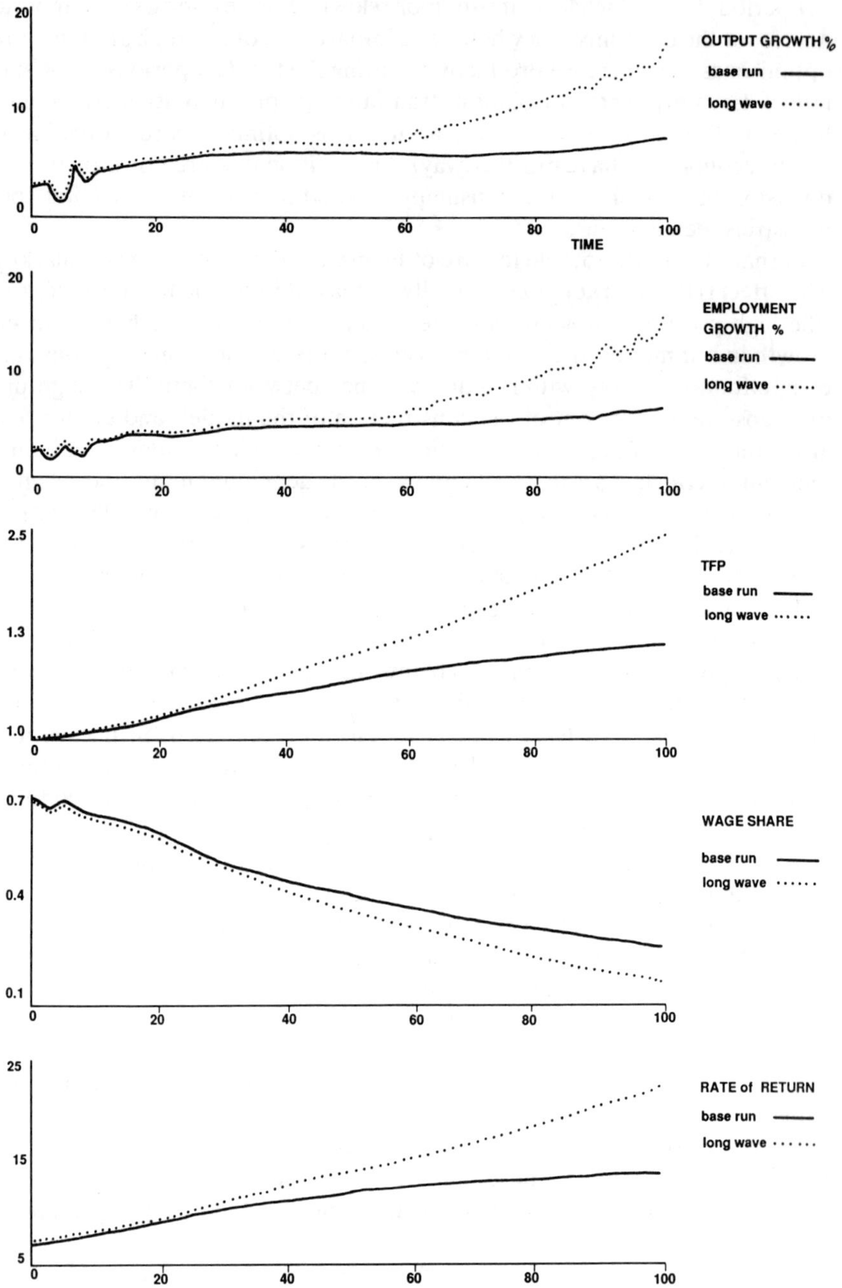

Figure 9. Performance of the economy: long wave.

The major innovation generates a continuing acceleration in growth rate. The extra growth which stems from innovations, is supplemented by the reduction in capital required to install a unit of capacity. An additional factor generating growth is the reduction in the share of wages. Together these effects create a continued surge in growth.

The potential for a major innovation to generate the upswing of an economic long wave is evident in Figure 9. However within our model there is nothing to generate the downswing of the wave since, as described above, their are no constraints on growth. We can obtain similar results if the major innovation occurs in just one capital goods industry.

4.2 *Selection and diffusion*

Selection and diffusion are vital elements of an evolutionary economy. To describe these phenomena in this section we innvestigate three scenarios analysed by Iwai (1984): selection with no imitation and no search, and with these two elements sequentially introduced. In order to generate results in a controlled way a very simple setting for technical change within an industry was devised. There are a number of exogenously determined basic processes to be discovered, each using 25% less of each input than its predecessor. Firms discover these when their search outcome passes a given point. To keep the environment as fixed as possible technical progress occurs in only one industry, where it is confined to implementing the basic processes themselves, with no incremental improvements.

No imitation or search. At the outset of the simulation firm 1 in industry 2 is given a 25% cost advantage over the other firms in that industry, with all the firms in the industry identical in all other respects. This equivalent to an unanticipated and disembodied improvement in productivity for the selected firm. There is no subsequent success by any firm in finding new processes. The only source of technical progress is from the relative growth of the advantaged firm.

Figure 10 shows results from this scenario. The graph shows firm 1's market share gradually rising (after the initial period of instability). After about period 40, when it has about 50% of the market, the rate of growth in share decreases, and obviously eventually approaches zero as monopoly is approached.

The rate of increase in firm 1's market share is limited by the ability to acquire new capacity. The need to limit the increase in market share is reflected in the price markup being consistently around 20%, which gives the firm only a very small price advantage over its competitors, who are forced to run at minimal price markups, covering their operating costs, but unable to fund much new capacity. This also gives firm 1 a higher profit margin and consequently both more borrowing and faster growth. The other firms in the industry are forced

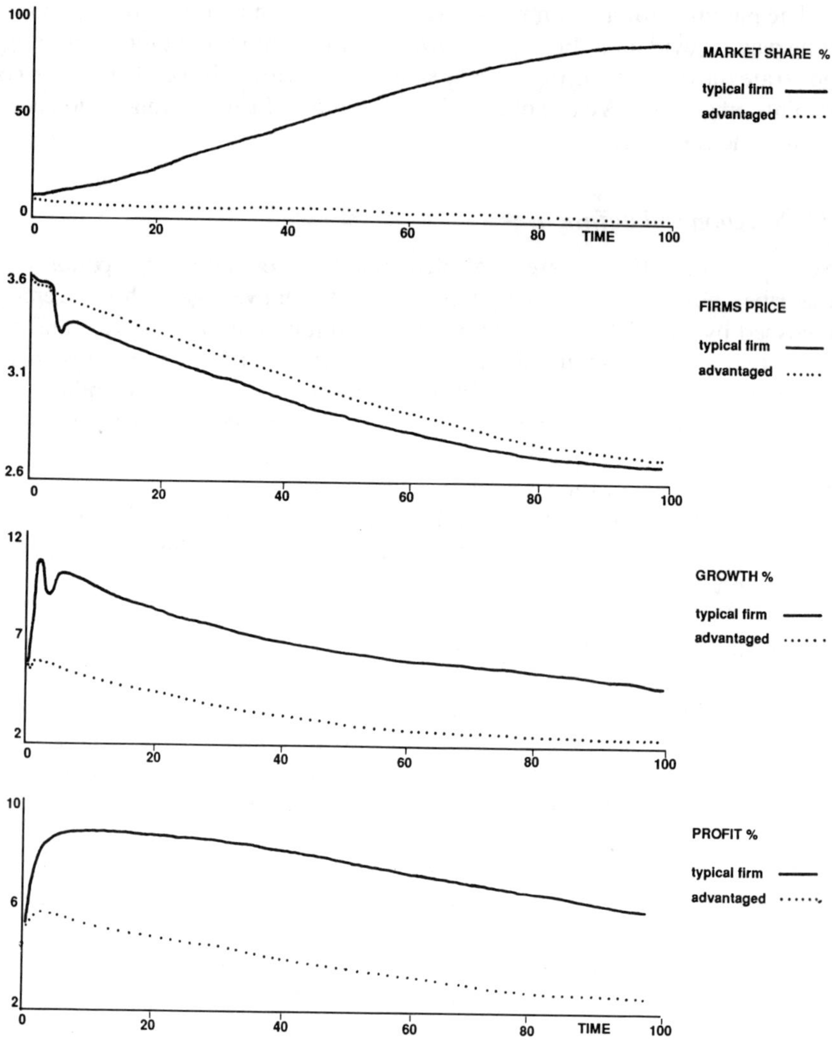

Figure 10. No search or imitation.

to lend their limited investment funds to the advantaged firm, thus further slowing their growth rate. Firm 1's growth of capacity does not follow a similar pattern to growth in market share, and is in fact fairly continuously decreasing, as it becomes more and more limited by the growth of the market. The growth rate of the firm is independent of its initial size, and the final size distribution of firms is determined solely by access to the superior technology, as in Iwai's results.

Imitation only. The scenario of the previous section is shown as the base run in Figure 11. It is shown in comparison to the same situation except that now the other firms can successfully imitate the new process. The bottom two graphs of Figure 11 show the means of two distributions used as indicators of progress; the proportion of firms and the proportion of capacity with better than a given level of revealed performance (using the second basic process). For the imitation allowed scenario, we see that in the early periods of the simulation a firm discovers the new process every few periods and that by period 15 five firms are using the second process. Recalling Figure 10, this is fortunate for those firms, since if the initially advantaged firm is allowed to gain too much ground it will force down the profit margins of the other firms and borrow their investment funds. This explains why the rate of inter-firm diffusion slows after period 15, and we note that only 8 of the firms actually succeed in imitating the innovator by the end of the simulation run.

The rate at which the second process comes into use, as shown by its percentage of total industry capacity, is speeded by the introduction of imitation, as expected. The graphs show a sigmoid diffusion path such as is typically found in empirical studies of diffusion. This pattern is reflected in the average cost of production in the industry. As well as mean preformance we can also examine the variance. The second graph in Figure 11 shows the variance in average production cost plotted against time. We see that the imitation case achieves its more rapid change will a lesser variance than the non-imitation case. This is because the zero rate of interfirm diffusion if the base run means that more investment takes place in the inferior process.

The third graphs illustrate Fisher's equation of evolutionary change; that rate of advance (fall in average cost) is proportional to variance. We see that the graphs are downward sloping as expected. In the imitation allowed case, any given degree of variance causes a much more rapid level of cost reduction. The loop shapes in the graphs seem to be explained by differential levels of fitness. By virtue of their ability to borrow, high profit firms also have 'fitness', in that they are adept at converting profit into growth. The level of cost reduction is less for any level of variance once it has passed its peak in both the runs shown, that is when 50% of capacity is in the second process. The dominant firms are by then reducing their prices and their profits, hence reducing their fitness.

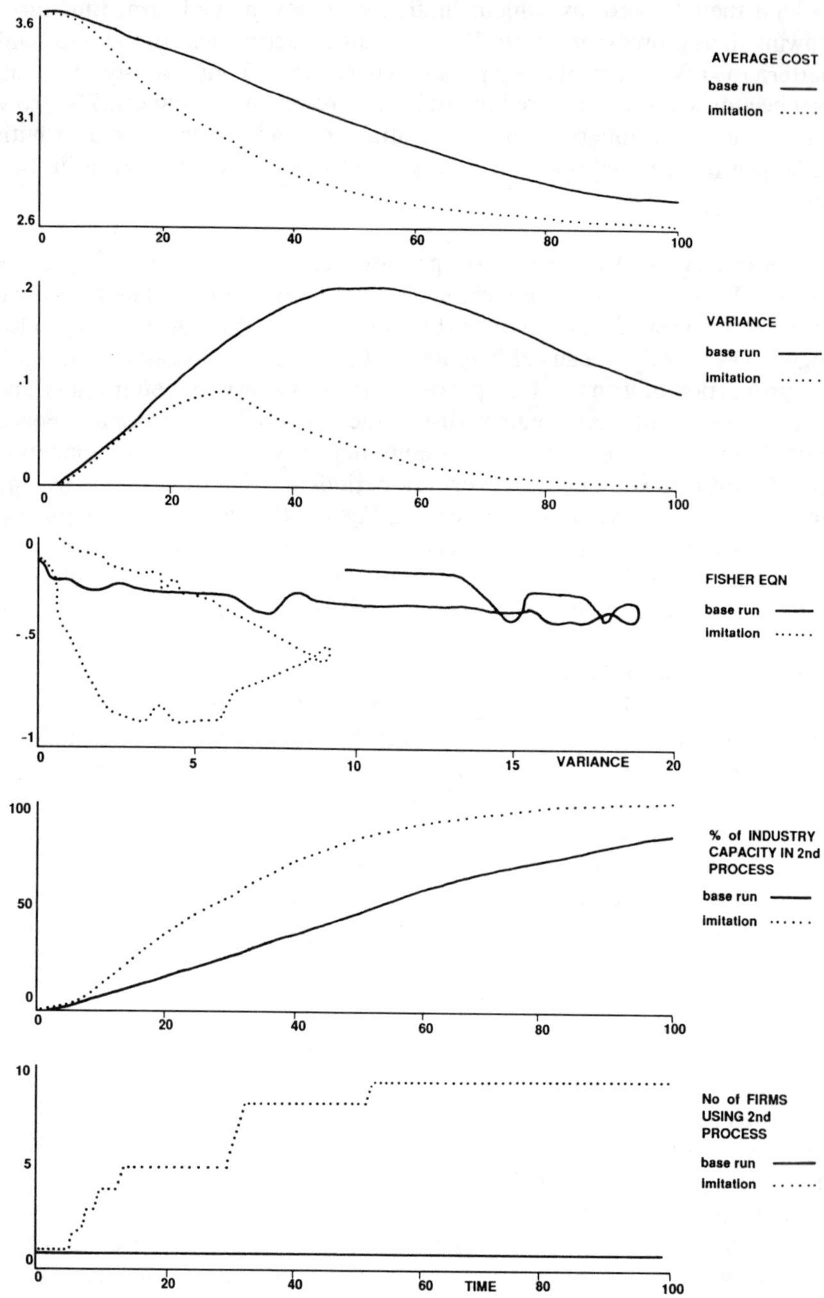

Figure 11. Industry performance and imitation.

Search and Imitation. To investigate the effects of search, a scenario similar to that used in the previous section was devised. The only differences are that there are now three new basic processes (2, 3 and 4) to be discovered, and no firm is given an initial advantage.

In Figure 12 we see that the rate of diffusion, in terms of the % of industry capacity devoted to each process, increases as we discover ever more advanced technology, but that the converse is true for the number of firms. This is attributable to the large market share of the early users of process 3 and then process 4, as illustrated in Figure 13. The graphs show that the best firm, as defined by its market share at the end of the simulation, is the first to discover process 3 and the first to discover 4 also, whilst the worst firm is the last to find process 2 and gets no further. We observe also in Figure 13 the surge in growth rate and borrowing as the best firm discovers each new process, and how these are reduced as imitators come in, reinforcing the picture of dynamic competition described above. Continued success for a firm depends on a stream of innovations or imitation. Failure to keep up with current best practice will ultimately cause a firm to be eliminated from the market.

Turning again to Figure 12, we see that each new innovation creates an upswing in variance, which eventually dies out and becomes a fall. Only in the case analysed by Iwai (op cit), of a constant exogenous rate of technical advance in best practice, would we expect to see variance constant over time, and then only if firms all have an equal chance of being innovators. In our example, where innovators are increasingly monopolistic, an accelerating level of technical advance in best practice would be required for constant variance.

Industry performance. In this section we consider the industry in two cases of a changing environment. The results are averaged over 10 simulation runs. In the first case, described by Figure 14, we show the result of increasing the level of 'static' competition. The parameter which governs the rate at which firms gain or lose market share within each period, was increased. The Base Run is the scenario used to generate Figure 12.

Figure 14 shows that the difference in competition only begins to make a major impact on performance in the later periods of the simulation run. However this is the culmination of less evident changes in earlier periods. Disadvantaged firms find they must match the price of more productive firms that much more closely if they are to maintain their market share. This means cutting price markups, reducing both their growth and search. Prices are marginally higher as a result of these tensions. Search is also marginally higher when 'static' competition is increased, reflecting the high propensity to search of larger firms. These two effects jointly mean that the fourth process is discovered somewhat earlier when 'static' competition is increased, but by a smaller number of firms. Within the confines of our model, increased 'static' competition, paradoxically, tends to increase the concentration within the industry, through allowing advanced firms to increase their dominance of search.

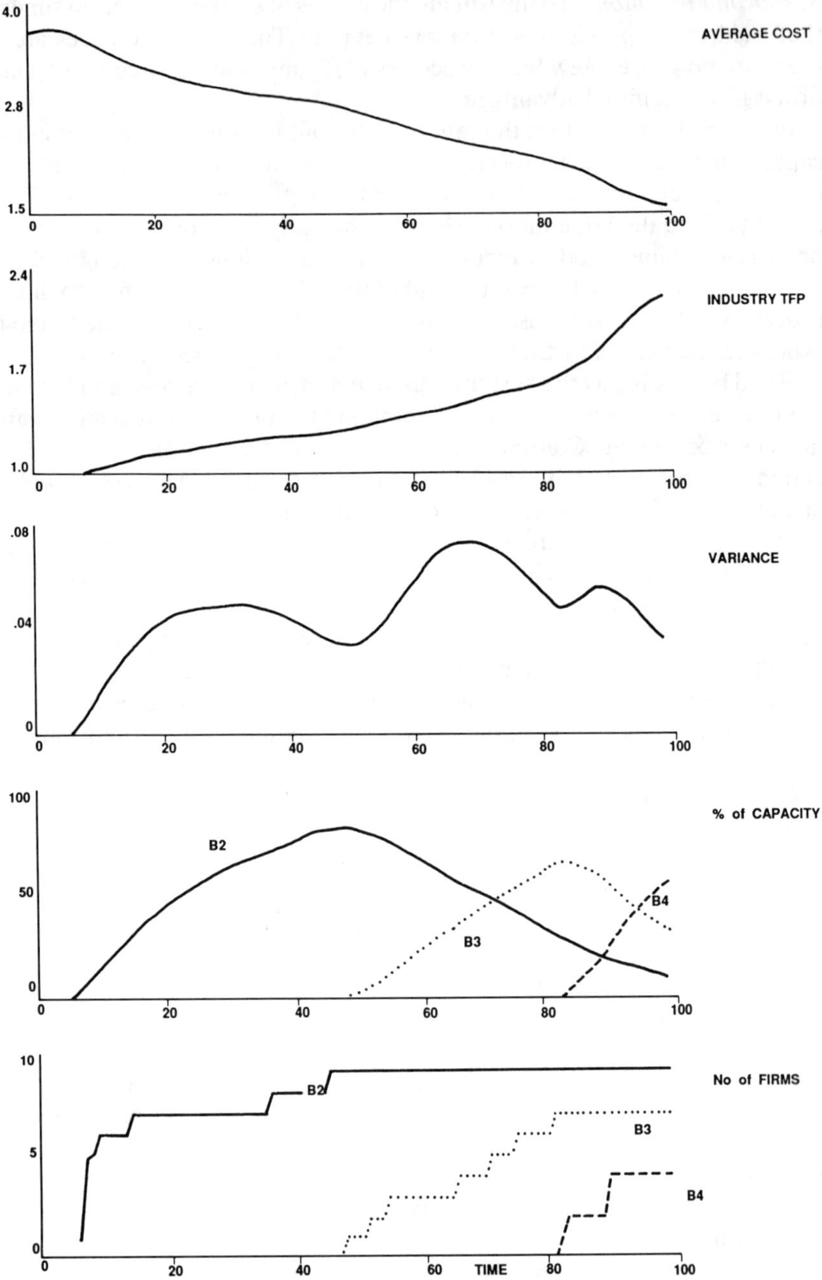

Figure 12. Industry performance: search.

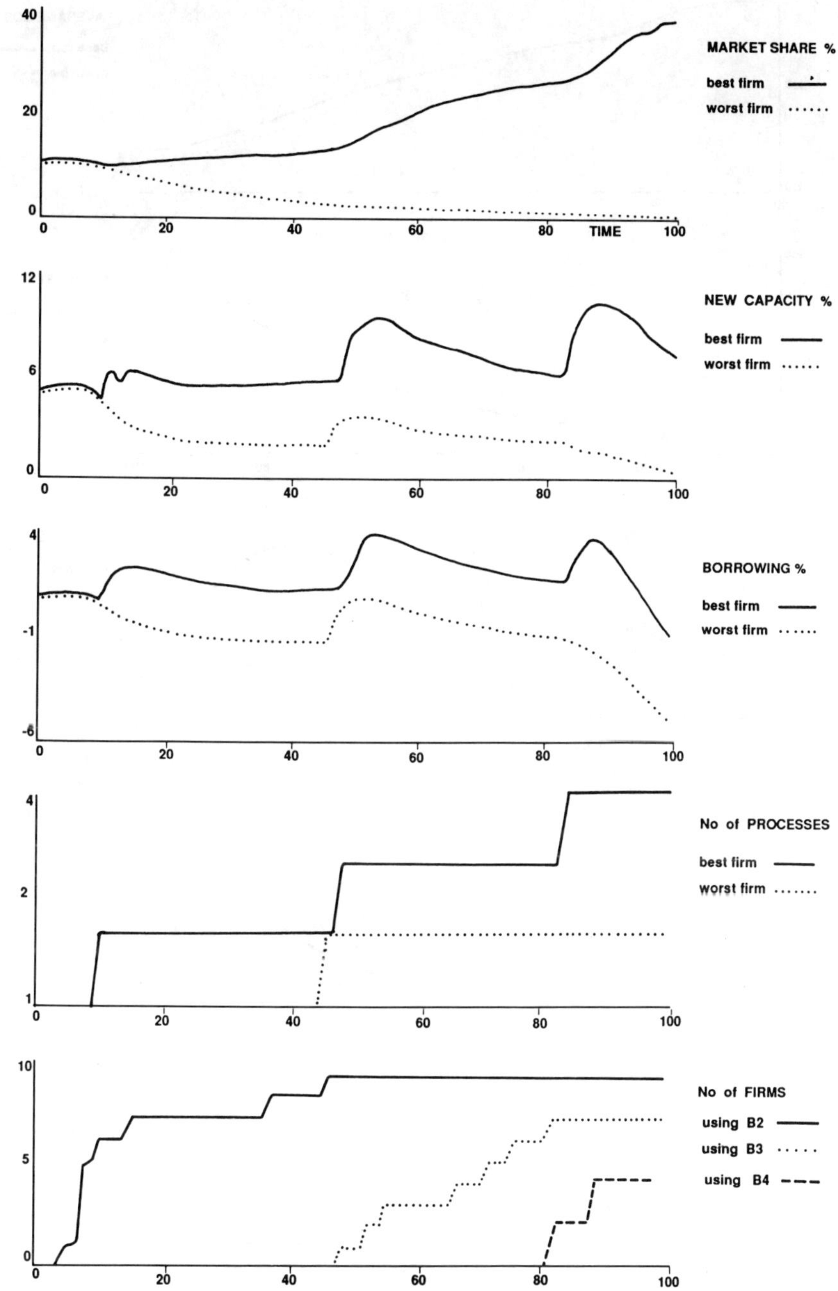

Figure 13. Best and worst firms: search and imitation case.

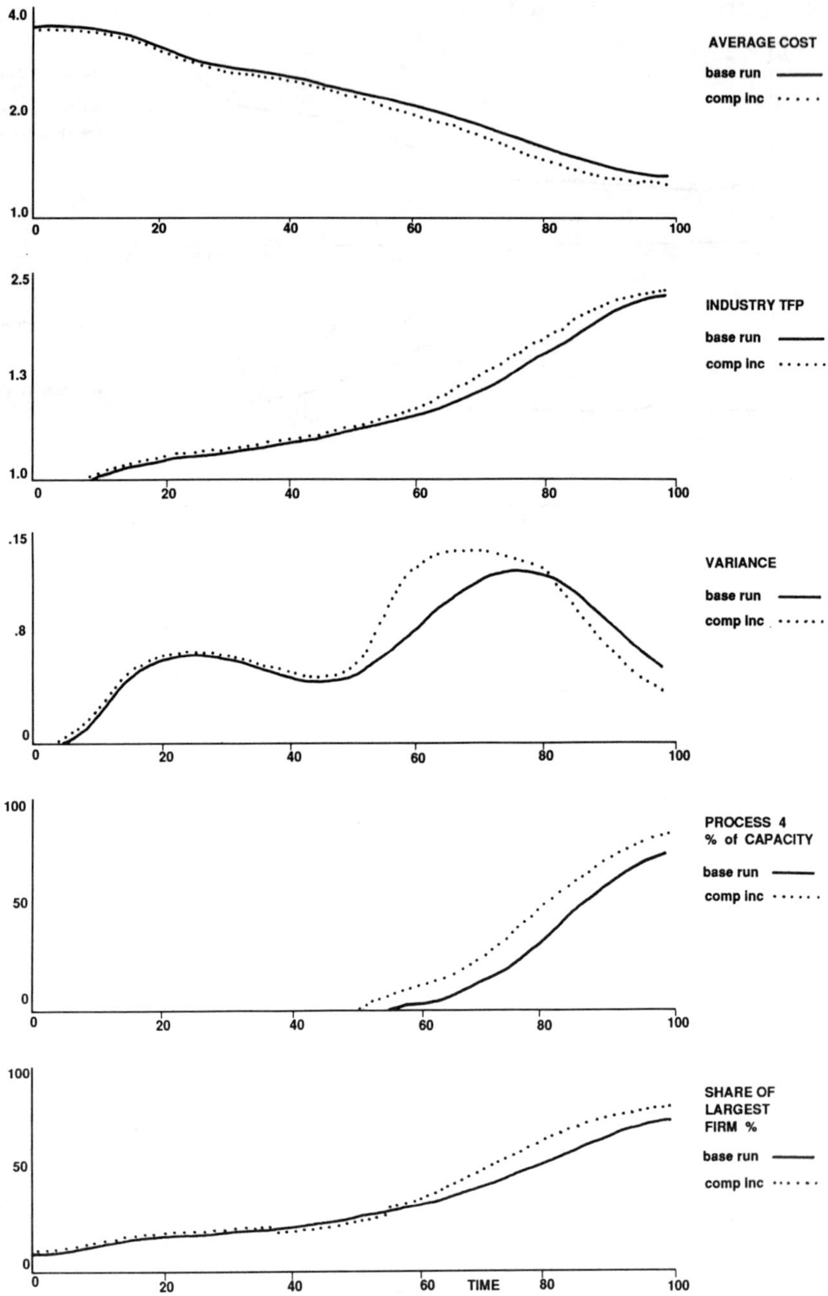

Figure 14. Industry performance and competition.

Figure 15 shows the effect of a simple system of patent rights, as described above. The patents last for 25 periods. The base run is as for Figure 12. The effect of the patent system is to speed the diffusion of new innovations. This is confirmed by the bottom graph of Figure 15 which shows more rapid diffusion of process 3, and the middle graph which shows that the patent system achieves diffusion with a much lower level of variance.

Figure 16 illustrates an interesting paradox of the patent system. Above we saw that increasing 'static' competitiveness lead to a greater degree of market concentration. Here we see that a policy designed to advantage a particular firm results in a lower level of industry concentration, both in the share of the top firm and of the top 3 firms. This lesser degree of concentration is the reason why, towards the end of the simulation run, productivity is lower in the patent on scenario; economies of scale in search are being less exploited with patents in operation because firms tend to remain more equal in size. Clearly this conclusion would not apply if the patent holder refused to license its discovery. Figure 4 illustrates that within a 25 year period the single advantaged firm attained a 40% market share.

5. CONCLUSIONS

The key elements of the evolutionary system are diversity of behaviour and economic selection on the basis of revealed performance. Competitive selection and a stream of new innovations create a constantly changing economic structure. Diversity of behaviour at the micro level must generate a functioning market economy, in which the transition of states at the macro level is relatively smooth and orderly.

The results from our simulation model demonstrate these features. In the simulation described in section 4.1, the economy undergoes a technological development which increases productivity by 60 percent. Performance at the economy level shows a continuously growing and stable macro economy, with technical progress seen to be a powerful source of that growth. Yet behind this appearance of harmonious developmment of the market economy, we see, at the micro economic level, the forces of dynamic competition at work.

Our illustrations of firms' behaviour show that simple decision rules, involving very limited calculations of profitability, allow firms to function effectively in the market place. Whilst it will eventually be eliminated, poor revealed performance can persist over long periods. The forces of induced innovation and competitive selection result in the dominance of successive technologies, together with the firms that can make best use of them. Firms which fail to adopt new technologies are eventually eliminated. Our results on competition and patents show that the short run effects of policy changes may end up having the exact opposite result from that intended, as a result of system

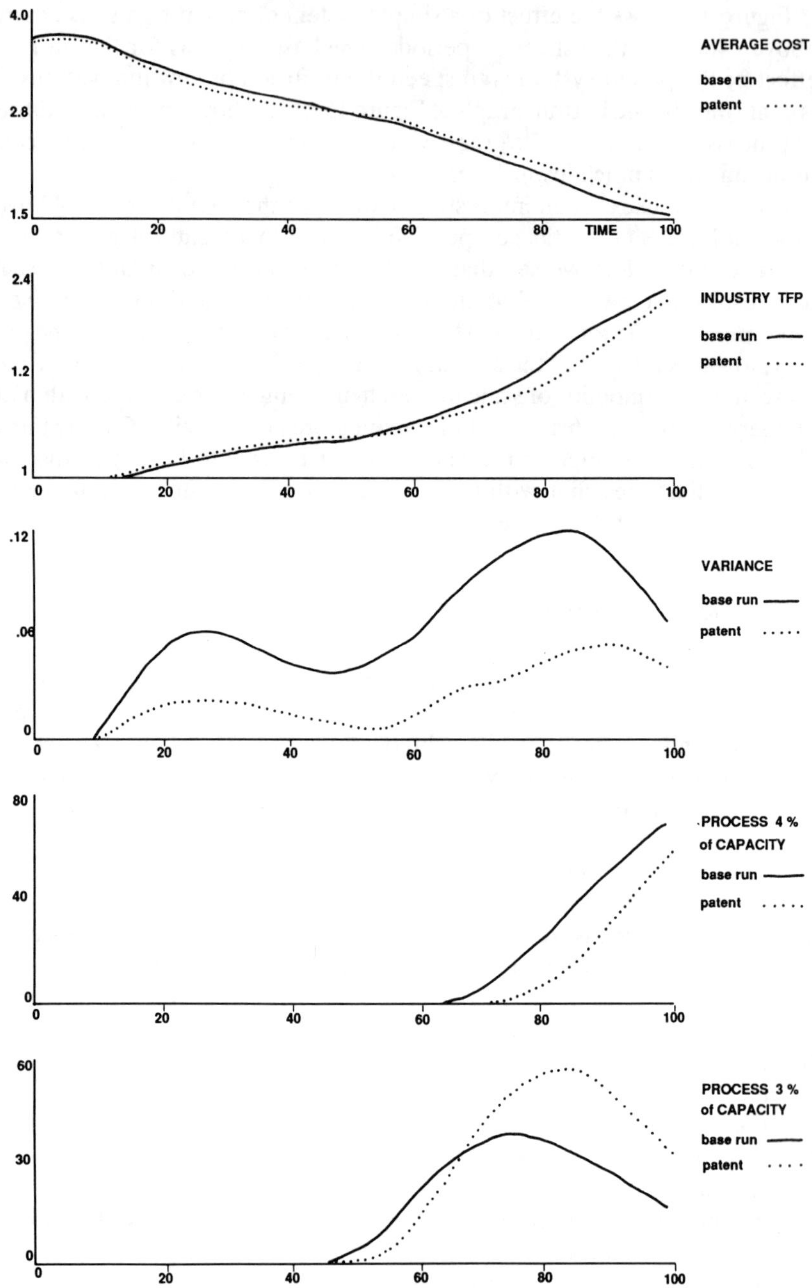

Figure 15. Industry performance 1: effect of patents.

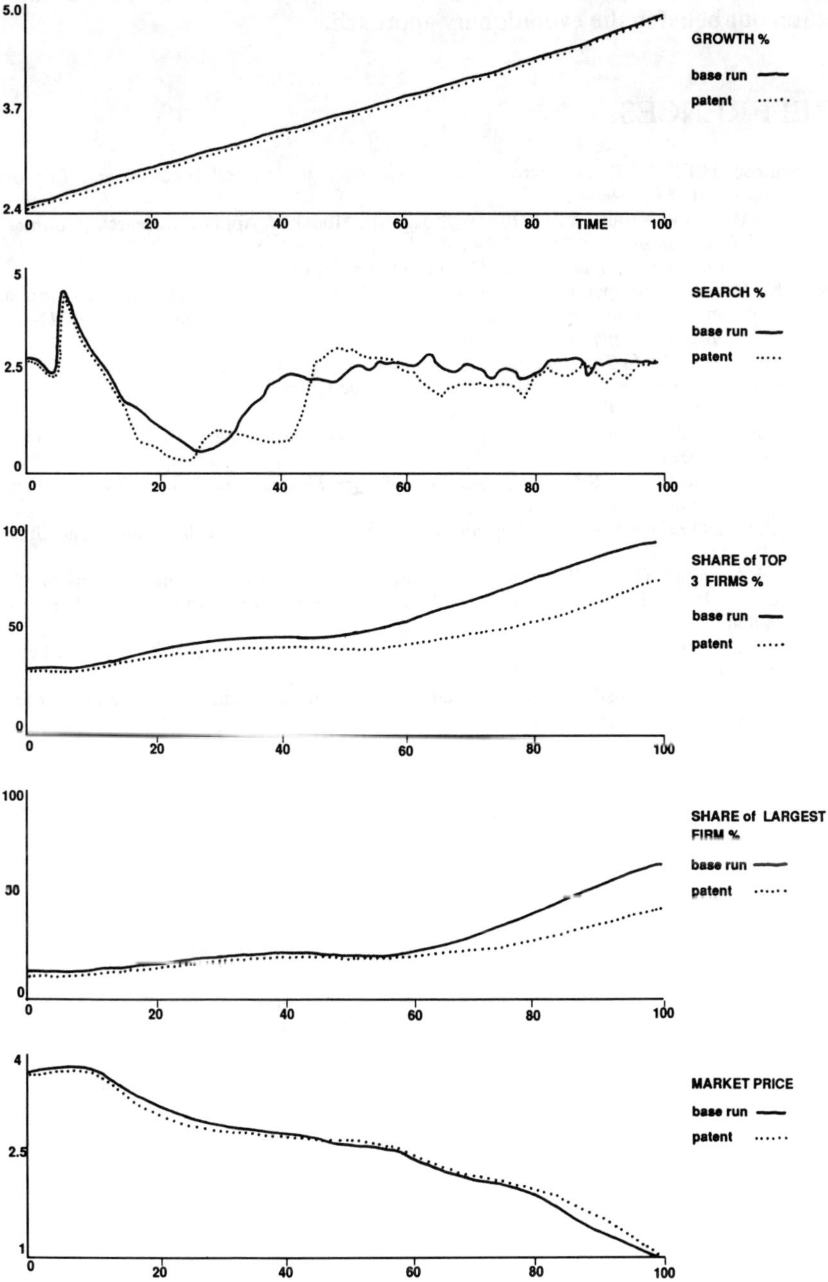

Figure 16. Industry performance 2: effect of patents.

nondecomposability and the dynamics of change. The results from our model affirm our belief in the evolutionary approach.

REFERENCES

Binswanger, H.P. (1974) 'A Microeconomic Approach to Induced Innovationn', *Economic Journal* vol **84** pp. 940–958.

Evenson, R.E. and Kislev, Y. (1976) 'A Stochastic Model of Applied Research', *Journal of Political Economy*, vol **84** pp. 265–281.

Hicks, J.R. (1965) *Capital and Growth*. Oxford: Oxford University Press.

Iwai, K. (1984) 'Schumpeterian Dynamics, Part 2: Technical Progress, Firm Growth and Economic Selection', *Journal of Economic Behaviour and Organisation*, vol **5** pp. 321–351.

Kay, N.M. (1984) *The Emergent Firm*, London: Macmillan.

McCloskey, D.N. (1965) *The Rhetoric of Economics*, Hemel Hempstead: Harvester.

Metcalfe, J.S. (1984) 'Technological Innovation and the Competitive Process', *Greek Economic Review*, vol **6** pp. 287–316.

Nelson, R.R. and Winter, S.G. (1974) 'Neoclassical vs. Evolutionary Theories of Economic Growth' *Economic Journal*, vol **84** pp. 886–905.

Nelson, R.R. and Winter, S.G. (1982) *An Evolutionary Theory of Economic Change,* Change, Mass: Harvard University Press.

Pasinetti, L.L. (1981) *Structural Change and Economic Growth*, Cambridge: Cambridge University Press.

Simon, H.A. (1986) 'On the Behavioural and Rational Foundations of Economic Dynamics'. In Day, R.H. and Eliasson. G. (eds.) *The Dynamics of Market Economies*. Amsterdam: North Holland.

Soete, L. and Turner, R. (1984) 'Technology Diffusion and the Rate of Technical Change', *Economic Journal*, vol **94** pp. 612–623.

Winter, S.G. (1971) 'Satisficing, Selection and the Innovating Remnant', *Quarterly Journal of Economics*, vol **85** pp. 237–261.

Econometric Methodology in an Environment of Evolutionary Change

JOHN FOSTER
Department of Economics, University of Queensland,
St Lucia, Queensland 4072, Australia.

1. INTRODUCTION

In the 1980s, econometric methodology has become an area of lively debate and disagreement. This controversy has not arisen primarily from debates amongst econometric theorists but from those of applied econometricians faced with difficulties encountered in econometric estimation. The 1970s was a decade when economists discovered that it was not straightforward to test hypotheses, suggested by economic theory, without encountering problems of observational equivalence and instability of estimated parameters. It was the decade of 'Goodhart's Law' where, one after another, estimated specifications in macroeconomics, when called upon to aid policymakers, promptly broke down.

The promise of instrumentalism, as promoted by Milton Friedman and other monetarists, and widely adopted in the 1970s soon gave way to new Classical

I would like to thank the participants of the Workshop on Economic Theories of Economic and Technological Change, held by the Department of Science and Technology Policy, University of Manchester, 21–2 March 1989, for their valuable comments on a preliminary draft of this paper.

macroeconomics which relied much more upon the force of expectational logic rather than econometric results to guide policymakers towards a non-interventionist stance. Instrumentalism was turned on its head with the emphasis being upon econometric results that revealed 'rationality' in various time series. The emphasis was on reduced-forms rather than structural equations, such as the demand for money function so emphasised by monetarists. Indeed, Robert Lucas (1976) dismissed the estimation of structural equations, for the purpose of constructing econometric models for forcasting purposes, as a misguided exercise.

From the perspective of the 'Lucas critique', the observed breakdowns in econometric results were viewed as entirely understandable: they were due to structural shifts in expectations in the wake of attempts to apply policy innovations, guided by parameter estimates derived from the past. This point concerning the possibility of structural change originating in expectations resulted in some disarray in econometric modelling circles. However, given that the problem was one of structural change rather narrowly defined, the result was not a new econometric methodology which could deal with structural change in a general sense. Instead, some modifications were made to existing practice to ensure that the Lucas critique, concerning the effect of rational expectations was, in some sense, met. However, even this led to controversy and disagreement as to the most appropriate strategy.

The purpose of this paper is to enquire into what an econometric methodology, which takes structural change fully into account, might look like. By structural change we do not merely mean the shifting of expectations of econometric agents who have adopted a static (or steady state) view of the long run. We mean structural change in the sense of technological and institutional change, which alters the parameteric configurations across the economic system in continuous and discontinuous ways.

In section 2, the onset of crisis in econometric methodology, in the late 1970s, will be discussed. In Section 3, we shall evaluate the methodological debate which has developed in the wake of widespread dissatisfaction with 'classical' econometric methodology. Section 4 examines how the econometric specifications and results yielded by new methodological approaches can be interpreted from the perspective of evolutionary change in the economic system. Section 5 offers suggestions as to how such change could be dealt with, explicity, in econometric methodology. Section 6 contains the conclusions.

2. THE ONSET OF CRISIS IN CLASSICAL ECONOMETRIC METHODOLOGY

In what follows, attention will be focused only upon time series econometric methodology since structural change is temporal in character. However, there will be no attempt to provide an extended review of the methodological debate which has take place in time series econometrics over the past decade. This has been done with great thoroughness and clarity by Pagan (1987) and Gilbert (1988). Our discussions will be confined to issues which have a bearing upon the question of structural change and its econometric manifestations.

The extent of dissatisfaction with econometric methodology at the present time can be gauged by the recent comments of Durbin (1988) who calls for an eclectic approach to dealing with time series statistical models and a significant downgrading of economic theory, in the guise of 'true' models which are presumed to be parameterised in econometric estimation. There are serious implications which follow from such a view, in terms of the scientific status of orthodox economics. This has caused some distinguished econometicians, such as Malinvaud (1989), to defend classical econometrics as the only appropriate methodology, however imperfect, for verifying economic hypotheses.

Dissatisfaction with classical econometrics began to be voiced in the late 1970s as practitioners encountered difficulties with their econometric results, in terms of both parametric stability and theoretical interpretation. The optimism concerning the use of econometric methods for the purposes of verifying hypotheses, which was widespread at the beginning of the decade, had begun to disappear. The problem was that, in order to operationalise a static equilibrium hypothesis in historical time, it was necessary to introduce a 'translating' hypothesis. Two types were used extensively: the backward-looking partial adjustment (PAH) and the forward-looking adaptive expective expectations hypothesis (AEH). The former appealed to a disequilibrium rationale whereas the latter emphasised an information deficiency problem.

Although, PAH and AEH could both be justified in terms of some 'special case' mathematical formalisation involving adjustment costs and information costs, respectively, their principle attraction was that they both rationalised the inclusion of lagged independent variables and/or a lagged dependent variable in estimable specifications in a manner which permitted estimated coefficients to be solved for estimated 'equilibrium' parameters. Furthermore, the inclusion of lagged variables invariably resulted in superior fits of models to historical data.

However, what seemed, initially, to be useful ways of converting hypotheses into specifications in historical time, ran into problems of observational equivalence when certain hypotheses were compared. For example, it became very difficult to know if a 'dynamic' specification of the consumption function revealed permanent income, generated by AEH, or current income, subject to PAH, as the relevant theoretical variable. In a scientific sense, this was hardly surprising given that one equation was being used to test both an equilibrium and a translating hypothesis.

The instrumentalist methodology promoted by Milton Friedman (1953) had become popular in the 1970s, particularly amongst monetarists. In the face of observational equivalence, instrumentalists argued that the simplest hypothesis, derivable from economic theory, should be preferred, provided that the estimated specification exhibited a high degree of predictiveness in forcasting. Specifications appealing to PAH due to factors outside orthodox economic theory, such as habit persistence and money illusion, were to be rejected unless they could be shown to predict *better* than specifications grounded in orthodox economic theory. Of course, observational equivalence almost guaranteed that this could not be the case.

Trading realism for simple theoretical consistency and predictiveness soon ran into trouble in the 1970s as the empirical specification upon which macroeconomic instrumentalists lavished most of their attention, namely, the demand for money function, refused to exhibit parametric stability. However, we did not witness an abandonment of the simple demand for money hypothesis and associated monetarist ideas. Instead, a more exhaustive search was conducted to discover a dynamic formulation which could generate a parametrically stable result. Courakis (1978) argued strongly that this promoted the predictiveness maxim to such an extent that what was left was 'measurement without theory'.

The *ad hoc* translating hypothesis had rendered the equilibrium one undiscoverable so that, instead of verifying a simple hypothesis, no economic hypothesis was discernable at all, only a statistical forecasting equation of unknown pedigree. This was a quite general feature of the structural equations estimated by macroeconometric modellers in the 1970s. Modellers were in the business of generating good forecasts and, therefore, were not particularly interested in verifying hypotheses but, instead, they drew upon whatever theoretical inspiration they could find to improve their modelling efforts. The links between their essentially statistical modelling tasks and economic analysis were rather tenuous, a point made most vividly by Sims (1980). It became almost a question of taste as to whether a macroeconomic model was labelled 'Keynesian' or 'monetarist'.

Instrumentalism had been indissolubly linked to forcasting 'usefulness' and had failed in the sense that abstract theory and econometric specification could not be decisively connected. The issue concerning the validity of instrumentalism was, therefore, left unsettled because econometrics had proved inadequate for instrumentalist objectives. More worrying was the vast flow of published material which purported to use econometrics to offer verification of some hypothesis with no useful forcasting application in mind. The instrumentalist could raise two objections to this activity. First, there was no effort to simply propose a hypothesis and test it, econometric search was used to discover the 'appropriate' hypothesis as well. Second, there was little attempt to subject discovered formulations to stringent forecasting and parameter constancy tests.

Model testing/estimation and model discovery had become mixed up, this time not to obtain a better forecast of a variable but, more seriously, to claim support for some 'revealed' hypothesis. The early 1980s witnessed a chorus of objections, most notably by Mayer (1980), Blaug (1980) and Leamer (1983). It was clear that the association of an equilibrium hypothesis, drawn from economic theory in abstract time, and a translating hypothesis containing unknowable features, could be juggled into an infinite number of inferential situations. This juxtaposition of two disctinct types of hypothesis, not surprisingly, became the focus of attention in time series econometrics for the remainder of the 1980s.

3. THE NEW ECONOMETRIC METHODOLOGIES

Keeping orthodox economic theory intact, in the sense of proposing abstract time (or long-run) equilibrium casual relationships, which are parameter invariant (static), involves three possibilities:

(i) Accept that model estimation and model discovery are mixed up and develop empirical procedures to deal with the situation;
(ii) Abandon attempts to estimate individual structural equations;
(iii) Continue along classical lines but do not rely on classical statistical inference to argue that parameter estimates are robust and, therefore, supportive of hypothetical detail.

All three of these routes have been pursued, respectively, by the LSE (London School of Economics), VAR (Vector Auto-Regression) and EBA (Extreme Bounds Analysis) schools of econometric methodology. The position of each of these schools has been explained and compared in Gilbert (1988), whose work is drawn on extensively here, so only summary descriptions are necessary. Emphasis is upon the ability of each approach to offer a way of incorporating evolutionary processes.

(i) LSE

The LSE tradition (see Hendry (1983) (1987)) seeks to reverse the procedure whereby a simple hypothesis is continually augmented in response to equation diagnostics. Instead, a 'general to specific' procedure is adopted whereby a general specification of all variables, expressed in distributed lags, are entered into the initial specification which is then 'tested down' to a parsimonious representation. Such representations are usually restricted so that long run theoretical consistency is assured, leaving short run disequilibrium dynamics to be generated by the data. The parsimonious representation is accepted only after stringent diagnostic tests have been applied. Parameter constancy and forecasting tests are assigned a high priority. Superiority over other competing specifications is settled by encompassing tests which involve standard F tests when the explanatory variables contained in each hypothesis are omitted, in turn, from a joint hypothesis.

The LSE methodology represents a pragmatic compromise, with the needs of a macroeconometric model-builder in mind. The emphasis is upon the instrumentalist's desire for parsomony and forecasting power, but the role of economic theory is confined to guidance as to the variables which should enter the initial general specification and the form of long run retrictions on the parsimonious representation. As such, it has been a methodology which has improved, considerably, the performance of macroeconometric models where it has been used but, not unexpectedly, it has proved more controversial as a discriminator of competing hypotheses.

As a methodology which encapsulates technological and structural change, its general reliance on invariant long run equilibrium parameters and a disequilibrium depiction of the short run suggests that it is unpromising. Indeed Hendry (1989) p. 103 states that:

> 'inherent non-stationarity due to innovative human behaviour or natural processes, which as yet we do not know how to model . . . raises a number of issued as yet unexplored'.

However, he goes on to ask:

> 'How well can learning and innovation themselves be modelled by constant parameter processes? Theoretical analyses of R & D, technical change, financial innovation etc. seem to have progressed, so a constant 'meta-parameterisation' in a high dimensional non-linear mechanism cannot be excluded *a priori*.'

These comments indicate both an interest in accommodating structural change in econometric methodology and an acknowledgement that it may be possible that constant parameter processes might well be able to pick up structural change which is of a steady, developmental character. Indeed, in Baba and Hendry (1987), financial innovation is successfully introduced into money stock specifications for the US. However, such an effect is found to be relatively marginal. Furthermore, when Hendry and Ericsson (1988) estimated money stock determination equations for the UK using annual data, no such innovation effect was included. If we are to view structural change as a pervasive force over time series which now span over a quarter of a century, it is necessary to explain why LSE constant parameter specifications have been so successful.

(ii) VAR

The VAR methodology grew out of Sims (1980) complaints that the structural equations contained in conventional macroeconometric models are simplified to an extent which renders them 'incredible'. The VAR answer is to focus attention upon reduced-forms which are estimated using unrestricted lag structures, which include lagged dependent variables, for all non-deterministic variables. Unlike univariate ARMA models, the history of all other independent variables is used as well as that for the variable in question. An appeal is made to the theory of rational expectations to justify modelling in such an interdependent way. Reduction of the model size is guided by Granger causality experiments to attempt to establish weak exogeneity. Typically, VAR method, as outlined, provides relatively poor forecasts, although they might seem quite good to some given that they contain little or no direct input from economic theory. The problem can be due to serious collinearity between lagged variables and Doan, Litterman and Sims (1984), once again appealing to the theory of rational expectations, argued that the problem could be solved by imposing a random walk prior on the distributed lag coefficients. The introduction of this Bayesian dimension to the VAR methodology proved successful with both Litterman (1986) and McNees (1986) claiming that the resultant models outperformed their conventional competitors.

The VAR methodology is explicitly a statistical forecasting methodology, in contrast to the LSE methodology which admits specific economic theory priors rather than a universal random walk prior. This, of course, has led to widespread criticism by economists that the VAR methodology is 'atheoretical'. However, this seems to hang upon each economist's view of what they regard as theory. Some (see Sims (1987)) contend that the influence of the theory of rational expectations on the development of the VAR methodology, in fact, implies a very strong role for theory.

From our perspective of enquiring into the process of structural change, the VAR methodology is interesting in two respects.

First, the VAR method is evolutionary in the sense that its recursive approach permits estimated parameters to change systematically without being 'out of equilibrium' in any non-random sense. The VAR response to the Lucas critique is to argue that parameters will shift but this will occur over significant periods of time and, therefore, can be captured by the VAR method. Unlike the LSE method, no specific static, long run economic prior is imposed.

Second, an evolutionary economist has a different view of theory to a conventional economist. Systemic thinking leads to the theoretical conclusion that the economy will exhibit a high degree of *continuity* over time and that this will be punctuated by short periods of substitution, realignment and structural discontinuity. The latter will be different to predict from the past. This implies a reversal of theoretical emphasis. Conventional economists tend to adopt substitution hypotheses as their long run theoretical anchors and to view 'dynamic structure' as outside theory. Evolutionary theorists see the latter as theoretically predictable and the former as the force which disturbs predictable evolutionary development.

From the perspective of the evolutionary economist we can view the VAR method as a way of describing the dynamic structure of an evolving economy in statistical terms. Furthermore, the random walk prior can be viewed as a crude characterisation of the impact of underlying adaptive or selection mechanisms. This is the view of Geweke (1989) who argues that the Doan *et al* (1984) approach to forecasting and signal extraction can be replaced by a chaotic systems approach where deterministic, but non-linear, processes give rise to illusory stochasticness. Thus VAR methodology may be viewed as atheoretical, or derived from an unacceptable theoretical basis, by conventional economists but that is largely a reflection of the static, timeless nature of what they regard as economic theory.

In this regard, VAR methodology seems more appropriate to capturing evolutionary processes than the LSE methodology. Indeed, Sims (1980) comments, with regard to one of his own estimated VAR models, that '. . . the infinitely long run behaviour of this system is nonsensical, though over any reasonable forecasting horizon the system is quite well-behaved'. This accords well with the expectations of an evolutionary modeller concerning the robustness of predictions derived from extrapolations of past historical data.

The disadvantage of VAR, compared with LSE, methodology is its lack of detail concerning structure in an evolving system. The LSE approach can pinpoint which part of the system is structurally changing, given the priority accorded to a parametrically stable specification. If the disequilibrium/equilibrium dichotomy could be suitably interpreted in evolutionary terms, then the LSE approach could prove the more insightful methodology.

(iii) EBA

The EBA methodology, inspired by Leamer (1983), is different to the other two in that its primary concern is the fragility of econometric results. Although the Bayesian approach adopted is more flexible than that adopted in VAR methodology, it does not confront the Lucas critique and, therefore, the evolutionary possibilities do not present themselves. It is not the methodology of a forecaster but of an economist concerned with valid hypothesis testing. Indeed, it is applied to cross-section econometrics as well as time series econometrics (see Leamer (1984)). Gilbert (1988) argues that in the latter context the emcompassing approach in the LSE methodology seems to offer as much.

However, the evolutionary modeller can draw one lesson from EBA and that is a healthy scepticism for hypothesis testing based on classical statistical inference using only t-values on estimated parameters. Almon (1988) has argued that, once we leave dubious arena of testing static economic hypotheses and focus, as the forecaster does, on statistical processes, we should adopt different ways of evaluating the contribution of variables to explanatory power.

He offers marginal explanatory value (mexval) which is a measure related to the fall in the residual sum of squares (root mean squared error is the favoured measure) when a variable is added to a specification. It is likely to be the case that, as we move from testing conventional economic theories to evolutionary modelling, new diagnostics will be required to evaluate results from this different perspective.

4. AN EVOLUTIONARY RE-INTERPRETATION OF ERROR CORRECTION MODELS

It has been argued that the LSE and VAR methodologies are capable of re-interpretation from an evolutionary perspective. In this section we shall attempt such a re-interpretation by focussing upon the LSE approach. The latter frequently adopts an implicit prior, in the form of an error correction model (ECM) which is often applied to variables which are integrated of order I(1). Since this is equivalent to a random walk, the LSE model is often quite similar to the corresponding VAR model. The only substantive difference in such cases is in the way that parsimonious representations are arrived at. The VAR modeller tends to argue that the LSE method does not face the Lucas critique directly enough in considering model reduction as an empirical question and, in turn, the LSE modeller is concerned that the over-parameterisation and variability of parameters, inherent in random coefficient VAR models, distract attention from model misspecification.

From the perspective of the evolutionary modeller, these are not issues of pivotal importance so let us focus on the ECM, which has proved so successful in the 1980s. The derivation of an ECM is well known (see, for example Hendry (1983)) and if we consider two variables x and y, which are I(1) co-integrated (see Engle and Granger (1987)), we can write:

$$y_t - y_{t-1} = b_0 + b_1(x_t - x_{t-1}) - b_2(y_{t-1} - x_{t-1}) + u_t \qquad (1)$$

Co-integration ensures that if, for example, economic theory argues for $y = ax$ in the long run, then this holds over the data sample. However, it is not easy, in practice, to know if such co-integratedness exists because available tests tend to be weak. Despite the fact that lack of co-integration can imply structural change and invalidate the ECM model there is a tendency in the literature for modellers to proceed on the grounds that any problem ought to show up in equation diagnostics. However, this is further blurred by the fact that, because of collinearity between the constant and the error correction term (ECT), the constant is often suppressed in estimation.

Suppose the 'true' relationship was one which omitted the ECT, then we would have a structural change formulation with b_0 performing the same role as technical progress term did in Solow's (1970) growth model, provided x and y are specified in natural logarithms. Thus, there is a danger that the ECM approach will be interpreted in a disequilibruim way when it is, in fact, picking up structural change. This kind of difficulty was highlighted some time ago by Nelson and Plosser (1982) who argued that trending could either be viewed as deterministic with stationary deviations or subject to non-stationary drift. More recently Brock and Dechert (1988) have depicted the latter in terms of non-linear or 'chaotic' dynamics.

Furthermore, the form of the linear specification may also admit a structural change interpretation. If x and y are natural logarithms and both grow over time, then it is possible that b_1 could deviate from, say, its 'true' value of unity due to learning, experience and other evolutionary effects (see Conlisk (1989) for discussion along these lines). Again, this equivalence was demonstated in the growth literature, albeit from the opposite perspective, by Black (1962) in his critique of Kaldor's (1957) estimated 'technical progress function'.

For example, in Foster and Malley (1988) it was argued that the particular logarithmic x and y relevant to their study *logically* had a b_1 of unity, allowing evolutionary interpretations of their estimated equations. The ECT was never significant in the presence of a time trend (ie b_0) which is what would be expected when lack of co-integration yields an ECT which is non constant over the data period. This was a study which used annual time series data and it does seem that the ECT term tends to have a stronger role to play in much shorter quarterly time series studies where, indeed, 'disequilibrium' fluctuations are likely to be observed. In other words, in such studies, we would expect the ECT to statistically dominate the constant term because it can pick up both structural change and short term random walk type oscillations.

However, interpretation of coefficients becomes difficult. What are viewed as 'disequilibrium' terms in ECM, which are set to zero to calculate 'long run' coefficients, turn out to be more difficult to interpret and the constant b_0 can be viewed differently. Indeed, it may be these problems which have led to to the computation of implausible long run elasticities in some ECM studies. From our evolutionary perspective, things look different. The 'ECT' is the measured y/x ratio in the previous period. A tendency for this to change systematically will qualify the relationship between the x and y change variables in a more 'intelligent' way than a simple constant such as b_O. Any tendency, for example, for technical progress to speed up or slow down, because of learning curve effects, will be captured in some proximate way. Furthermore, business cycle shift effects may also be proxied by the inclusion of the lagged y/x ratio.

This way of interpreting the ECM specification suggests that it is able to capture, in a Roximate way, a non-linear dynamic relationship between x and y as well as homeostatic correction in the face of shocks.

We know from the non-linear dynamic evolution (NLDE) literature that non-linear specifications are necessary to deal with structural change in time series economic data (see Barratt and Chen (1988). The non-linearity in the ECM specification depends, of course, upon the values of b_1 and b_2. It is not the objective here to examine parametric configurations in detail, but it is intuitively obvious that, if the sign on b_2 can assume positive as well as negative values, a wide range of functional forms are possible including approximations to the logistic curve shape so popular in the NLDE literature.

It is uncommon in macroeconometrics to have the ECM dominating other explanatory variables. Indeed, in many cases the ECM becomes insignificant when we extend the observation period from a quarter to a year (see Malley, Foster and Bell (1988) for examples using annual data). At high degrees of aggregation it is often the case that linear relationships perform well, despite the presence of non-linearity at the sub-aggregate level (see Rostow (1989) p.9 for the example of industrial production in the UK where the logistic shape is so shallow that a linear approximation over a period as long as a century is very close).

Seeing ECM as a rather ingenious and simple way of capturing the non-linear character of an evolving system, described in terms of time series statistics, implies a different perspective on economic behaviour. In discussing VAR methodology, in the previous section, it was argued that explanation in terms of lagged variables, particularly lagged dependent variables, reflected the commitment which characterises evolving systems (David (1988) calls it 'path dependence'). Instead of autocorrelation being a nuisance, it is the backbone of statistical description of system behaviour over time. ECM is a method which offers a way of capturing some of the non-linear character of dynamics, which is difficult to discern from distributed lag specifications expressed in terms of levels of variables.

The excellent record of LSE modellers in their application of ECM has not only been because the world is characterised by disequilibrium but because ECM can capture evolutionary development. Indeed, evolutionary analysis renders the equilibrium/disequilibrium dichotomy meaningless because true adaptive behaviour involves target revision not only in response to current changes in stimuli, but also revision of targets aimed at in response to previous stimuli. In such circumstances, the conventional dichotomy breaks down and we are in a world where there may be a number of long run attractors. Kostelich and Yorke (1989) provide general examples in natural science applications and Geweke (1988) provides evidence that, in the face of non-linear dynamics of the simplest type, standard time series methods fail to identify the correct long run parameters.

So, although we can argue that ECM models allow for non-linearity, once we abandon the view that we are seeing the dynamics of disequilibrium in a world of unique, static long run equilibrium, it is unlikely that ECM could forecast successfully through a rapid bifurcation. Indeed, it is the presence of such bifurcations which permits us to envision smooth drift and enables us to relax the requirement that an econometric specification must have a long run static equilibrium solution. In the absence of the latter we are confronted with the possibility of a range of attractors. Unfortunately, the isolation of attractor characteristics typically requires much longer sets of time series data than we are accustomed to.

However, the importance of this potential disruption, due to structural breakdown depends on the level of aggregation at which enquiry is taking place. Microeconomic catastrophies aggregate into smooth macroeconomic change. This is particularly the case when heterongenious activity is aggregated through the homogenising medium of monetary valuation. Thus, the damage inflicted by the existence of non-linear dynamics may be much less in some macroeconimic applications than implied in the natural science applications widely cited. The evidence reviewed by Chen (1988) suggests that phase transitions from periodic to chaotic motion do occur in, for example, monetary data of very short periodicity. However, it is likely that an ECM model using quarterly or annual data could cope with the dynamics discovered. This possibility remains to be explored more fully.

5. THE EXPLICIT DEVELOPMENT OF AN EVOLUTIONARY METHODOLOGY FOR AN ENVIRONMENT OF EVOLUTIONARY CHANGE

An attempt has been made to interpret current econometric methodologies in terms of the extent to which they can capture evolutionary change. However, there is a limit to the extent to which this can be done given that the methods considered were developed with a different perspective in mind. In this section we shall consider how econometric methods could be applied with evolutionary change explicitly in mind. No attempt will be made to discuss, in any detail, what constitutes an evolutionary approach to economics (see Foster (1987) and David (1988) for extended discussion which is compatible with what follows). An evolving system, encapsulated in a money-valued aggregation can be described as follows:

$$X_t = X_{t-1} - Z_t + W_t \tag{2}$$

where:

X_t: the output flow from X-structure over period t;

Z_t: that part of X-structure lost through entropic decay or ephemeral effects over period t;

W_t: that part of X-structure which is new due to developmental activity and, again, ephemeral effects over period t.

If we assume that ephemeral effects (z and w) are random we get:

$$Z_t = Z^*_t + z \tag{3}$$

$$W_t = W^*_t + w \tag{4}$$

then

$$X_t = X_{t-1} - Z^*_t + W^*_t + (z + w) \tag{5}$$

Z^*_t depends upon the commitment horizon of X-structure, embodied in the residual life of productive capital, outstanding contractual agreements and other factors which induce temporal bonding. If no development were to take place ($W^*_t = 0$) then the commitment horizon would shorten and Z^*_t would rise as a proportion of X_t as X-structure eroded away. Homeostatic balance is maintained through replenishment, maintenance and developmental activity in W^*_t. For an evolving X-structure the main distinction between maintenance and development is that the former is directed at slowing down Z^*, by defending existing structure, and the latter, by extending existing structure and commitment in a way that permits Z^* to occur as obsolete structure is ejected.

In an economic structure, where monetary valuation is used to measure all flows, it is difficult to disentangle production, maintenance and development in monetary flows, particularly at higher levels of aggregation. One of the reasons that macroeconomists have had so much difficulty in dealing with structural change may be that time series statistics have not been collated with such an evolutionary division in mind. Investment expenditure can be looked on as developmental and consumption of domestic goods as reflecting production characteristics, but maintenance activity disappears into both types of expenditure. If an economy increased its tendency to engage in maintenance, in preference to development, it could be difficult to spot this ossifying trend in macroeconomic statistics.

Even at the level of an individual firm, with published sets of detailed accounts, it is difficult for financial analysts to gauge its evolutionary condition, so crucial to its future prospects. However, the economist, unlike the financial analyst, is not interested in discovering differences in the condition of economic units such as firms but, instead, in understanding aggregate behaviour. The crucial question is: what is the correct behavioural aggregate? As economic evolution proceeds, this will change. Institutions of a particular type will follow their logistic development paths over time as they are superceded.

For example, financial innovation has led to changes in the availablity of financial instruments for a particular purpose. Without ongoing statistical revision, in line with such evolution, the macroeconometrician will obtain a misleading impression of monetary relationships in the economic system. Detailed institutional study becomes an essential preliminary to the construction and use of statistical magnitudes. If this is not done then non-linearities and structural shifts will be recorded, which should be attributed to the institutional and technological change that is occurring and not to orthodox economic factors.

Such a suggestion, of course, runs counter to the conventional preference of applied econometricians not to 'tamper with' the data. However, if the data does not take account of evolutionary change, it is difficult to see how tampering can be avoided. If a definitionally rigid approach cannot be circumvented, then the econometric modeller can minimise the damage by starting from a homeostatic flow taxonomy, as contained in eq (5), rather than from an equilibrium/disequilibrum standpoint. Although the resultant specification may not be radically different than before, the interpretation of the estimated parameters will.

There is a fundamental implication of eq (5). The lagged dependent variable enters, *not* for causal reasons but for intertemporal bonding reasons. It represents the continuity of the system, or even the existence of the system, and therefore it cannot be said to 'explain' anything in conventional, hypothesis testing terms, despite its capacity to predict. This can also apply to variables which enter via Z^* and W^*. Their importance may be due to their *systemic* link rather than a causal association. In conventional terms we view this as simultaneity, but in a non-causal, systemic context it is the hierarchical bonding between components of a system that gives rise to such associations.

Looking at econometric results from such a perspective means that the breakdown of a relationship no longer is a 'nuisance' but rather a useful indication of structural breakdown. An econometric model then assumes the role of a monitoring device, analogous to an X-ray machine. It can point to problems that would not be otherwise visible and assists in pointing towards the sector of a system which requires detailed institutional evaluation. It may be that mutation has rendered the data obsolete or it may be that genuine structural breakdown has taken place.

Such a role for econometric modelling is, in essence, that which forecasters adopt at the present time and, indeed, we have discussed how VAR modellers have accepted a particular type of evolutionary representation in their modelling. However, there continues to be a desire to pay lip-service to rationales derived from orthodox economic theory to ensure that models are viewed as 'methodologically sound'. Yet, if it is accepted that history is characterised by evolutionary change, then reliance on static, equilibrium theoretical shadows would seem to imply the reverse.

This suggest that the problem in econometric methodology is not primarily what modellers do but, instead, the economics from which they choose to draw theoretical priors and depictions of dynamic mechanisms. It is this adherence to orthodox economic theory which makes if difficult for modellers to incorporate the structural change, taking place in historical time, as quickly as they might otherwise do.

For example, from an evolutionary perspective, it makes a difference whether the data to be analysed relates to a stock or a flow. The 'money stock' usually relates to a particular definition which means that an institutional 'fix' exists. In the face of financial innovation we would expect to observe typical logistic behaviour as other institutions come into being. When we are dealing with monetary flows, such as total expenditure, evolutionary change can take place within the statistical measures used. Thus, the econometric modeller can assess, from the outset, the likely time series configuration of the particular data examined.

6. CONCLUSION

In this paper an attempt has been made to argue for the development of an econometric methodology with incorporates an explicit evolutionary change dimension. It has been argued that existing methodological positions have been influenced in their development by a desire to interface econometric estimation with a body of economic theory which is static, timeless and equilibrium in nature. However, to obtain satisfactory forecasts, applied econometricians have been forced to model dynamics which can be explained much more easily from an evolutionary perspective on economic behaviour. What this means in practice is that some of the successful methodologies, which have become popular over the past decade, have evolutionary interpretations and, therefore. remain useful to the evolutionary economist.

There has been a tendency amongst some evolutionary economists to argue that time series econometrics is unhelpful in analysing a structurally changing economy. The conclusion here is that such a judgement is over hasty. Time series econometrics can offer an excellent statistical description of how an economic system evolves over time and it can offer timely warnings when developmental phases begin to turn into periods of structural fragmentation. Furthermore, there remains great potential for the 're-focussing' of econometric methods to address the evolutionary character of economic processes more directly (see, for example, Kalaba and Testfatsion's (1989) attempt to develop a dynamic estimation method better suited to estimating models which are evolutionary in character). Econometric methodology, up until recently, has been largely constrained by a hypothesis testing, or a hypothesis acknowledgement, aspiration which has been non-evolutionary in construction.

There are signs in the applied litereature that structural change is becoming accepted as a force which results in non-constant parameters over time rather than something to be 'dummied' into specifications (see, for example, Landesmann and Snell (1989)). However, there is little sign of a general preference for straighforward evolutionary rationalisations over complex attempts to use orthodox theory to guide such applied work. A revolution in econometric methodology cannot occur before there is a revolution in the economics to which it is subservient.

REFERENCES

Almon, C. (1988) *The Craft of Economic Modelling*, Ginn Press, Needham Heights, Mass.

Baba,Y., D.F. Hendry and R.M. Starr (1987) *US money demand 1960–84*, University of Oxford (Mimeo).

Barnett, W.A. and Chen P. (1988) 'The aggregation-theoretic monetary aggregates are chaotic and have strange attractors: an econometric application of mathematical chaos'. In *Dynamic Econometric Modelling*, eds. W.A. Barnett, E. Berndt and H. White., Cambridge University Press, Cambridge.

Black, J. (1962) 'The technical progress function and the production function', *Economica*, vol **29**, pp. 166–170.

Blaug, M. (1980) *The Methodology of Economics*, Cambridge University press, Cambridge.

Brock, W.A. and W. Dechert (1988) 'Theorems on distinguishing deterministic from random systems'. In Barnett *et al, op cit.*

Chen, P. (1988) 'Empirical and theoretical evidence of economic chaos', *System Dynamics Review*, vol **4**, pp. 81–108.

Conlisk, J. (1989) 'Optimisation, adaptation and random innovations in a model of economic growth', Department of Economics, University of California, San Diego (Mimeo, March).

Courakis, A.S. (1978) 'Serial correlation and a Bank of England study of the demand for money: an exercise in measurement without theory', *Economic Journal*, vol **88**, pp. 537–548.

David, P.A. (1988) 'Path dependence: putting the past into the future of economics', Institute for Mathematical Studies in the Social Sciences, Stanford University, Technical Report No 533.

Doan, T.,R. Litterman and C.A. Sims (1984) 'Forecasting and conditional projections using realistic prior distributions', *Econometric Reviews*, vol **3**, pp. 1–100.

Durbin, J. (1988) 'The ET interview by P.C.B. Phillips', *Econometric Theory*, vol **4**, pp. 125–157.

Foster, J. (1987) *Evolutionary Macroeconomics*, Allen & Unwin, London.

Foster, J. and J. Malley (1988) 'The domestic and foreign-owned sectors of Scottish manufacturing: a macroeconomic approach to their relative prospects and performance', *Scottish Journal of Political Economy*, vol **35**, pp. 250–265.

Friedman, M (1953) 'The methodology of positive economics'. In *Essays in Positive Economics*, University of Chicago Press, Chicago.

Geweke, J. (1989) *Inference and forecasting for chaotic nonlinear timer series*, Institute of Statistics and Decision Sciences, Duke University, Durham, NC 27706, USA (Mimeo, February).

Gilbert, C.L. (1988) *Alternative approaches to time series methodology in econometrics*, Institute of Economics and Statistics, Oxford (Mimeo, May 1988).

Hendry, D.F. (1983) 'Econometric modelling: the consumption function in retrospect', *Scottish Journal of Political Economy*, vol **30**, pp. 193–220.

Hendry, D.F. (1987) 'Econometric methodology: a personal perspective'. In Bewley, T.F., *Advances in Econometrics*, Econometric Society, Cambridge, Mass.

Hendry, D.F. and N.R. Ericcson (1988) 'An econometric analysis of UK money demand' in *Monetary Trends in the US and the UK* by Milton Friedman and Anna J. Schwartz, (Mimeo, November 1988).

Kalaba, R. and Testfatsion, L. (1989) 'Time varying linear regression via flexible least squares', *Computers and Mathematics with Applications: Special Issue on System-Theoretic Methods in Economic Modelling* (forthcoming).

Kaldor, N. (1957) 'A model of economic growth' *Economic Journal*, vol **67**, pp. 591–624.

Kostelich, E.J. and Yorke, J.A. (1989) 'The analysis of experimental data using time delay embedding methods', Institute for Physical Science and Technology, University of Maryland (Mimeo, January 30th).

Landesmann M. and A. Snell (1989) 'The consequences of Mrs Thatcher for UK manufacturing exports', *Economic Journal*, vol **99**, pp. 1–27.

Leamer, E.E. (1983) 'Let's take the con out of econometrics', *American Economic Review*, vol. **73**, pp. 31–43.

Leamer, E.E. (1984) *Sources of Comparative Advantage*, MIT Press, Cambridge Mass.

Litterman, R.B. (1986) 'Forecasting with Bayesian vector autoregressions – five years of experience' *Journal of Business and Economic Statistics*, vol **4**, pp. 25–38.

Lucas, R.E. (1976) 'Econometric policy evaluation: a critique'. In Brunner, K. and A.H. Meltzer (eds) *The Phillips curve and Labour Markets*, North Holland, Amsterdam.

McNees, S.K. (1986) 'Forecasting accuracy of alternative techniques: a comparison of US macroeconomic forecasts' *Journal of Business and Economics*, vol **4**, pp. 5–15.

Malinvaud, E. (1988) 'Econometric methodology at the Cowles Commission: rise and maturity', *Econometric Theory*, vol **4**, pp. 187–209.

Malley, J., J. Foster and D. Bell (1988) 'US budgetary policy and the World economy: the estimation of a small multicountry model' University of Glasgow Discussion Paper in Economics No 8813.

Mayer, T. (1980) 'Economics as a hard science: realistic goal or wishful thinking?' *Economic Inquiry*, vol **18**, pp. 165–178.

Nelson, C.R. and C.I. Plosser (1982) 'Trends and random walks in macroeconomic time series: some evidence and implications', *Journal of Monetary Economics*, vol **10**, pp. 139–162.

Pagan, A. (1987) 'Three econometric methodologies: a critical appraisal', *Journal of Economic Surveys*, vol **1**, pp. 3–24.

Rostow, W. (1989) *Non-linear dynamics: implications for economics in historical perspective*. Paper presented at the International Symposium on Evolutionary Economics and Nonlinear Dynamics, IC^2 Institute, Austin, Texas, April 16–19.

Sims, C.A. (1980) 'Macroeconomics and reality' *Econometrica*, vol **48**, pp. 1–48.

Sims, C.A. (1987) 'Making economics credible'. In Bewley, T.F. (ed.) *Advances in Econometrics*, Cambridge, Mass.

Solow, R.M. (1970) *Growth Theory: an Exposition*, Oxford University Press, Oxford.

Innovation Policy in an Evolutionary Context

KEITH SMITH
Department of Economics and Management Science, University of Keele
and
Resource Policy Group, Oslo

1. INTRODUCTION

This paper deals with the conceptual foundations – in economic analysis – of technology and innovation policy. Its point of departure is the fact that there is now a sharp contrast between the usual approach to the theoretical rationale for such policy, which is based on neoclassical welfare economics, and the most important current research programme on the innovation process and its economic effects: the latter is in large part based on evolutionary modelling, and is often influenced by the legacy of Joseph Schumpeter. Evolutionary analysis of innovation issues is in a state of flux, but even so its concepts, methods and results are reasonably well defined and are by no means in accord with neoclassical ideas. Does this matter? Do we need a general conceptual basis for innovation policy – as opposed to ad hoc or partial decision-making procedures – and if so, how well suited is evolutionary modelling to define such foundations, and what are its policy implications?

The question of the rationale for technology policy is a significant one in a number of respects. Against the background of a general move towards less interventionist policy stances within market economies, research and innovation policy is one of the few areas where anything like a consensus on the need for public support remains; yet the basis for this consensus is often unclear.

At the same time, this is not a marginal area of intervention: for example, public expenditure on R&D within the OECD in 1989 was in the region of $250 billion – the major economies are spending in excess of one percent of GDP each, from public funds, on R&D. We also know, from a major effort in empirical research in the economics of technological change, that technological change is the fundamental contributing factor in economic growth, that it is central in explaining inter-economy growth rate differences, that inter-industry R&D differences are highly correlated with productivity and growth rate differences, that the fastest growing industries in output and trade are R&D and innovation-intensive, and that trade shares are closely correlated with variations in innovation performance. Much of the public R&D expenditure referred to above is aimed, directly or indirectly, at improving growth, trade and productivity performance through improved scientific knowledge and faster technological innovation. Yet the links between policy and performance are unclear, and it is often the case that innovation policy is made with little reference to the actual characteristics of innovation processes, and with little importance attached to the relation between innovation activities and other arenas of economic policy (such as stabilisation policy of financial policy).

Of course it does not follow from the scale of the problem that it should be approached in any particular way. What are the arguments for thinking in terms of a *general* conceptual framework for policy? The basic reason is that although policy decisions are usually concerned with matters of detail, the economy is a system, and it is necessary to relate both partial decisions, and overall regulation or allocation decisions, to systemic inter-relations and their outcomes. It is clear that this requires detailed knowledges of some complexity, but these in turn depend on general conceptual bases. As an epistemological point, scientific understanding of complex systems requires fundamental organising concepts – specific problems can only be addressed within the context of general concepts related to the nature of the system, its structure, its dynamics and so on. These general concepts also define the scope and methods of analysis, domains of evidence, canons of proof and so on. The real issue is not whether we need such general conceptual bases, but how we can choose between the adequacy of competing conceptual systems.[1] The systemic aspects of economic activity mean that any rationale for technology policy requires such a conceptual framework; moreover, as I shall show in a later section, some policy issues turn quite directly on these general conceptual questions.

2. NEOCLASSICAL ANALYSIS OF TECHNOLOGY POLICY

There can be little doubt that neoclassical approaches have dominated the theoretical analysis of microeconomic policy.[2] Applications to research and innovation policy are complicated by the fact that, as Dasgupta has recently remarked, when it comes to scientific and technological change, 'modern resource allocation theory pretty much single-mindedly ignores the phenomenon'.[3] But in general, the case for any form of industrial intervention derives from analysis of the welfare properties of competitive systems. Perhaps the most important result rests on the existence of competitive equilibria. With a number of well-known assumptions (strict convexity, constant returns, complete markets, perfect information, absence of externalities and public goods etc), Arrow and Debreu showed that a vector of equilibrium prices exists, that an intertemporal equilibrium exists, that the Walrasian general equilibrium which results is Pareto optimal, that perfect competition is a sufficient condition for such an optimum, and that with an appropriate distribution of initial resources any Pareto optimum can be achieved as a competitive equilibrium.

Modern public sector economics is in large part based on this result in the sense that it is concerned with the extreme restrictiveness of the assumptions necessary to achieve it. If these assumptions do not hold, we have market failure in the sense that trading does not deliver an optimum equilibrium solution. Then, in principle, the public sector can intervene to overcome such failure: 'one may judge allocations of resources using this efficiency yardstick and if an existing allocation is *not* Pareto-efficient that it can be improved upon, and as such there may be a role for using the appropriate instruments of technology policy . . . thus for example, in some situations, transferring resources from consumption into R&D may generate a Pareto improvement and increase welfare'.[4] This approach, based on the identification and correction of market failure, is widely used in the analysis of industrial policy, and in research policy discussion itself.[5] And in general it can be said that within economic theory, policy measures are normally seen in terms of adjustment towards such competitive equilibria, and that this is as true of macroeconomic policy – both Keynesian and New Classical – as it is of microeconomic approaches. It should be noted that within this framework there is nothing particularly important about technology policy: technological change is just one of many phenomena whose properties involve some degree of market failure. It has no more intrinsic significance than, for example, a case of imperfect competition.

It seems to me that this theoretical system is unsatisfactory as a point of departure for policy analysis. This is not necessarily because of limitations in neo-classical theory itself. Of course, many critics have argued that the neo-classical system is fundamentally unsuited to analysis of problems of technological change; in practice this sometimes boils down to the non-sequitur that because neoclassical analyses of technological change have not been produced, then they cannot be produced. When it comes to details, most critiques of neoclassical theory tend to extend the argument that Pareto-optimal general equilibria are unattainable, particularly in dynamic contexts. Thus Corricelli and Dosi argue that aggregation from representative agents is internally inconsistent, that rational expectations models – since they involve optimising with respect to a 'correct' model – assume the ex ante consistency of plans rather than proving it (and are therefore tautological), that equilibria are not unique in the absence of representative agents and the presence of differing beliefs and conjectures, and that adaptive processes cannot converge in general to rational expectations equilibria.[6]

These arguments, although convincing, do not in themselves demonstrate that a neo-classical approach (of the type sketched in fn 2 above) could not in principle be applied to technical change phenomena. But nonetheless it does seem to me that the Corricelli and Dosi critique leads to the wider questions of whether the restrictive assumptions of neo-classicism are in some sense unmodifiable: that a theory based on rational otimising equilibria simply could not be extended to a world of radical uncertainty (especially in the face of different forms of relevant knowledge), of incomplete choice sets and non-convexities, of dynamics, of genuine hetergeneity of agents, and so on. Whether this is so must remain an open question, although if neoclassical approaches were to be suitably adapted to deal with technological change phenomena we would certainly end up with something radically different from current practice, and possibly quite close to the evolutionary models described below.[7]

However the real issue for policy is not neo-classical theory itself, and its possibilities and impossibilities, but the status of the conceptualisation of policy action which we actually have. This conceptualisation follows from the assumption, rather than the proof, that the general equilibrium framework is an adequate abstract descriptive point of departure for the analysis of a market economy. The question is, how adequate is the model, and what follows from accepting this assumption?

The main limitations of the neoclassical policy framework derive from the underlying concepts of production and competition within the neoclassical model. Production in neoclassical models is essentially a process of combination of factors, rather than a technical process of transformation of inputs. This simplifies the nature of the choice set and thus, at least implicitly, production techniques are both well-defined and well known. This makes it possible for the firm to separate two decisions, namely what to produce and how to produce it; it also makes the dual decision of the firm, at any point, path independent. Competition is simply the two processes of rational maximising calculation which correspond to the above decisions: one calculation optimally allocates capital in response to exogenous shifts in preferences or environmental conditions, and the other chooses an optimal capital-labour ratio. Information in this context is not technological knowledge; it is price information which enables calculation to occur. What all this leads to is a conceptualisation of the economy as a deterministic system in which the fundamental internal processes are those which adapt to exogenous change, rather than those which endogenously produce change. The extensive research programme in search processes undertaken in recent years within the neo-classical framework does not invalidate this view, since search is primarily seen as a method of generating the optimal level of information for adaptive choice where there is uncertainty or dispersed information. It is not seen in terms of search for new methods or techniques producing endogenous change.

Market failure is anything which prevents optimal allocation. Since the role of policy is to overcome market failure by promoting adaptation towards equilibrium, it explicitly cannot be concerned with any internal transformations which may change the nature of the competitive equilibrium state itself. That is to say, policy can only be concerned with the choice among a series of states characterised by given techniques (and hence given levels of income) rather than with, for example, processes of search for new states. A serious problem for policy analysis in this framework is that it is not clear how rectifying a perceived market failure actually promotes adjustment towards equilibrium, since the theory is by and large concerned with existence theorems for equilibrium states, not with an account of how those states are achieved. If there is no adequate theory of adjustment processes, then there is no guarantee that changing an environmental condition (such as inappropriability of technical information) in a disequilibrium state will in fact produce an equilibrium outcome. A further, more serious, difficulty is that setting up the problem in terms of this kind of market failure makes it extremely difficult to identify any legitimate role for policy. This is because although it is not in principle difficult to identify market failure, it is very difficult to assess whether the effects of any policy intervention are Pareto-superior. This has two aspects: firstly, once the basic assumptions of the neoclassical model are accepted, it is not particularly difficult to show that policy failure is at least as likely as market failure.[8] Secondly, even where a solution which empirical analysis suggests is superior is produced, it rarely rests on a general equilibrium framework; and there is no guarantee that any partial solution is an overall improvement.

Such problems ultimately derive from the fact that competitive equilibrium states, as noted above, are based on concepts of production and competition which are not abstractions but misrepresentations. This in turn is a result of an inadequate conceptualisation of (1) the agents who make up the elements of the system, whose capabilities in terms of technological knowledge and calculation are (to put it mildly) overstated, and (2) the availability and nature of information itself. A better abstract conceptualisation in these two fundamental areas should be constructed on the basis of, firstly, the heterogeneity of agents, the constraints on their acquisition of technological and economic information, and hence the, at best, bounded rationality of their calculation processes. Secondly, these characteristics generate diverse behaviour and technological structures, and hence non-price competition. Agents and behaviour which are conceptualised in this way have characteristics which are not deviations generating market failure. Rather, they are the essential features of the system: they shape its dynamics and produce its characteristic phenomena. The fundamental interest of evolutionary modelling in economics springs from the fact that it does indeed make these conceptual modifications, and that it analyses the systemic implications.

3. CHARACTERISTICS OF EVOLUTIONARY ECONOMIC SYSTEMS

This section describes some main characteristics of evolutionary models, before moving to a discussion of policy issues.[9]

Quite apart from whether evolutionary models present a more plausible theory of the economic system, an important reason for their potential relevance to technology policy is that they are constructed in large part around an account of why and how competing technologies develop. Underpinning evolutionary models is diversity among profit-seeking economic agents. For our purposes here the important agent is the firm, an entity with shifting boundaries and heterogeneous internal functions. This diversity is not simply an assumption of evolutionary models: it follows firstly from a prior view, sometimes explicit, about the nature of information in economic activity. Technical information is not a general category which can take the form of a free good, or of a good which is uniformly accessible. It is localised, highly specific, and often tacit. Having a particular productive capability involves a hierarchy of knowledges, at different levels of abstraction and with different functional characteristics. This means that the problem for the firm is not to gather information, but to produce it; information and capabilities are therefore endogenously generated within firms. For this purpose, firms must engage in search activity. Since technologies are complex and multifaceted, and since search strategies must be directed and focused, variation in knowledge bases results. This knowledge tends to cumulate along a search path. On the one hand this implies diversity in firm structures and capabilities, in the characteristics of products, and in the types of innovations which are introduced. On the other it implies path dependence of firm capabilities.

A further impulse towards technological diversity is the character of the selection environment. Markets are clearly the fundamental selection mechanism in evolutionary models: demand takes the form of requirements for combinations of prices and characteristics, and firms survive and grow according to the extent to which they can generate these combinations. So far, the concept of demand is shared with neo-classicism. But the prices/attributes combinations which succeed are always uncertain, and in general changing.

This produces what Metcalfe calls 'innovative competition', which is 'based upon the notion of differentiation as the chief means by which firms gain market advantages relative to their rivals'.[10] This notion of competition runs throughout evolutionary models, and is central to their dynamics; it derives of course from Schumpeter's well-known rejection of competition in terms of 'a rigid pattern of invariant conditions, methods of production and forms of industrial organisation'. By contrast, Schumpeter insisted that

> '. . . in capitalist reality as distinguished from its text book picture, it is not the kind of competition which counts but the competition from the new commodity, the new technology, the new source of supply, the new type of organisation . . . competition which commands a decisive cost or quality advantage and which strikes not at the margins of the profits and outputs of existing firms but at their foundations and their very lives.'[11]

The point which follows from this is that the system as a whole generates diversity not just because of informational factors, but because it positively selects for diversity: the nature of the competitive process generates conscious strategies aimed at producing differentiation. A significant problem for firms is to find mechanisms for generating variety on the basis of constrained informational capabilities.[12]

Market selection generates an evolution towards particular patterns of products and processes in two ways. Firstly, there is a survival and growth mechanism: a successful technology generates resources for investment and hence for growth of the firm. This relationship between selection and the investment function is a main determinant of industrial structure within an evolutionary framework. Secondly there is active learning and imitation, which introduces a Lamarckian element into economic and general cultural versions of evolutionary theory. Firms have an incentive to adapt towards selected technical attributes, and it is this which limits or constrains the overall process of technological variation. Hence there is convergence towards sets of technical principles which are linked to the configuration of demand: this underpins the emergence of so-called 'technological paradigms', or 'technological trajectories'.

Variations of the above concepts appear in most evolution-influenced models.[13] Georghiou et al on the one hand distinguish a 'technological regime' which

> may be represented in terms of a set of design parameters which embody the principles which will generate both the physical configuration of the product and the process and materials from which it is to be constructed. The basic design parameters are the heart of the technological regime, and they constitute a framework of knowledge which is *shared* by the firms in the industry.[14]

On the other hand, within the technological regime, firms choose a 'design configuration' which defines the particular process/product characteristics they are initiating or developing. This is a continuous process, for two reasons. Firstly, innovations usually possess only marginal advantages over competing technologies, and therefore require an often long process of post-innovation improvement.[15] Secondly, in the presence of search and innovation even successful innovations cannot deliver a permanent competitive advantage, and maintaining success therefore implies a process of continuous development and improvement within design configurations.[16] It ought to be noted that such points about the character and evolution of a technology also define technological conditions governing entry to an industry or a market: the entry decision implies the ability to place the firm within a technological regime or trajectory. What firms do is in part a matter of perceived opportunities for profit, but rather more a matter of technological capabilities which are path dependent and cumulative. This has implications for the operation of the 'market mechanism', and also has policy implications which will be referred to below.

A central problem for firms within this framework is to integrate and coordinate two quite different types of behaviour. Firstly, there is the construction and operation of routine production, which requires relatively stable relationships between the components of the firm. Secondly, there is the need to change and innovate, which itself must be carried through on a quasi-routine basis. In some evolutionary models, notably Nelson and Winter's work, when the first type of routine becomes non-viable in the face of enhanced competition, increased search is initiated for better techniques and routines; this motivates learning and imitation.

In summary, then, what does it mean to speak of 'evolutionary' models, and what is their *differentia specifica*, especially in terms of the neoclassical framework? Firstly, there is the emphasis on variation and the competitive mechanisms which continually generate it; here there is a sharp contrast with neoclassical competitive equilibrium which implies one optimal technique and set of firm behaviours, with rapid selection, and hence decreasing variation towards the optimal form which characterises equilibrium.[17] Secondly, there is the selection mechanism, which is also an inducement mechanism toward variety and hence technological innovation. Selection operates relatively slowly, so that 'losing' enterprises can engage in learning and imitation. Thirdly there is the place of time within the economic process; the selection mechanism, and the processes of differential growth and learning which it sets in train, obviously occur through time. But time is central in another way, closely analogous to its role in evolutionary biology; the technological state of the economy at any point in time is the result of sequences of changes which – however radical or incremental they are in the short run – are predicated on cumulation through time.[18] States are path dependent, as are firms capabilities; they are not the result of a globaloptimising choice. The latter implies reversibility, whereas in the longer term the evolutionary approach implies irreversibility, since the capability of using an old technology tends to disappear (although it can be reconstructed and even improved with further learning). As noted above, this implies a major asymmetry between firm exit from an industry, and the entry decision which implies complex technological preconditions.

4. POLICY IN AN EVOLUTIONARY CONTEXT

It seems immediately plain that a concept of technology policy based on the repair of a defective co-ordination mechanism has little relevance in an evolutionary context. A familiar way of looking at this is in terms of the so-called 'Schumpeterian trade-off', the contrast between static co-ordination efficiency and dynamic growth with imperfect competition.[19] However since two essential points about evolutionary models are firstly, that information does not take a form which permits optimisation to occur, and secondly that the firm is not an entity which under most circumstances could actually make optimising choices, this contrast is perhaps more rhetorical than analytical. In fact it is not clear what the welfare properties of market co-ordination really are in an evolutionary framework; in a system of quantity djustments, learning and quasi-continuous innovation, there are co-ordination concepts available, but they seem to have no clear static interpretation. This means that in an evolutionary context welfare and policy problems are intrinsically concerned with the nature and effects of dynamic paths, and co-ordination problems are an issue only insofar as they affect these paths.

It is also plain that from a welfare point of view there is nothing particulary special about any co-ordination pattern which does emerge: within a system of bounded rationality (based on the informational and firm-structure points noted above) whatever market solution emerges is necessarily imperfect. It is *always* possible to do better.[20] What it is not possible to do, even in outline, is to identify the characteristics of an equilibrium situation, or to distinguish between patterns of firm behaviour. This does not mean that firms do not engage in the kinds of substitution behaviour central to the neo-classical approach. Within the evolutionary framework, firms will respond to exogenous price shifts: in fact evolutionary approaches give a significantly better account of this, since they can explain why substitution takes the form of technological search and change rather than movements along isoquants.[21] The question of whether firm actions are adjustments based on the data of the equilibrium situation, or whether they are attempts to disrupt equilibria within an evolving technological trajectory is based on a false contrast. All adjustment behaviour occurs in a dynamic disequilibrium context, and the question then becomes whether a standard of system performance exists (whether there is an evolutionary equivalent of the Pareto-optimal Walrasian equilibrium). If firms can always do better, from a policy perspective one question is whether the routes towards improved performance are identifiable and whether they have any consistent pattern.

It seems, as with co-ordination issues, difficult to define what a normative description (as a basis for developing policy principles) of evolutionary system performance might look like; any such investigation is likely to be considerably more complex even than the Arrow-Debreu work which established a performance concept within neo-classical theory. What can be said as a point of departure is that such a description could not take the form of a suitably transformed optimum concept. Since a central theme of economic evolution, in common with evolutionary biology, is that future courses of events in principle cannot be identified, the idea that policy analysis is a matter either of optimising behaviour, or of the achievement of optima, makes little sense.[22] Even if it is argued that adaptive variation generates results which are in some sense optimal, these results are based on a prior diversity which is both non-optimal and 'blind' to what the optimal solution might be.

However an evolutionary normative framework might be structured around the following issues: firstly, the operation of economic and social mechanisms which generate experiment and search generally; secondly, the operation of selection mechanisms at different levels; and thirdly, the interaction between institutional structures and strategic behaviour.

In an evolutionary context, growth depends on search for and development of new techniques which can follow either from existing knowledge bases, or from learning. In either case there are general problems concerning the capacity to undertake such change. These include, for example, issues concerning the provision of competence, which relate to the operation of the education and training system. On the other, there are issues concerning the regulation of firm behaviour, which involve such matters as the attitude of capital markets to R&D expenditure and technological development generally. Many of the problems in such areas bear on or depend in large part on public policy, the point here being that from an evolutionary perspective these are integral to technology and innovation policy.

In terms of the selection mechanism, a key issue for growth concerned the rapidity and 'force' with which it operates. Given the fact that many new technologies have long gestation periods, and in particular that they often undergo long periods of post-innovation improvement before becoming genuinely viable, the speed and finality of selection can be critical to whether certain new technologies are developed at all. But the operation of the selection mechanism is, in many respects, a policy variable, affected sharply by fiscal and monetary policy, by procurement policies, by trade policy and so on. This too should be seen in the context of its technological implications. From a neo-classical perspective, strengthening the operation of the selection mechanism improves efficiency. The evolutionary argument is that selection is myopic: a population which is too well adapted to its existing circumstances may not generate variation, and hence may not adapt to change. Applied to firms, this suggests that those selected on the basis of current, short-term, efficiency need not be particularly able when it is a matter of generating new technologies nor be technologically adaptable as circumstances change.

Next, policy-making institutions, a topic which neo-classical analysis usually either neglects or incorporates within a political analysis based on vote-maximising, are important to the analysis in various respects. One key aspect of this, as Nelson and Winter emphasize, is that 'knowledge of *how* decisions are arrived at in business firms may tell us something about *what* decisions will be reached . . . this is even more strikingly true regarding government decision-making'. The point here is that once we dispense with the idea of agents as unified decision-making entities capable of optimisation, then the question becomes one of how organisations integrate different types of learning into flexible behaviour:

> Public policies and programs, like private activities, are embedded in and carried out by organisations . . . the design of a good policy is, to a considerable extent, the design of an organisational structure capable of learning and adjusting behaviour in response to what is learned.[23]

Nelson and Winter see this primarily in terms of how closely legislation should specify policy measures and instruments, the argument being that they should not be delimited within legislation but should be open to exploration within processes of implementation. But there are further aspects of the question of institutions which will be explored below; in particular there is the issue of the allocation of responsibilities among agencies, and relative importance of different functions within overall economic policy.

The question of strategy also arises in different forms. In evolutionary environments, the impossibility of optimisation imposes strategic behaviour on organisations, but objectives and characteristics of such behaviour may vary significantly among organisations. Firms are seeking profits, often at each others' expense. We can even say that they are seeking to maximise profits subject to certain constraints (although some of the constraints can be endogenously determined); what firms cannot do is achieve this through a global maximising decision. Technological uncertainty, economic uncertainty (about price structures and the nature of demand), and uncertainty about the actions of competitors all generate the need to construct competitive strategies; the interesting policy problems in this context concern the factors which limit or facilitate the achievement of adequate firm strategies. But an obvious problem here is that government agencies also have strategies (in the sense of decision rules for the attainment of objectives in an uncertain environment) which firstly, may conflict or harmonise badly with those of enterprises, and secondly, may conflict with those of other government agencies. The latter conflict follows from the fact that there is no clearly defined 'public interest', let alone a clear method to achieve it.

5. SOME GENERAL POLICY IMPLICATIONS OF EVOLUTIONARY MODELLING

An immediate conclusion from the evolutionary approach concerns the hierarchy of priorities within economic policy. Macroeconomic performance in the model is not the result of aggregation from a complex set of optimising decisions, but is shaped by the ability of firms to sustain activities within an evolving technological trajectory. These abilities shape, in the first instance, the growth rate and the external balance. Technology and innovation policy is not therefore an activity of equivalent importance to other arenas of market failure. Technological performance has quite direct effects, unlike any other category of economic activity, on the long-run productivity growth rate, on the balance of payments, and on the tax base (and hence on public finance). For that reason, it ought to be a central concern of economic policy. Evolutionary models imply shifting the focus away from a targets/instruments framework which is aimed essentially at setting indirect environmental conditions which will generate desired equilibria, towards an approach which looks much more directly at the underlying production conditions of the growth of welfare. It can be noted that if economic welfare derives from long-run dynamic technological performance, then the equity-efficiency trade-off takes on a very different aspect: firstly, there is no longer any static choice, and secondly, it may be that certain equity decisions (particularly in education and training provision) are not alternatives to output maximisation, but are underlying preconditions for the growth of output.

Such points do not of course imply that technology policy as such ought to be at the centre of economic policy; rather, it is technological performance which becomes a fundamental object of policy in and evolutionary framework. The innovation activity of firms responds to a wide range of policy measures, from the general macroeconomic stance to such areas as contract law and accounting regulations; the problem is to investigate whether these arenas of policy can be integrated with innovation objectives in a consistent way.

There nonetheless remain a number of arguments for definite technology and innovation policies. Firstly, even if an attempt is made to relate other policy arenas to innovation performance in a consistent way, a key problem is that innovation processes vary across firms and industries; one role for technology policy therefore is to regulate the differential impacts of other policy actions. Secondly, there are roles related to developing and maintaining the ability of firms to operate within technological trajectories. For a firm, the trajectory is determined by large-scale, often global, trends in demand and technological opportunity which are usually uncertain and at best involve risk for the firm; firm decision-making largely takes the form of conjectures. Since decisions by the firm involve focusing on a specific set of design and technology choices within an uncertainly evolving trajectory, where outcomes are unknown and where reasonable assessments of the situation can be sharply divergent, there is a high probability not just that mistakes will be made but that successive mistakes will be made. In such cases the ability of firms to regain the trajectory may depend heavily on policy support. Thirdly, there is the widely recognised problem that certain potentially viable avenues of pre-competitive search are simply too risky for individual firms to undertake. Public activity in such areas not only widens subsequent technological horizons for firms, it also entails external benefits. Within an evolutionary framework, such policy intervention is not related to market failure, but is a matter of shaping the potential technological trajectory, and providing access to it for firms; since policy-makers face the same uncertainties as firms, overall success in such R&D is likely to result not from individual projects but from portfolios.

6. OBJECTIVES OF PUBLIC POLICY AND PUBLIC–PRIVATE INTERACTION

An initial point here is that public policy in general has economic objectives which are distinct from those of economic agents. These differences arise from the fact that the economic system has 'boundaries' which are different from those of policy-making; the latter are usually, although not completely, national. That is, they are defined not by economic considerations but by political factors. Objectives such as balance of payments equilibrium, employment, public expenditure targets, or the achievement of a particular growth rate have at best a tangential connection with those of agents in general and particularly the large multinational enterprises who produce a substantial proportion of manufacturing output and perform a large share of national R&D. Now to some extent the objectives of policy-makers are directly linked with the success of enterprises; the balance of payments constraints on fiscal or interest rate policy, for example, are linked with shares of enterprises in world trade. But these objectives are shared only insofar as the activities of firms take place within particular national borders. From the point of view of consumers and producers it matters little whether domestic demand is satisfied from domestic or foreign sources. From a policy point of view (and ultimately for consumers) it may matter a great deal whether manufactured products (which for the most part are traded goods) are domestically produced or imported. However large enterprises – and, it might be argued, most manufacturing enterprises – have a wide variety of international activities, through trade, joint ventures, transnational finance sources, technology licensing and so on; their interest in location decisions is not wholly economic, but is certainly only partly political. It follows from this that policy decisions relate to two types of strategic conflict: one between firms, which is decided by their relative capabilities, and one with competing groups of policy-makers, which is decided by the location of successful firms.

However this is not a straightforward conflict, for there are also shared interest. The evolving economic system is global in its relationships, and some national policy objectives rely on a general increase in technological capabilities; another way of putting this would be to say that policy-makers have interests in nationally-located companies reaching the technological frontier, but they also have a joint interest in pushing the frontier outward. Here the evolutionary approach has important implications for technology policy strategies, in particular for levels and compositions of publicly-funded R&D. Evolutionary systems imply that economic effects depend on the dynamics of technological trajectories. Now trajectory dynamics, pushing the frontier out, relies in part on science policy and so-called 'basic' research, an issue which has on the whole been neglected in discussion of technology policy.[24] The key point about such research is not simply that it is difficult to appropriate, but that is is produced within a system which places a positive value on disclosure, and in which the transmission costs of information are very low. The problem which this produces is exactly that described by Arrow in his classic 1962 paper, except that it is transposed onto an international scene.[25] Fundamental research is in effect a pure public good. Such considerations mean that competitive national policies aimed at supporting or providing short-run or least-cost R&D improvements for industrial applications will tend towards free-riding; a prisoner's dilemma solution is a likely result, with general under-provision of fundamental research. This implies that an important element of technology policy ought to be the development of co-operative strategies aimed at collective provision of scientific research. As Kindleberger has pointed out, the international economy already operates with collectively provided public goods (such as the trade and payments system).[26] The evolutionary framework suggests that this should be extended to the scientific system.

What about public-private interaction and national policy objectives within the technological frontier? Here the achievement of policy objectives depend on the capabilities of firms within the national boundaries, and a number of issues arise. Firstly, there is the determination of the competitive environment itself, through competition policy, regulation, and industrial policy. What is being decided is the degree of competition versus co-operation between firms, and the degree of public-private co-operation. From an evolutionary viewpoint the issue is whether competitive strategies are superior to co-operative strategies. In a neoclassical framework the answer is unambiguous: competitive strategies are superior, usually even in the presence of market failure. In an evolutionary approach, things are less clear. Firstly, innovation-based competition means that we need to treat the actual characteristics of the innovation process carefully. In many cases, especially where radical innovations are involved, innovation processes entail the development of generic technologies with quasi-public good aspects. In such cases, inter-firm co-operation can be risk reducing and R&D enhancing. Secondly, the path dependence of firm capabilities means that a firm which is in competitive crises will not necessarily be replaced by domestic production if it fails; path dependence within technological trajectories provides a basis for understanding cumulative causation, but also a rationale for policy support. If firm survival is a policy objective, then it will firstly require the provision of R&D and/or financial support to take the firm out of downward cumulation, and secondly, such support would have to be provided to a degree sufficient to reach the investment, R&D and innovation capabilities of the current state of the technological trajectory. When firms enter crisis, they may respond – as in Nelson and Winter's model – with increased search, but they do so in a situation in which their capacity to sustain search is severely depleted. Thirdly, the problem of crisis for firms is in a sense systemic, since the performance of R&D generates continuous technological uncertainty for the system. From time to time the overall science and technology R&D system throws out more or less completely unexpected results which reshape the technological environment and redirect trajectories. Hence, of course, the endogeneity and unpredictability of change; but an implication of this, particularly for smaller economies, may be the provision of collective support in access to new generic technologies.

7. THE COMPOSITION OF PUBLIC R&D

In this section I discuss the allocation of public R&D between civil and military applications. This problem is of interest for an essentially empirical reason, which is that – among the five largest OECD economies who perform most of the world's R&D – there appears to be a long-term inverse relation between the share of military R&D in total R&D, and the long-run growth rates of output and productivity.

An evolutionary approach to the question of the civil-military R&D split might begin by linking growth rates to trade performance, which in turn is related to the innovative underpinnings of competitive behaviour described in Section 3 above. Now trade performance generally entails strong technological performance: the fastest-growing industries in world trade are R&D- and innovation-intensive, are essentially civilian, and are characterised by more or less exactly the competitive situation analysed in evolutionary models.[27] The problem then becomes one of asking how military innovation activity might affect the civilian technological trajectory. In the first place, we can note that although military and civilian technologies share a common knowledge base and common generic technologies, there are significant technological differences between them at the level of product performance. Moreover military R&D efforts are almost entirely related to the development and testing of products; the proportion of basic and generic research is typically low.

Why should this matter? One reason might be that competition in evolutionary environments in fact involves a complex mixture of struggle and co-operation. Co-operative elements within an ostensibly competitive framework are very well established empirically in the economics of innovation.[28] There is a biological analogy of this, namely 'reciprocal altruism', which does not imply a lack of self-interest but rather aid with the probability of help in return.[29] It is not difficult to see that this can generate a process of external benefits in innovation behaviour which could affect system performance: technological corridors are defined by a number of agents, and the pace at which they develop will presumably reflect the distribution or allocation of technological priorities by those agents. In this context, large volumes of military R&D could systematically deflect civilian trajectory development.

8. VARIATION, DUPLICATION AND THE LEVEL AND COMPOSITION OF R&D

Economic analysis of the level and composition of R&D has placed heavy emphasis on two countervailing trends. On the one hand there are the arguments deriving from Arrow and Nelson which suggest that scientific and technological knowledges have economic and other characteristics which inhibit appropriability; this combined with non-insurance against risk generates under-provision of R&D in a system of decentralised competition.[30] On the other hand, there are analyses based on effects of the patent system: firms race towards patentable information, with one winner appropriating all benefits. Dasgupta and Stiglitz showed that with free entry and an absence of sunk costs in R&D, an equilibrium with excessive firms and over-provision R&D will result; a policy conclusion follows, namely an R&D tax.[31] In sub sequent work Dasgupta and Maskin extended this to argue that the patent system generates not only an excessive volume of R&D, but a sub-optimal composition of R&D, with excessively risky projects.[32] Central to these analyses is that these search races involve different routes to similar functional results, and therefore that society must necessarily be indifferent to who wins; it is this which sustains the argument of overprovision.

An evolutionary perspective looks rather different. Firstly, there are no well defined ends, let alone routes towards them. Diversity in search paths is thus a social mechanism for the reduction of uncertainty, and the central danger is that the paths may be closed off too soon. Here the problem is twofold. Firstly, noted above, innovation is not a clearly defined phase, and successful innovation may require post-innovation improvements over a long period. Secondly, as Businaro remarks, 'the ultimate fate of an invention is not decided at the moment when it appears, so that more favourable future conditions could govern the ultimate selection'.[33] The possibility is therefore of being 'locked-in' to a technology which is in performance terms unsatisfactory. Arthur has shown just how likely this is, and why, with a number of empirical cases. His 'increasing returns to adoption' model depends on learning by using, scale economies in production and information, and various forms of technological inter-relatedness and externalities. His policy conclusions are very different from those of Dasgupta, Stiglitz and Maskin:

> 'It may sometimes be desirable as a policy option to keep more than one technology "alive", to avoid monopoly problems (if the technology is marketed), or to retain "requisite variety" as a hedge against shifts in the economic environment or against future "Chernobyl" revelations that the technology is unsafe.[34]

Here the policy conclusion is for subsidies to prevent the adoption/diffusion process selecting out technologies with long-run potential. The point is not the particular merits of this policy proposal, however. It is that an analytical framework based on an evolutionary conceptualisation of firms and the innovation process is likely to generate very different policy results, not just at a general level, but when it comes to specifics.

9. CONCLUSION

Research, technology and innovation policy are increasingly important areas within economic and social policy. Yet they lack a secure conceptual foundation, and this can have very significant effects. Anyone with a reasonable familiarity with policy making in these areas can readily give examples, from a range of countries, of decisions ranging from individual policy actions to sweeping rearrangements of entire research and institutional systems, which are taken in the more or less complete absence of an analytical framework other than that of the 'market failure' model. Can an alternative framework be developed? This paper has argued that the answer is yes, and that the evolutionary model of economic and technological change can provide such a framework. What has been presented here is, of necessity, rather tentative, but it suggests that there does indeed exist a body of theory which can in time provide a much-improved analytical basis for this critical policy arena.

NOTES

1. Some epistemological issues relevant to the problems addressed in this paper are explored in N. Clark and C. Juma, (1987), Part 1, and (1988), pp. 197–218.
2. The term 'neoclassical' here follows the general sense in which it is used by Hahn, to denote theories which are (a) reductionist in that they 'attempt to locate explanations in the actions of individual agents', (b) theorise about agents in terms of axioms of rationality, and (c) 'hold

that some notion of equilibrium is required and that the study of equilibrium states is useful'. But I would add (as Hahn might not) an emphasis on market clearing under competitive conditions, and that rationality is understood in terms of optimising choice, (1984), pp. 1–3.
3. P. Dasgupta, (1988), p.66.
4. Paul Stoneman, (1987), pp. 17–18. For similar approach, see P.H. Hall, (1986), pp. 3–7. A related approach, which does not necessarily rely on general equilibrium ideas and which is probably compatible with a range of theoretical perspectives, examines technology policy in terms of market failure in the production of information; the first work was Arrow's classic paper, but a recent analysis is P. Dasgupta and P. Stoneman, (1987), pp. 7–23.
5. For an application in the context of general industrial policy, see John Kay and David Thompson, (187). For examples of this approach in policy discussion, see K. Smith, (1989).
6. F. Corricelli and G. Dosi, (1988), pp. 124–147.
7. Richard Nelson has suggested that 'if one wants to expand the set of models, that one understands as neoclassical, to include those admitting substantial diversity of expectations as well as performance and friction, and to define maximisation and equilibrium so as to be consistent with this, then we are talking about the same kind of analytic animal'. But since this would probably have to imply highly asymmetric information and an absence of rational optimising behaviour, it is tantamount to doing away with the core of neoclassical modelling. R. Nelson, (1987), p.16.
8. For an example, see C. Wolf, (1986), pp. 43–70.
9. The outline here is broadly consistent with the approaches in, for example, R. Nelson and S. Winter, (1982); R.R. Nelson, (1987); J.S.Metcalfe, (1986), pp. 35–64; J.S. Metcalfe, (1988); L. Gheorghiou et al (1986), Chs. 2–4; G. Dosi and L. Orsenigo, (1988), pp. 13–37; C. De Bresson, (1987), pp. 751–762.
10. Metcalfe 1986, p.36.
11. J. Schumpeter, (1987), p.84.
12. This problem is analysed in P.P. Saviotti, (1988), pp. 89–103.
13. See, for example, G. Dosi, (1982), pp. 147–162; Nelson and Winter (1982), Ch.11.
14. L. Georghiou *et al*, 1986, p.32.
15. This is a major influence on diffusion paths: N. Rosenberg, (1977), pp. 189–210.
16. This is the central theme of Georghiou *et al* (1986).
17. This neoclassical version of natural selection suppresses the need for any 'optimising choice' model of firm behaviour, but it relies on a static notion of competition in terms of cost minimisation, and thus suppresses the competitive forces generating diversity and, in general, all other elements of a genuinely evolutionary theory. For an excellent discussion of Milton Friedman's use of the natural selection metaphor, see Nelson and Winter (1982), pp. 139–141.
18. Kim B. Clark, (1985), pp. 235 251.
19. The best analysis is found in Nelson and Winter (1982), Part V.
20. This point and its implications for firm behaviour, particularly search activity in response to crisis, are fully discussed in F. Sejersted, (1984).
21. Within the neo-classical system it is very difficult to make the steps from increasing real wages to labour saving technological change: certainly factor substitution will occur, but why has this been consistently labour-saving in the long-run? See N. Rosenberg, (1977). For an account of systematic substitution behaviour involving technological search, consistent with evolutionary approaches, see F. Lichtenberg, (1986), 67–76.
22. Nelson and Winter (1982), pp. 382–383.
23. Nelson and Winter (1982), pp.384.
24. An exception is P. Dasgupta, (1987), pp. 9–12.
25. K.J. Arrow, (1962), pp. 609–625.
26. C.P. Kindleberger, (1986). For a relevant discussion of institutions and externalities in an evolutionary framework, see R.C.O. Matthews, (1984), pp. 91–117.
27. In general, evolutionary models have not been extended to open economy contexts. However, an important recent evolution-influenced approach combines a Kaldorian growth model in which the balance of payments constraint is the primary determinant of the growth rate, with a Schumpeterian model of competition based on technological performance; the model is tested using patent and R&D data as proxies for innovation performance; and explains levels and patterns of trade and growth well. See J. Fagerberg, (1988).

28. For a detailed treatment, see E. von Hippel, (1988), especially Ch. 6, 'Co-operation between rivals: the informal trading of know-how'.
29. 'Animals aid others, including non-relatives, because there are the possibilities of reciprocating benefits. Thus altruism evolves . . . not because of some group of other supposed benefit, but because of returns or potentiality of returns to the individual actor'; M. Ruse, (1987).
30. Arrow 1962 *op. cit.*; R. Nelson, (1959).
31. P. Dasgupta and J. Stiglitz, (1980), pp. 266–293.
32. P. Dasgupta and E. Maskin, (1987).
33. U. Businaro, (1988), p.603.
34. W.B. Arthur, (1988), p.603.

REFERENCES

W.B. Arthur, 'Competing technologies: an overview' in G. Dosi *et al* (eds) *Technical Change and Economic Theory* (London: Frances Pinter, 1988).

K.J. Arrow, 'Economic welfare and the allocation of resources for invention', in National Bureau of Economic Research, *The Rate and Direction of Inventive Activity*, (1962).

Kim B. Clark, 'The interaction of design hierarchies and market concepts in technological evolution', *Research Policy,* **14** (1985).

N. Clark and C. Juma, 'Evolutionary theories in economic thought', in G. Dosi *et al* (eds) *Technical Change and Economic Theory*, (London: Frances Pinter, 1988).

N. Clark and C. Juma, *Long-Run Economics: An Evolutionary Approach to Economic Growth* (London: Frances Pinter, 1987), Part I.

F. Corricelli and G. Dosi, 'Co-ordination and order in economic change and the interpretive power of economic theory' in G. Dosi *et al*, *Technical Change and Economic Theory*, (London, 1988), pp. 124–147.

P. Dasgupta and J. Stiglitz, 'Industrial structure and the nature of innovative activity', *Economic Journal,* **90**, (1980).

P. Dasgupta and E. Maskin, 'The simple economics of research portfolios', *Economic Journal,* **97**, (1987).

P. Dasgupta and P. Stoneman (eds) *Economic Policy and Technological Performance* (Cambridge, 1987).

P. Dasgupta, 'Patents, priority and imitation or, the economics of races and waiting games', *Economic Journal*, Vol. **98**, (March 1988), p. 66.

P. Dasgupta and P. Stoneman (eds) *Economic Policy and Technological Performance* (Cambridge, 1987).

C. De Bresson, 'The evolutionary paradigm and the economics of technological change', *Journal of Economic Issues,* **21**, 2, (1987).

U. Businaro, 'Applying the biological evolution metaphor to technological innovation', in C. Freeman (ed) *Design, Innovation and Long Cycles in Economic Development*, (London: Frances Pinter, 1986).

G. Dosi, 'Technological paradigms and technological trajectories: a suggested interpretation of the determinants and directions of technical change', *Research Policy,* **11**, (1982), pp. 147–162.

G. Dosi and L. Orsenigo, 'Co-ordination and transformation: an overview of structures, behaviours and change in evolutionary environments', in G. Dosi *et al* (eds) *Technical Change and Economic Theory* (Frances Pinter: London, 1987).

J. Fagerberg, 'International competitiveness', *Economic Journal*, (1988).

L. Georghiou *et al, Post Innovation Performance*, (London: Macmillan, 1986).

P.H. Hall, 'The theory and practice of innovation policy: an overview' in P.H. Hall (ed) *Technology, Innovation and Economic Policy*, Oxford: Philip Allan, (1986).

F. Hahn, *Equilibrium and Macroeconomics*, Oxford, Oxford University Press, (1984).

E. von Hippel, *Sources of Innovation* (Oxford: OUP, 1988).

C.P. Kindleberger, 'International public goods without international government', *American Economic Review*, (1986).

F. Lichtenberg, 'Energy prices and induced innovation', *Research Policy,* **15**, (1986).

R.C.O. Matthews, 'Darwinism and economic change', *Oxford Economic Papers*, Vol. **36** (Supplement), (1984).

J.S. Metcalfe, 'Technological innovation and the competitive process', in P. Hall, *Technology, Innovation and Economic Policy* (Philip Allan: Oxford, 1986), pp. 35–64.

J.S. Metcalfe, 'Evolutionary and economic change', paper presented to the Economics of Technological Change Conference, Manchester, (1988).

R. Nelson, 'The simple economics of basic scientific research', *American Economic Review*, (1959).

R. Nelson, *Understanding Technical Change as an Evolutionary Process*, (Amsterdam: North-Holland, 1987).

R. Nelson and S. Winter, *An Evolutionary Theory of Economic Change* (Harvard, 1982).

J. Kay and D. Thompson, 'Policy for industry' in R. Dornbusch and R. Layard, *The Performance of the British Economy* (Oxford, 1987).

N. Rosenberg, 'Factors affecting the diffusion of technology', *Perspectives on Technology* (Cambridge: CUP, 1977).

N. Rosenberg, 'Problems in the economist's conceptualisation of technological change', *Perspectives on Technology* (Cambridge: CUP, 1977).

M. Ruse, 'Evolutionary models and social theory. Prospects and problems', in M. Schmid and F. Wuketits, *Evolutionary Theory in Social Science*, (Dordrecht: eidel/Kluwer, 1987).

P.P. Saviotti, 'Information, variety and entropy in technoeconomic development', *Research Policy*, **17** (1988), 89–103.

J. Schumpeter, *Capitalism, Socialism and Democracy*, (Unwin: London, 1987).

F. Sejersted, 'Routine and choice. On various concepts of the firm and their implications for economic development', *Technology History and Project Paper No 2*, University of Oslo, (1984).

K. Smith, 'Civil R&D in the UK: limitations of recent policy debate', *Research Policy*, **18**, (2), (1989).

Paul Stoneman, *The Economic Analysis of Technology Policy*, (Oxford, 1987).

C. Wolf, 'Market and non-market failures: comparison and assessment', *Journal of Public Policy*, **7**, 1, (1986).

Index

K

L

M

N

O

P

R

S

T

V

W